THE ELEGANT UNIVERSE

Brian Greene is Professor of Physics and Mathematics at
Columbia University and Cornell University.

D0332607

Brian Greene

THE ELEGANT UNIVERSE

Superstrings,
Hidden Dimensions,
and the Quest for the
Ultimate Theory

V

VINTAGE

Published by Vintage 2000

21

First published in Great Britain in 1999
by Jonathan Cape

Vintage
Random House, 20 Vauxhall Bridge Road,
London SW1V 2SA

www.vintage-books.co.uk

Addresses for companies within The Random House Group
Limited can be found at: www.randomhouse.co.uk/offices.htm

The Random House Group Limited Reg. No. 954009

A CIP catalogue record for this book
is available from the British Library

ISBN 9780099289920

The Random House Group Limited makes every effort to
ensure that the papers used in its books are made from
trees that have been legally sourced from well-managed and
credibly certified forests. Our paper procurement policy
can be found at: www.randomhouse.co.uk/paper.htm

Printed in the UK by CPI Bookmarque, Croydon, CR0 4TD

To my mother and the memory of my father,
with love and gratitude

Contents

Part IV: String Theory and the Fabric of Spacetime

Part V: Unification in the Twenty-First Century

Preface

During the last thirty years of his life, Albert Einstein sought relentlessly for a so-called unified field theory—a theory capable of describing nature's forces within a single, all-encompassing, coherent framework. Einstein was not motivated by the things we often associate with scientific undertakings, such as trying to explain this or that piece of experimental data. Instead, he was driven by a passionate belief that the deepest understanding of the universe would reveal its truest wonder: the simplicity and power of the principles on which it is based. Einstein wanted to illuminate the workings of the universe with a clarity never before achieved, allowing us all to stand in awe of its sheer beauty and elegance.

Einstein never realized this dream, in large part because the deck was stacked against him: In his day, a number of essential features of matter and the forces of nature were either unknown or, at best, poorly understood. But during the past half-century, physicists of each new generation—through fits and starts, and diversions down blind alleys—have been building steadily on the discoveries of their predecessors to piece together an ever fuller understanding of how the universe works. And now, long after Einstein articulated his quest for a unified theory but came up empty-handed, physicists believe they have finally found a framework for stitching these insights together into a seamless whole—a single theory that, in principle, is capable of describing all physical phenomena. The theory, *superstring theory*, is the subject of this book.

I wrote *The Elegant Universe* in an attempt to make the remarkable insights emerging from the forefront of physics research accessible to a broad spectrum of readers, especially those with no training in mathematics or physics. Through public lectures on superstring theory I have given over the past few years, I have witnessed a widespread yearning to understand what current research says about the fundamental laws of the universe, how these laws require a monumental restructuring of our conception of the cosmos, and what challenges lie ahead in the ongoing quest for the ultimate theory. I hope that, by explaining the major achievements of physics going back to Einstein and Heisenberg, and describing how their discoveries have grandly flowered through the breakthroughs of our age, this book will both enrich and satisfy this curiosity.

I also hope that *The Elegant Universe* will be of interest to readers who do have some scientific background. For science students and teachers, I hope this book will crystallize some of the foundational material of modern physics, such as special relativity, general relativity, and quantum mechanics, while conveying the contagious excitement of researchers closing in on the long-sought unified theory. For the avid reader of popular science, I have tried to explain many of the exhilarating advances in our understanding of the cosmos that have come to light during the last decade. And for my colleagues in other scientific disciplines, I hope this book will give an honest and balanced sense of why string theorists are so enthusiastic about the progress being made in the search for the ultimate theory of nature.

Superstring theory casts a wide net. It is a broad and deep subject that draws on many of the central discoveries in physics. Since the theory unifies the laws of the large and of the small, laws that govern physics out to the farthest reaches of the cosmos and down to the smallest speck of matter, there are many avenues by which one can approach the subject. I have chosen to focus on our evolving understanding of space and time. I find this to be an especially gripping developmental path, one that cuts a rich and fascinating swath through the essential new insights. Einstein showed the world that space and time behave in astoundingly unfamiliar ways. Now, cutting-edge research has integrated his discoveries into a quantum universe with numerous hidden dimensions coiled into the fabric of the cosmos—dimensions whose lavishly entwined geometry may well hold the key to some of the most profound questions ever posed. Although some of these concepts are subtle, we will see that they can be grasped through down-to-earth analogies. And when these ideas are understood, they provide a startling and revolutionary perspective on the universe.

Throughout this book, I have tried to stay close to the science while giving the reader an intuitive understanding—often through analogy and metaphor—of how scientists have reached the current conception of the cosmos. Although I avoid technical language and equations, because of the radically new concepts involved the reader may need to pause now and then, to mull over a section here or ponder an explanation there, in order to follow the progression of ideas fully. A few sections of Part IV (focusing on the most recent developments) are a bit more abstract than the rest; I have taken care to forewarn the reader about these sections and to structure the text so that they can be skimmed or skipped with minimal impact on the book's logical flow. I have included a glossary of scientific terms for an easy and accessible reminder of ideas introduced in the main text. Although the more casual reader may wish to skip the endnotes completely, the more diligent reader will find in the notes amplifications of points made in the text, clarifications of ideas that have been simplified in the text, as well as a few technical excursions for those with mathematical training.

I owe thanks to many people for their help during the writing of this book. David Steinhardt read the manuscript with great care and generously provided sharp editorial insights and invaluable encouragement. David Morrison, Ken Vineberg, Raphael Kasper, Nicholas Boles, Steven Carlip, Arthur Greenspoon, David Mermin, Michael Popowits, and Shani Offen read the manuscript closely and offered detailed reactions and suggestions that greatly enhanced the presentation. Others who read all or part of the manuscript and offered advice and encouragement are Paul Aspinwall, Persis Drell, Michael Duff, Kurt Gottfried, Joshua Greene, Teddy Jefferson, Marc Kamionkowski, Yakov Kanter, Andras Kovacs, David Lee, Megan McEwen, Nari Mistry, Hasan Padamsee, Ronen Plesser, Massimo Poratti, Fred Sherry, Lars Straeter, Steven Strogatz, Andrew Strominger, Henry Tye, Cumrun Vafa, and Gabriele Veneziano. I owe special thanks to Raphael Gunner for, among many other things, his insightful criticisms at an early stage of writing that helped to shape the overall form of the book, and to Robert Malley for his gentle but persistent encouragement to go beyond thinking about the book and to put "pen to paper." Steven Weinberg and Sidney Coleman offered valuable advice and assistance, and it is a pleasure to acknowledge many helpful interactions with Carol Archer, Vicky Carstens, David Cassel, Anne Coyle, Michael Duncan, Jane Forman, Wendy Greene, Susan Greene, Erik Jendresen, Gary Kass, Shiva Kumar, Robert Mawhinney, Pam Morehouse, Pierre Ramond, Amanda Salles, and Eero Simoncelli. I am indebted to Costas Efthimiou for his help in fact-checking and reference-finding, and for turning my initial sketches into line drawings from which Tom Rockwell created—with the patience of a saint and a masterful artistic eye—the figures that illustrate the text. I also thank Andrew Hanson and Jim Sethna for their help in preparing a few of the specialized figures.

For agreeing to be interviewed and to lend their personal perspectives on various topics covered I thank Howard Georgi, Sheldon Glashow, Michael Green, John Schwarz, John Wheeler, Edward Witten, and, again, Andrew Strominger, Cumrun Vafa, and Gabriele Veneziano.

I am happy to acknowledge the penetrating insights and invaluable suggestions of Angela Von der Lippe and the sharp sensitivity to detail of Traci Nagle, my editors at W. W. Norton, both of whom significantly enhanced the clarity of the presentation. I also thank my literary agents, John Brockman and Katinka Matson, for their expert guidance in shepherding the book from inception to publication.

For generously supporting my research in theoretical physics for more than a decade and a half, I gratefully acknowledge the National Science Foundation, the Alfred P. Sloan Foundation, and the U.S. Department of Energy. It is perhaps not surprising that my own research has focused on the impact superstring theory has on our conception of space and time, and in a couple of the later chapters I describe some of the discoveries in which I had the fortune to take part. Although I hope the reader will enjoy reading these "inside" accounts, I realize that they may leave an exaggerated impression of the role I have played in the development of superstring theory. So let me take this opportunity to acknowledge the more than one thousand physicists around the world who are crucial and dedicated participants in the effort to fashion the ultimate theory of the universe. I apologize to all whose work is not included in this account; this merely reflects the thematic perspective I have chosen and the length limitations of a general presentation.

Finally, I owe heartfelt thanks to Ellen Archer for her unwavering love and support, without which this book would not have been written.

The Elegant
Universe

Part I

The Edge of
Knowledge

Chapter 1

Tied Up with String

C alling it a cover-up would be far too dramatic. But for more than half
a century—even in the midst of some of the greatest scientific
achievements in history—physicists have been quietly aware of a dark
cloud looming on a distant horizon. The problem is this: There are two
foundational pillars upon which modern physics rests. One is Albert Einstein's general relativity, which provides a theoretical framework for understanding the universe on the largest of scales: stars, galaxies, clusters
of galaxies, and beyond to the immense expanse of the universe itself.
The other is quantum mechanics, which provides a theoretical framework
for understanding the universe on the smallest of scales: molecules, atoms,
and all the way down to subatomic particles like electrons and quarks.
Through years of research, physicists have experimentally confirmed to almost unimaginable accuracy virtually all predictions made by each of
these theories. But these same theoretical tools inexorably lead to another
disturbing conclusion: As they are currently formulated, general relativity
and quantum mechanics *cannot both be right*. The two theories underlying the tremendous progress of physics during the last hundred years—
progress that has explained the expansion of the heavens and the
fundamental structure of matter—are mutually incompatible.

If you have not heard previously about this ferocious antagonism you
may be wondering why. The answer is not hard to come by. In all but the
most extreme situations, physicists study things that are either small and

light (like atoms and their constituents) or things that are huge and heavy (like stars and galaxies), but not both. This means that they need use only quantum mechanics *or* only general relativity and can, with a furtive glance, shrug off the barking admonition of the other. For fifty years this approach has not been quite as blissful as ignorance, but it has been pretty close.

But the universe *can* be extreme. In the central depths of a black hole an enormous mass is crushed to a minuscule size. At the moment of the big bang the whole of the universe erupted from a microscopic nugget whose size makes a grain of sand look colossal. These are realms that are tiny and yet incredibly massive, therefore requiring that both quantum mechanics and general relativity simultaneously be brought to bear. For reasons that will become increasingly clear as we proceed, the equations of general relativity and quantum mechanics, when combined, begin to shake, rattle, and gush with steam like a red-lined automobile. Put less figuratively, well-posed physical questions elicit nonsensical answers from the unhappy amalgam of these two theories. Even if you are willing to keep the deep interior of a black hole and the beginning of the universe shrouded in mystery, you can't help feeling that the hostility between quantum mechanics and general relativity cries out for a deeper level of understanding. Can it really be that the universe at its most fundamental level is divided, requiring one set of laws when things are large and a different, incompatible set when things are small?

Superstring theory, a young upstart compared with the venerable edifices of quantum mechanics and general relativity, answers with a resounding no. Intense research over the past decade by physicists and mathematicians around the world has revealed that this new approach to describing matter at its most fundamental level resolves the tension between general relativity and quantum mechanics. In fact, superstring theory shows more: Within this new framework, general relativity and quantum mechanics *require one another* for the theory to make sense. According to superstring theory, the marriage of the laws of the large and the small is not only happy but inevitable.

That's part of the good news. But superstring theory—string theory, for short—takes this union one giant step further. For three decades, Einstein sought a unified theory of physics, one that would interweave all of nature's forces and material constituents within a single theoretical tapestry.

He failed. Now, at the dawn of the new millennium, proponents of string theory claim that the threads of this elusive unified tapestry finally have been revealed. String theory has the potential to show that all of the wondrous happenings in the universe—from the frantic dance of subatomic quarks to the stately waltz of orbiting binary stars, from the primordial fireball of the big bang to the majestic swirl of heavenly galaxies—are reflections of one grand physical principle, one master equation.

Because these features of string theory require that we drastically change our understanding of space, time, and matter, they will take some time to get used to, to sink in at a comfortable level. But as shall become clear, when seen in its proper context, string theory emerges as a dramatic yet natural outgrowth of the revolutionary discoveries of physics during the past hundred years. In fact, we shall see that the conflict between general relativity and quantum mechanics is actually not the first, but the third in a sequence of pivotal conflicts encountered during the past century, each of whose resolution has resulted in a stunning revision of our understanding of the universe.

The Three Conflicts

The first conflict, recognized as far back as the late 1800s, concerns puzzling properties of the motion of light. Briefly put, according to Isaac Newton's laws of motion, if you run fast enough you can catch up with a departing beam of light, whereas according to James Clerk Maxwell's laws of electromagnetism, you can't. As we will discuss in Chapter 2, Einstein resolved this conflict through his theory of special relativity, and in so doing completely overturned our understanding of space and time. According to special relativity, no longer can space and time be thought of as universal concepts set in stone, experienced identically by everyone. Rather, space and time emerged from Einstein's reworking as malleable constructs whose form and appearance depend on one's state of motion.

The development of special relativity immediately set the stage for the second conflict. One conclusion of Einstein's work is that no object—in fact, no influence or disturbance of any sort—can travel faster than the speed of light. But, as we shall discuss in Chapter 3, Newton's experimentally successful and intuitively pleasing universal theory of gravitation involves influences that are transmitted over vast distances of space *instantaneously*. It was Einstein, again, who stepped in and resolved the conflict by offering a new conception of gravity with his 1915 general theory of relativity. Just as special relativity overturned previous conceptions of space and time, so too did general relativity. Not only are space and time influenced by one's state of motion, but they can warp and curve in response to the presence of matter or energy. Such distortions to the fabric of space and time, as we shall see, transmit the force of gravity from one place to another. Space and time, therefore, can no longer to be thought of as an inert backdrop on which the events of the universe play themselves out; rather, through special and then general relativity, they are intimate players in the events themselves.

Once again the pattern repeated itself: The discovery of general relativity, while resolving one conflict, led to another. Over the course of the three decades beginning in 1900, physicists developed quantum mechanics (discussed in Chapter 4) in response to a number of glaring problems that arose when nineteenth-century conceptions of physics were applied to the microscopic world. And as mentioned above, the third and deepest conflict arises from the incompatibility between quantum mechanics and general relativity. As we will see in Chapter 5, the gently curving geometrical form of space emerging from general relativity is at loggerheads with the frantic, roiling, microscopic behavior of the universe implied by quantum mechanics. As it was not until the mid-1980s that string theory offered a resolution, this conflict is rightly called the central problem of modern physics. Moreover, building on special and general relativity, string theory requires its own severe revamping of our conceptions of space and time. For example, most of us take for granted that our universe has three spatial dimensions. But this is not so according to string theory, which claims that our universe has many more dimensions than meet the eye—dimensions that are tightly curled into the folded fabric of the cosmos. So central are these remarkable insights into the nature of space and time that we shall use them as a guiding theme in all that follows. String theory, in a real sense, is the story of space and time since Einstein.

To appreciate what string theory actually is, we need to take a step back and briefly describe what we have learned during the last century about the microscopic structure of the universe.

The Universe at Its Smallest: What We Know about Matter

The ancient Greeks surmised that the stuff of the universe was made up of tiny "uncuttable" ingredients that they called *atoms*. Just as the enormous number of words in an alphabetic language is built up from the wealth of combinations of a small number of letters, they guessed that the vast range of material objects might also result from combinations of a small number of distinct, elementary building blocks. It was a prescient guess. More than 2,000 years later we still believe it to be true, although the identity of the most fundamental units has gone through numerous revisions. In the nineteenth century scientists showed that many familiar substances such as oxygen and carbon had a smallest recognizable constituent; following in the tradition laid down by the Greeks, they called them *atoms*. The name stuck, but history has shown it to be a misnomer, since atoms surely are "cuttable." By the early 1930s the collective works of J. J. Thomson, Ernest Rutherford, Niels Bohr, and James Chadwick had established the solar system–like atomic model with which most of us are familiar. Far from being the most elementary material constituent, atoms consist of a nucleus, containing protons and neutrons, that is surrounded by a swarm of orbiting electrons.

For a while many physicists thought that protons, neutrons, and electrons were the Greeks' "atoms." But in 1968 experimenters at the Stanford Linear Accelerator Center, making use of the increased capacity of technology to probe the microscopic depths of matter, found that protons and neutrons are not fundamental, either. Instead they showed that each consists of three smaller particles, called *quarks*—a whimsical name taken from a passage in James Joyce's *Finnegan's Wake* by the theoretical physicist Murray Gell-Mann, who previously had surmised their existence. The experimenters confirmed that quarks themselves come in two varieties, which were named, a bit less creatively, *up* and *down*. A proton consists of two up-quarks and a down-quark; a neutron consists of two down-quarks and an up-quark.

Everything you see in the terrestrial world and the heavens above appears to be made from combinations of electrons, up-quarks, and down-quarks. No experimental evidence indicates that any of these three particles is built up from something smaller. But a great deal of evidence indicates that the universe itself has additional particulate ingredients. In the mid-1950s, Frederick Reines and Clyde Cowan found conclusive experimental evidence for a fourth kind of fundamental particle called a *neutrino*—a particle whose existence was predicted in the early 1930s by Wolfgang Pauli. Neutrinos proved very difficult to find because they are ghostly particles that only rarely interact with other matter: an average-energy neutrino can easily pass right through many trillion miles of lead without the slightest effect on its motion. This should give you significant relief, because right now as you read this, billions of neutrinos ejected into space by the sun are passing through your body and the earth as well, as part of their lonely journey through the cosmos. In the late 1930s, another particle called a *muon*—identical to an electron except that a muon is about 200 times heavier—was discovered by physicists studying cosmic rays (showers of particles that bombard earth from outer space). Because there was nothing in the cosmic order, no unsolved puzzle, no tailor-made niche, that necessitated the muon's existence, the Nobel Prize–winning particle physicist Isidor Isaac Rabi greeted the discovery of the muon with a less than enthusiastic "Who ordered that?" Nevertheless, there it was. And more was to follow.

Using ever more powerful technology, physicists have continued to slam bits of matter together with ever increasing energy, momentarily recreating conditions unseen since the big bang. In the debris they have searched for new fundamental ingredients to add to the growing list of particles. Here is what they have found: four more quarks—*charm, strange, bottom,* and *top*—and another even heavier cousin of the electron, called a *tau,* as well as two other particles with properties similar to the neutrino (called the *muon-neutrino* and *tau-neutrino* to distinguish them from the original neutrino, now called the *electron-neutrino*). These particles are produced through high-energy collisions and exist only ephemerally; they are not constituents of anything we typically encounter. But even this is not quite the end of the story. Each of these particles has an *antiparticle* partner—a particle of identical mass but opposite in certain other respects such as its electric charge (as well as its charges with respect to other forces discussed below). For instance, the antiparticle of an electron is called a *positron*—it has exactly the same mass as an electron, but its electric charge is +1 whereas the electric charge of the electron is −1. When in contact, matter and antimatter can annihilate one another to produce pure energy—that's why there is extremely little naturally occurring antimatter in the world around us.

Physicists have recognized a pattern among these particles, displayed in Table 1.1. The matter particles neatly fall into three groups, which are often called *families*. Each family contains two of the quarks, an electron or one of its cousins, and one of the neutrino species. The corresponding particle types across the three families have identical properties except for their mass, which grows larger in each successive family. The upshot is that physicists have now probed the structure of matter to scales of about a billionth of a billionth of a meter and shown that *everything* encountered to date—whether it occurs naturally or is produced artificially with giant atom-smashers—consists of some combination of particles from these three families and their antimatter partners.

A glance at Table 1.1 will no doubt leave you with an even stronger sense of Rabi's bewilderment at the discovery of the muon. The arrangement into families at least gives some semblance of order, but innumerable "whys" leap to the fore. Why are there so many fundamental particles, especially when it seems that the great majority of things in the world around us need only electrons, up-quarks, and down-quarks? Why are there three families? Why not one family or four families or any other number? Why do the particles have a seemingly random spread of masses—why, for in-

Family 1		Family 2		Family 3	
Particle	*Mass*	*Particle*	*Mass*	*Particle*	*Mass*
Electron	.00054	Muon	.11	Tau	1.9
Electron-neutrino	$< 10^{-8}$	Muon-neutrino	$< .0003$	Tau-neutrino	$< .033$
Up-quark	.0047	Charm Quark	1.6	Top Quark	189
Down-quark	.0074	Strange Quark	.16	Bottom Quark	5.2

Table 1.1 The three families of fundamental particles and their masses (in multiples of the proton mass). The values of the neutrino masses have so far eluded experimental determination.

stance, does the tau weigh about 3,520 times as much as an electron? Why does the top quark weigh about 40,200 times as much an up-quark? These are such strange, seemingly random numbers. Did they occur by chance, by some divine choice, or is there a comprehensible scientific explanation for these fundamental features of our universe?

The Forces, or, Where's the Photon?

Things only become more complicated when we consider the forces of nature. The world around us is replete with means of exerting influence: balls can be hit with bats, bungee enthusiasts can throw themselves earthward from high platforms, magnets can keep superfast trains suspended just above metallic tracks, Geiger counters can tick in response to radioactive material, nuclear bombs can explode. We can influence objects by vigorously pushing, pulling, or shaking them; by hurling or firing other objects into them; by stretching, twisting, or crushing them; or by freezing, heating, or burning them. During the past hundred years physicists have accumulated mounting evidence that all of these interactions between various objects and materials, as well as any of the millions upon millions of others encountered daily, can be reduced to combinations of four fundamental forces. One of these is the *gravitational force.* The other three are the *electromagnetic force,* the *weak force,* and the *strong force.*

Gravity is the most familiar of the forces, being responsible for keeping us in orbit around the sun as well as for keeping our feet firmly planted on earth. The mass of an object measures how much gravitational force it can exert as well as feel. The electromagnetic force is the next most familiar of the four. It is the force driving all of the conveniences of modern life—lights, computers, TVs, telephones—and underlies the awesome might of lightning storms and the gentle touch of a human hand. Microscopically, the electric charge of a particle plays the same role for the electromagnetic force as mass does for gravity: it determines how strongly the particle can exert as well as respond electromagnetically.

The strong and the weak forces are less familiar because their strength rapidly diminishes over all but subatomic distance scales; they are the nuclear forces. This is why these two forces were discovered only much more recently. The strong force is responsible for keeping quarks "glued" together inside of protons and neutrons and keeping protons and neutrons tightly crammed together inside atomic nuclei. The weak force is best known as the force responsible for the radioactive decay of substances such as uranium and cobalt.

During the past century, physicists have found two features common to all these forces. First, as we will discuss in Chapter 5, at a microscopic level all the forces have an associated particle that you can think of as being the smallest packet or bundle of the force. If you fire a laser beam—an "electromagnetic ray gun"—you are firing a stream of *photons,* the smallest bundles of the electromagnetic force. Similarly, the smallest constituents of weak and strong force fields are particles called *weak gauge bosons* and *gluons.* (The name *gluon* is particularly descriptive: You can think of gluons as the microscopic ingredient in the strong glue holding atomic nuclei together.) By 1984 experimenters had definitively established the existence and the detailed properties of these three kinds of force particles, recorded in Table 1.2. Physicists believe that the gravitational force also has an associated particle—the graviton—but its existence has yet to be confirmed experimentally.

The second common feature of the forces is that just as mass determines how gravity affects a particle, and electric charge determines how the electromagnetic force affects it, particles are endowed with certain amounts of "strong charge" and "weak charge" that determine how they are affected by the strong and weak forces. (These properties are detailed in

Force	Force particle	Mass
Strong	Gluon	0
Electromagnetic	Photon	0
Weak	Weak gauge bosons	86, 97
Gravity	Graviton	0

Table 1.2 The four forces of nature, together with their associated force particles and their masses in multiples of the proton mass. (The weak force particles come in varieties with the two possible masses listed. Theoretical studies show that the graviton should be massless.)

the table in the endnotes to this chapter.[1]) But as with particle masses, beyond the fact that experimental physicists have carefully measured these properties, no one has any explanation of *why* our universe is composed of these particular particles, with these particular masses and force charges.

Notwithstanding their common features, an examination of the fundamental forces themselves serves only to compound the questions. Why, for instance, are there four fundamental forces? Why not five or three or perhaps only one? Why do the forces have such different properties? Why are the strong and weak forces confined to operate on microscopic scales while gravity and the electromagnetic force have an unlimited range of influence? And why is there such an enormous spread in the intrinsic strength of these forces?

To appreciate this last question, imagine holding an electron in your left hand and another electron in your right hand and bringing these two identical electrically charged particles close together. Their mutual gravitational attraction will favor their getting closer while their electromagnetic repulsion will try to drive them apart. Which is stronger? There is no contest: The electromagnetic repulsion is about a million billion billion billion billion (10^{42}) times stronger! If your right bicep represents the strength of the gravitational force, then your left bicep would have to extend beyond the edge of the known universe to represent the strength of the electromagnetic force. The only reason the electromagnetic force does not completely overwhelm gravity in the world around us is that most things are composed of an equal amount of positive and negative electric charges whose forces cancel each other out. On the other hand, since gravity is always attractive, there are no analogous cancellations—more stuff means greater gravitational force. But fundamentally speaking, gravity is an extremely feeble force. (This fact accounts for the difficulty in experimentally confirming the existence of the graviton. Searching for the smallest bundle of the feeblest force is quite a challenge.) Experiments also have shown that the strong force is about one hundred times as strong as the electromagnetic force and about one hundred thousand times as strong as the weak force. But where is the rationale—the raison d'être—for our universe having these features?

This is not a question borne of idle philosophizing about why certain details happen to be one way instead of another; the universe would be a vastly different place if the properties of the matter and force particles were even moderately changed. For example, the existence of the stable nuclei forming the hundred or so elements of the periodic table hinges delicately on the ratio between the strengths of the strong and electromagnetic forces. The protons crammed together in atomic nuclei all repel one another electromagnetically; the strong force acting among their constituent quarks, thankfully, overcomes this repulsion and tethers the protons tightly together. But a rather small change in the relative strengths of these two forces would easily disrupt the balance between them, and would cause most atomic nuclei to disintegrate. Furthermore, were the mass of the electron a few times greater than it is, electrons and protons would tend to combine to form neutrons, gobbling up the nuclei of hydrogen (the simplest element in the cosmos, with a nucleus containing a single proton) and, again, disrupting the production of more complex elements. Stars rely upon fusion between stable nuclei and would not form with such alterations to fundamental physics. The strength of the gravitational force also plays a formative role. The crushing density of matter in a star's central core powers its nuclear furnace and underlies the resulting blaze of starlight. If the strength of the gravitational force were increased, the stellar clump would bind more strongly, causing a significant increase in the rate of nuclear reactions. But just as a brilliant flare exhausts its fuel much faster than a slow-burning candle, an increase in the nuclear reaction rate would cause stars like the sun to burn out far more quickly, having a devastating effect on the formation of life as we know it. On the other hand, were the strength of the gravitational force significantly decreased, matter would not clump together at all, thereby preventing the formation of stars and galaxies.

We could go on, but the idea is clear: the universe is the way it is because the matter and the force particles have the properties they do. But is there a scientific explanation for *why* they have these properties?

String Theory: The Basic Idea

String theory offers a powerful conceptual paradigm in which, for the first time, a framework for answering these questions has emerged. Let's first get the basic idea.

The particles in Table 1.1 are the "letters" of all matter. Just like their linguistic counterparts, they appear to have no further internal substructure. String theory proclaims otherwise. According to string theory, if we could examine these particles with even greater precision—a precision many orders of magnitude beyond our present technological capacity—we would find that each is not pointlike, but instead consists of a tiny one-dimensional *loop*. Like an infinitely thin rubber band, each particle contains a vibrating, oscillating, dancing filament that physicists, lacking Gell-Mann's literary flair, have named a *string*. In Figure 1.1 we illustrate this essential idea of string theory by starting with an ordinary piece of matter, an apple, and repeatedly magnifying its structure to reveal its ingredients on ever smaller scales. String theory adds the new microscopic layer of a vibrating loop to the previously known progression from atoms through protons, neutrons, electrons and quarks.[2]

Although it is by no means obvious, we will see in Chapter 6 that this simple replacement of point-particle material constituents with strings resolves the incompatibility between quantum mechanics and general relativity. String theory thereby unravels the central Gordian knot of contemporary theoretical physics. This is a tremendous achievement, but it is only part of the reason string theory has generated such excitement.

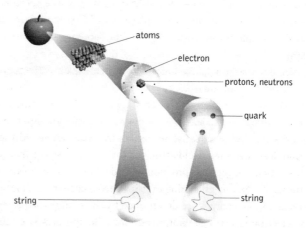

Figure 1.1 Matter is composed of atoms, which in turn are made from quarks and electrons. According to string theory, all such particles are actually tiny loops of vibrating string.

String Theory as the Unified Theory of Everything

In Einstein's day, the strong and the weak forces had not yet been discovered, but he found the existence of even two distinct forces—gravity and electromagnetism—deeply troubling. Einstein did not accept that nature is founded on such an extravagant design. This launched his thirty-year voyage in search of the so-called *unified field theory* that he hoped would show that these two forces are really manifestations of one grand underlying principle. This quixotic quest isolated Einstein from the mainstream of physics, which, understandably, was far more excited about delving into the newly emerging framework of quantum mechanics. He wrote to a friend in the early 1940s, "I have become a lonely old chap who is mainly known because he doesn't wear socks and who is exhibited as a curiosity on special occasions."[3]

Einstein was simply ahead of his time. More than half a century later, his dream of a unified theory has become the Holy Grail of modern physics. And a sizeable part of the physics and mathematics community is becoming increasingly convinced that string theory may provide the answer. From one principle—that everything at its most microscopic level consists of combinations of vibrating strands—string theory provides a single explanatory framework capable of encompassing all forces and all matter.

String theory proclaims, for instance, that the observed particle properties, the data summarized in Tables 1.1 and 1.2, are a reflection of the various ways in which a string can vibrate. Just as the strings on a violin or on a piano have resonant frequencies at which they prefer to vibrate—patterns that our ears sense as various musical notes and their higher harmonics—the same holds true for the loops of string theory. But we will see that, rather than producing musical notes, each of the preferred patterns of vibration of a string in string theory appears as a particle whose mass and force charges are determined by the string's oscillatory pattern. The electron is a string vibrating one way, the up-quark is a string vibrating another way, and so on. Far from being a collection of chaotic experimental facts, particle properties in string theory are the manifestation of one and the same physical feature: the resonant patterns of vibration—the

15

music, so to speak—of fundamental loops of string. The same idea applies to the forces of nature as well. We will see that force particles are also associated with particular patterns of string vibration and hence everything, all matter and all forces, is unified under the same rubric of microscopic string oscillations—the "notes" that strings can play.

For the first time in the history of physics we therefore have a framework with the capacity to explain every fundamental feature upon which the universe is constructed. For this reason string theory is sometimes described as possibly being the "theory of everything" (T.O.E.) or the "ultimate" or "final" theory. These grandiose descriptive terms are meant to signify the deepest possible theory of physics—a theory that underlies all others, one that does not require or even allow for a deeper explanatory base. In practice, many string theorists take a more down-to-earth approach and think of a T.O.E. in the more limited sense of a theory that can explain the properties of the fundamental particles and the properties of the forces by which they interact and influence one another. A staunch reductionist would claim that this is no limitation at all, and that in principle absolutely everything, from the big bang to daydreams, can be described in terms of underlying microscopic physical processes involving the fundamental constituents of matter. If you understand everything about the ingredients, the reductionist argues, you understand everything.

The reductionist philosophy easily ignites heated debate. Many find it fatuous and downright repugnant to claim that the wonders of life and the universe are mere reflections of microscopic particles engaged in a pointless dance fully choreographed by the laws of physics. Is it really the case that feelings of joy, sorrow, or boredom are nothing but chemical reactions in the brain—reactions between molecules and atoms that, even more microscopically, are reactions between some of the particles in Table 1.1, which are really just vibrating strings? In response to this line of criticism, Nobel laureate Steven Weinberg cautions in *Dreams of a Final Theory*,

At the other end of the spectrum are the opponents of reductionism who are appalled by what they feel to be the bleakness of modern science. To whatever extent they and their world can be reduced to a matter of particles or fields and their interactions, they feel diminished by that knowledge. . . . I would not try to answer these critics with a pep talk about the beauties of modern science. The reductionist worldview *is* chilling and impersonal. It has to be accepted as it is, not because we like it, but because that is the way the world works.[4]

Some agree with this stark view, some don't.

Others have tried to argue that developments such as chaos theory tell us that new kinds of laws come into play when the level of complexity of a system increases. Understanding the behavior of an electron or a quark is one thing; using this knowledge to understand the behavior of a tornado is quite another. On this point, most agree. But opinions diverge on whether the diverse and often unexpected phenomena that can occur in systems more complex than individual particles truly represent new physical principles at work, or whether the principles involved are derivative, relying, albeit in a terribly complicated way, on the physical principles governing the enormously large number of elementary constituents. My own feeling is that they do not represent new and independent laws of physics. Although it would be hard to explain the properties of a tornado in terms of the physics of electrons and quarks, I see this as a matter of calculational impasse, not an indicator of the need for new physical laws. But again, there are some who disagree with this view.

What is largely beyond question, and is of primary importance to the journey described in this book, is that even if one accepts the debatable reasoning of the staunch reductionist, principle is one thing and practice quite another. Almost everyone agrees that finding the T.O.E. would in no way mean that psychology, biology, geology, chemistry, or even physics had been solved or in some sense subsumed. The universe is such a wonderfully rich and complex place that the discovery of the final theory, in the sense we are describing here, would not spell the end of science. Quite the contrary: The discovery of the T.O.E.—the ultimate explanation of the universe at its most microscopic level, a theory that does not rely on any deeper explanation—would provide the firmest foundation on which to *build* our understanding of the world. Its discovery would mark a beginning, not an end. The ultimate theory would provide an unshakable pillar of coherence forever assuring us that the universe is a comprehensible place.

The State of String Theory

The central concern of this book is to explain the workings of the universe according to string theory, with a primary emphasis on the implications that these results have for our understanding of space and time. Unlike many other exposés of scientific developments, the one given here does not address itself to a theory that has been completely worked out, confirmed by vigorous experimental tests, and fully accepted by the scientific community. The reason for this, as we will discuss in subsequent chapters, is that string theory is such a deep and sophisticated theoretical structure that even with the impressive progress that has been made over the last two decades, we still have far to go before we can claim to have achieved full mastery.

And so string theory should be viewed as a work in progress whose partial completion has already revealed astonishing insights into the nature of space, time, and matter. The harmonious union of general relativity and quantum mechanics is a major success. Furthermore, unlike any previous theory, string theory has the capacity to answer primordial questions having to do with nature's most fundamental constituents and forces. Of equal importance, although somewhat harder to convey, is the remarkable elegance of both the answers and the framework for answers that string theory proposes. For instance, in string theory many aspects of nature that might appear to be arbitrary technical details—such as the number of distinct fundamental particle ingredients and their respective properties—are found to arise from essential and tangible aspects of the geometry of the universe. If string theory is right, the microscopic fabric of our universe is a richly intertwined multidimensional labyrinth within which the strings of the universe endlessly twist and vibrate, rhythmically beating out the laws of the cosmos. Far from being accidental details, the properties of nature's basic building blocks are deeply entwined with the fabric of space and time.

In the final analysis, though, nothing is a substitute for definitive, testable predictions that can determine whether string theory has truly lifted the veil of mystery hiding the deepest truths of our universe. It may be some time before our level of comprehension has reached sufficient depth to achieve this aim, although, as we will discuss in Chapter 9, experimental tests could provide strong circumstantial support for string theory within the next ten years or so. Moreover, in Chapter 13 we will see that string theory has recently solved a central puzzle concerning black holes, associated with the so-called Bekenstein-Hawking entropy, that has stubbornly resisted resolution by more conventional means for more than twenty-five years. This success has convinced many that string theory is in the process of giving us our deepest understanding of how the universe works.

Edward Witten, one of the pioneers and leading experts in string theory, summarizes the situation by saying that "string theory is a part of twenty-first-century physics that fell by chance into the twentieth century," an assessment first articulated by the celebrated Italian physicist Daniele Amati.[5] In a sense, then, it is as if our forebears in the late nineteenth century had been presented with a modern-day supercomputer, without the operating instructions. Through inventive trial and error, hints of the supercomputer's power would have become evident, but it would have taken vigorous and prolonged effort to gain true mastery. The hints of the computer's potential, like our glimpses of string theory's explanatory power, would have provided extremely strong motivation for obtaining complete facility. A similar motivation today energizes a generation of theoretical physicists to pursue a full and precise analytic understanding of string theory.

Witten's remark and those of other experts in the field indicate that it could be decades or even centuries before string theory is fully developed and understood. This may well be true. In fact, the mathematics of string theory is so complicated that, to date, no one even knows the exact equations of the theory. Instead, physicists know only approximations to these equations, and even the approximate equations are so complicated that they as yet have been only partially solved. Nevertheless, an inspiring set of breakthroughs in the latter half of the 1990s—breakthroughs that have answered theoretical questions of hitherto unimaginable difficulty—may well indicate that complete quantitative understanding of string theory is much closer than initially thought. Physicists worldwide are developing powerful new techniques to transcend the numerous approximate methods so far used, collectively piecing together disparate elements of the string theory puzzle at an exhilarating rate.

Surprisingly, these developments are providing new vantage points for reinterpreting some of the basic aspects of the theory that have been in place for some time. For instance, a natural question that may have occurred to you in looking at Figure 1.1 is, Why strings? Why not little frisbee disks? Or microscopic bloblike nuggets? Or a combination of all of these possibilities? As we shall see in Chapter 12, the most recent insights show that these other kinds of ingredients *do* have an important role in string theory, and have revealed that string theory is actually part of an even grander synthesis currently (and mysteriously) named M-theory. These latest developments will be the subject of the final chapters of this book.

Progress in science proceeds in fits and starts. Some periods are filled with great breakthroughs; at other times researchers experience dry spells. Scientists put forward results, both theoretical and experimental. The results are debated by the community, sometimes they are discarded, sometimes they are modified, and sometimes they provide inspirational jumping-off points for new and more accurate ways of understanding the physical universe. In other words, science proceeds along a zig-zag path toward what we hope will be ultimate truth, a path that began with humanity's earliest attempts to fathom the cosmos and whose end we cannot predict. Whether string theory is an incidental rest stop along this path, a landmark turning point, or in fact the final destination we do not know. But the last two decades of research by hundreds of dedicated physicists and mathematicians from numerous countries have given us well-founded hope that we are on the right and possibly final track.

It is a telling testament of the rich and far-reaching nature of string theory that even our present level of understanding has allowed us to gain striking new insights into the workings of the universe. A central thread in what follows will be those developments that carry forward the revolution in our understanding of space and time initiated by Einstein's special and general theories of relativity. We will see that if string theory is correct, the fabric of our universe has properties that would likely have dazzled even Einstein.

Part II

The Dilemma of Space, Time, and the Quanta

Chapter 2

Space, Time, and the Eye of the Beholder

In June 1905, twenty-six-year-old Albert Einstein submitted a technical article to the German *Annals of Physics* in which he came to grips with a paradox about light that had first troubled him as a teenager, some ten years earlier. Upon turning the final page of Einstein's manuscript, the editor of the journal, Max Planck, realized that the accepted scientific order had been overthrown. Without hoopla or fanfare, a patent clerk from Bern, Switzerland, had completely overturned the traditional notions of space and time and replaced them with a new conception whose properties fly in the face of everything we are familiar with from common experience.

The paradox that had troubled Einstein for a decade was this. In the mid-1800s, after a close study of the experimental work of the English physicist Michael Faraday, the Scottish physicist James Clerk Maxwell succeeded in uniting electricity and magnetism in the framework of the *electromagnetic field*. If you've ever been on a mountaintop just before a severe thunderstorm or stood close to a Van de Graaf generator, you have a visceral sense of what an electromagnetic field is, because you've felt it. In case you haven't, it is somewhat like a tide of electric and magnetic lines of force that permeate a region of space through which they pass. When you sprinkle iron filings near a magnet, for example, the orderly pattern they form traces out some of the invisible lines of magnetic force. When you take off a wool sweater on an especially dry day and hear a crackling sound and perhaps feel a momentary shock or two, you are witnessing evidence of electric lines of force generated by electric charges swept up by the fibers in your sweater. Beyond uniting these and all other electric and magnetic phenomena in one mathematical framework, Maxwell's theory showed—quite unexpectedly—that electromagnetic disturbances travel at a fixed and never-changing speed, a speed that turns out to equal that of light. From this, Maxwell realized that visible light itself is nothing but a particular kind of electromagnetic wave, one that is now understood to interact with chemicals in the retina, giving rise to the sensation of sight. Moreover (and this is crucial), Maxwell's theory also showed that all electromagnetic waves—visible light among them—are the epitome of the peripatetic traveler. They never stop. They never slow down. Light *always* travels at light speed.

All is well and good until we ask, as the sixteen-year-old Einstein did, What happens if we chase after a beam of light, at light speed? Intuitive reasoning, rooted in Newton's laws of motion, tells us that we will catch up with the light waves and so they will appear stationary; light will stand still. But according to Maxwell's theory, and all reliable observations, there is simply no such thing as stationary light: no one has ever held a stationary clump of light in the palm of his or her hand. Hence the problem. Luckily, Einstein was unaware that many of the world's leading physicists were struggling with this question (and were heading down many a spurious path) and pondered the paradox of Maxwell and Newton largely in the pristine privacy of his own thoughts.

In this chapter we discuss how Einstein resolved the conflict through his special theory of relativity, and in so doing forever changed our conceptions of space and time. It is perhaps surprising that the essential concern of special relativity is to understand precisely how the world appears to individuals, often called "observers," who are moving relative to one another. At first, this might seem to be an intellectual exercise of minimal importance. Quite the contrary: In the hands of Einstein, with his imaginings of observers chasing after light beams, there are profound implications to grasping fully how even the most mundane situations appear to individuals in relative motion.

Intuition and Its Flaws

Common experience highlights certain ways in which observations by such individuals differ. Trees alongside a highway, for example, appear to be moving from the viewpoint of a driver but appear stationary to a hitchhiker sitting on a guardrail. Similarly, the dashboard of the automobile does not appear to be moving from the viewpoint of the driver (one hopes!), but like the rest of the car, it does appear to be moving from the viewpoint of the hitchhiker. These are such basic and intuitive properties of how the world works that we hardly take note of them.

Special relativity, however, proclaims that the differences in observations between two such individuals are more subtle and profound. It makes the strange claim that observers in relative motion will have different perceptions of distance and of time. This means, as we shall see, that identical wristwatches worn by two individuals in relative motion will tick at *different rates* and hence will not agree on the amount of time that elapses between chosen events. Special relativity demonstrates that this statement does not slander the accuracy of the wristwatches involved; rather, it is a true statement about time itself.

Similarly, observers in relative motion carrying identical tape measures will not agree on the lengths of distances measured. Again, this is not due to inaccuracies in the measuring devices or to errors in how they are used. The most accurate measuring devices in the world confirm that space and time—as measured by distances and durations—are not experienced identically by everyone. In the precise way delineated by Einstein, special relativity resolves the conflict between our intuition about motion and the properties of light, but there is a price: individuals who are moving with respect to each other will not agree on their observations of either space or time.

It has been almost a century since Einstein informed the world of his dramatic discovery, yet most of us still see space and time in absolute terms. Special relativity is not in our bones—we do not feel it. Its implications are not a central part of our intuition. The reason for this is quite simple: The effects of special relativity depend upon how fast one moves, and at the speeds of cars, planes, or even space shuttles, these effects are minuscule. Differences in perceptions of space and of time between individuals planted on the earth and those traveling in cars or planes *do* occur, but they are so small that they go unnoticed. However, were one to take a trip in a futuristic space vehicle traveling at a substantial fraction of light speed, the effects of relativity would become plainly obvious. This, of course, is still in the realm of science fiction. Nevertheless, as we shall discuss in later sections, clever experiments allow clear and precise observation of the relative properties of space and time predicted by Einstein's theory.

To get a sense of the scales involved, imagine that the year is 1970 and big, fast cars are in. Slim, having just spent all his savings on a new Trans Am, goes with his brother Jim to the local drag strip to give the car the kind of test-drive forbidden by the dealer. After revving up the car, Slim streaks down the mile-long strip at 120 miles per hour while Jim stands on the sideline and times him. Wanting an independent confirmation, Slim also uses a stopwatch to determine how long it takes his new car to traverse the track. Prior to Einstein's work, no one would have questioned that if both Slim and Jim have properly functioning stopwatches, each will measure the identical elapsed time. But according to special relativity, while Jim will measure an elapsed time of 30 seconds, Slim's stopwatch will record an elapsed time of 29.99999999999952 seconds—*a tiny bit less*. Of course, this difference is so small that it could be detected only through a measurement whose accuracy is well beyond the capacity of hand-held stopwatches run by the press of a finger, Olympic-quality timing systems, or even the most precisely engineered atomic clocks. It is no wonder that our everyday experiences do not reveal the fact that the passage of time depends upon our state of motion.

There will be a similar disagreement on measurements of length. For example, on another test run Jim uses a clever trick to measure the length of Slim's new car: he starts his stopwatch just as the front of the car reaches him and he stops it just as the back of the car passes. Since Jim knows that Slim is speeding along at 120 miles per hour, he is able to figure out the length of the car by multiplying this speed by the elapsed time on his stopwatch. Again, prior to Einstein, no one would have questioned that the length Jim measures in this indirect way would agree *exactly* with the length Slim carefully measured when the car sat motionless on the showroom floor. Special relativity proclaims, on the contrary, that if Slim and Jim carry out precise measurements in this manner and Slim finds the car to be, say, exactly 16 feet long, then Jim's measurement will find the car to be 15.99999999999974 feet long—*a tiny bit less*. As with the measurement of time, this is such a minuscule difference that ordinary instruments are just not accurate enough to detect it.

Although the differences are extremely small, they show a fatal flaw in the commonly held conception of universal and immutable space and time. As the relative velocity of individuals such as Slim and Jim gets larger, this flaw becomes increasingly apparent. To achieve noticeable differences, the speeds involved must be a sizeable fraction of the maximum possible speed—that of light—which Maxwell's theory and experimental measurements show to be about 186,000 miles per second, or about 670 million miles per hour. This is fast enough to circle the earth more than seven times in a second. If Slim, for example, were to travel not at 120 miles per hour but at 580 million miles per hour (about 87 percent of light speed), the mathematics of special relativity predicts that Jim would measure the length of the car to be about eight feet, which is substantially different from Slim's measurement (as well as the specifications in the owner's manual). Similarly, the time to traverse the drag strip according to Jim will be about *twice* as long as the time measured by Slim.

Since such enormous speeds are far beyond anything currently attainable, the effects of "time dilation" and "Lorentz contraction," as these phenomena are technically called, are extremely small in day-to-day life. If we happened to live in a world in which things typically traveled at speeds close to that of light, these properties of space and time would be so completely intuitive—since we would experience them constantly—that they would deserve no more discussion than the apparent motion of trees on the side of the road mentioned at the outset of this chapter. But since we don't live in such a world, these features are unfamiliar. As we shall see, understanding and accepting them requires that we subject our worldview to a thorough makeover.

The Principle of Relativity

There are two simple yet deeply rooted structures that form the foundation of special relativity. As mentioned, one concerns properties of light; we shall discuss this more fully in the next section. The other is more abstract. It is concerned not with any specific physical law but rather with *all* physical laws, and is known as the *principle of relativity*. The principle of relativity rests on a simple fact: Whenever we discuss speed or velocity (an object's speed and its direction of motion), we must specify precisely who or what is doing the measuring. Understanding the meaning and importance of this statement is easily accomplished by contemplating the following situation.

Imagine that George, who is wearing a spacesuit with a small, red flashing light, is floating in the absolute darkness of completely empty space, far away from any planets, stars, or galaxies. From George's perspective, he is completely stationary, engulfed in the uniform, still blackness of the cosmos. Off in the distance, George catches sight of a tiny, green flashing light that appears to be coming closer and closer. Finally, it gets close enough for George to see that the light is attached to the spacesuit of another space-dweller, Gracie, who is slowly floating by. She waves as she passes, as does George, and she recedes into the distance. This story can be told with equal validity from Gracie's perspective. It begins in the same manner with Gracie completely alone in the immense still darkness of outer space. Off in the distance, Gracie sees a red flashing light, which appears to be coming closer and closer. Finally, it gets close enough for Gracie to see that it is attached to the spacesuit of another being, George, who is slowly floating by. He waves as he passes, as does Gracie, and he recedes into the distance.

The two stories describe one and the same situation from two distinct but equally valid points of view. Each observer feels stationary and perceives the other as moving. Each perspective is understandable and justifiable. As there is symmetry between the two space-dwellers, there is, on quite fundamental grounds, no way of saying one perspective is "right" and the other "wrong." Each perspective has an equal claim on truth.

This example captures the meaning of the principle of relativity: The concept of motion is relative. We can speak about the motion of an object, but only relative to or by comparison with another. There is thus no meaning to the statement "George is traveling at 10 miles per hour," as we have not specified any other object for comparison. There *is* meaning to the statement "George is traveling at 10 miles per hour past Gracie," as we have now specified Gracie as the benchmark. As our example shows, this last statement is completely equivalent to "Gracie is traveling at 10 miles per hour past George (in the opposite direction)." In other words, there is no "absolute" notion of motion. Motion is relative.

A key element of this story is that neither George nor Gracie is being pushed, pulled, or in any other way acted upon by a force or influence that could disturb their serene state of force-free, constant-velocity motion. Thus, a more precise statement is that *force-free* motion has meaning only by comparison with other objects. This is an important clarification, because if forces are involved, they cause changes in the velocity of the observers—changes to their speed and/or their direction of motion—and these changes can be felt. For instance, if George were wearing a jet-pack firing away from his back, he would definitely feel that he was moving. This feeling is intrinsic. If the jet-pack is firing away, George *knows* he is moving, even if his eyes are closed and therefore can make no comparisons with other objects. Even without such comparisons, he would no longer claim that he was stationary while "the rest of the world was moving by him." Constant-velocity motion is relative; not so for non-constant-velocity motion, or, equivalently, *accelerated motion*. (We will re-examine this statement in the next chapter when we take up accelerated motion and discuss Einstein's general theory of relativity.)

Setting these stories in the darkness of empty space aids understanding by removing such familiar things as streets and buildings, which we typically, although unjustifiably, accord the special status of "stationary." Nonetheless, the same principle applies to terrestrial settings, and in fact is commonly experienced.[1] For example, imagine that after you have fallen asleep on a train, you awake just as your train is passing another on adjacent parallel tracks. With your view through the window completely blocked by the other train, thereby preventing you from seeing any other objects, you may temporarily be uncertain as to whether your train is moving, the other train is moving, or both. Of course, if your train shakes or jostles, or if the train changes direction by rounding a bend, you can feel that you are moving. But if the ride is perfectly smooth—if the train's velocity remains constant—you will observe relative motion between the trains without being able to tell for certain which is moving.

Let's take this one step further. Imagine you are on such a train and that you pull down the shades so that the windows are fully covered. Without the ability to see anything outside your own compartment, and assuming that the train moves at absolutely constant velocity, there will be no way for you to determine your state of motion. The compartment around you will look *precisely* the same regardless of whether the train is sitting still on the tracks or moving at high speed. Einstein formalized this idea, one that actually goes back to insights of Galileo, by proclaiming that it is impossible for you or any fellow traveler to perform an experiment within the closed compartment that will determine whether or not the train is moving. This again captures the principle of relativity: since all force-free motion is relative, it has meaning only by comparison with other objects or individuals also undergoing force-free motion. There is no way for you to determine anything about your state of motion without making some direct or indirect comparison with "outside" objects. There simply is no notion of "absolute" constant-velocity motion; only comparisons have any physical meaning.

In fact, Einstein realized that the principle of relativity makes an even grander claim: the laws of physics—whatever they may be—must be absolutely identical for all observers undergoing constant-velocity motion. If George and Gracie are not just floating solo in space, but, rather, are each conducting the same set of experiments in their respective floating space-stations, the results they find will be identical. Once again, each is perfectly justified in believing that his or her station is at rest, even though the two stations are in relative motion. If all of their equipment is identical, there is nothing distinguishing the two experimental setups—they are completely symmetric. The laws of physics that each deduces from the experiments will likewise be identical. Neither they nor their experiments can feel—that is, depend upon in any way—constant-velocity travel. It is this simple concept that establishes complete symmetry between such observers; it is this concept that is embodied in the principle of relativity. We shall shortly make use of this principle to profound effect.

The Speed of Light

The second key ingredient in special relativity has to do with light and properties of its motion. Contrary to our claim that there is no meaning to the statement "George is traveling at 10 miles per hour" without a specified benchmark for comparison, almost a century of effort by a series of dedicated experimental physicists has shown that any and all observers will agree that light travels at 670 million miles per hour *regardless of benchmarks for comparison.*

This fact has required a revolution in our view of the universe. Let's first gain an understanding of its meaning by contrasting it with similar statements applied to more common objects. Imagine it's a nice, sunny day and you go outside to play a game of catch with a friend. For a while, you both leisurely throw the ball back and forth with a speed of, say, 20 feet per second, when suddenly an unexpected electrical storm stirs overhead, sending you both running for cover. After it passes, you rejoin to resume your game of catch but you notice that something has changed. Your friend's hair has become wild and spiky, and her eyes have grown severe and crazed. When you look at her hand, you are stunned to see that she is no longer planning to play catch with a baseball, but instead is about to toss you a hand grenade. Understandably, your enthusiasm for playing catch diminishes substantially; you turn to run. When your companion throws the grenade, it will still fly toward you, but because you are running, the speed with which it approaches you will be less than 20 feet per second. In fact, common experience tells us that if you can run at, say, 12 feet per second then the hand-grenade will approach you at (20 − 12 =) 8 feet per second. As another example, if you are in the mountains and an avalanche of snow is rumbling toward you, your inclination is to turn and run because this will cause the speed with which the snow approaches you to decrease—and this, generally, is a good thing. Again, a stationary individual perceives the speed of the approaching snow to be greater than that perceived by someone in retreat.

Now, let's compare these basic observations about baseballs, grenades, and avalanches to those about light. To make the comparisons tighter, think about a light beam as composed of tiny "packets" or "bundles" known as photons (a feature of light we will discuss more fully in Chapter 4). When we turn on a flashlight or a laser beam we are, in effect, shooting a stream of photons in whatever direction we point the device. As we did for grenades and avalanches, let's consider how the motion of a photon appears to someone who is moving. Imagine that your crazed friend has swapped her grenade for a powerful laser. If she fires the laser toward you—and if you had the appropriate measuring equipment—you would find that the speed of approach of the photons in the beam is 670 million miles per hour. But what if you run away, as you did when faced with the prospect of playing catch with a hand grenade? What speed will you now measure for the approaching photons? To make things more compelling, imagine that you can hitch a ride on the starship *Enterprise* and zip away from your friend at, say, 100 million miles per hour. Following the reasoning based on the traditional Newtonian worldview, since you are now speeding away, you would expect to measure a *slower* speed for the oncoming photons. Specifically, you would expect to find them approaching you at (670 million miles per hour − 100 million miles per hour =) 570 million miles per hour.

Mounting evidence from a variety of experiments dating back as far as the 1880s, as well as careful analysis and interpretation of Maxwell's electromagnetic theory of light, slowly convinced the scientific community that, in fact, this is *not* what you will see. *Even though you are retreating, you will still measure the speed of the approaching photons as 670 million miles per hour, not a bit less.* Although at first it sounds completely ridiculous, unlike what happens if one runs from an oncoming baseball, grenade, or avalanche, the speed of approaching photons is always 670 million miles per hour. The same is true if you run toward oncoming photons or chase after them—their speed of approach or recession is completely unchanged; they still appear to travel at 670 million miles per hour. Regardless of relative motion between the source of photons and the observer, the speed of light is always the same.[2]

Technological limitations are such that the "experiments" with light, as described, cannot actually be carried out. However, comparable experiments can. For instance, in 1913 the Dutch physicist Willem de Sitter suggested that fast-moving binary stars (two stars that orbit one another) could be used to measure the effect of a moving source on the speed of light. Various experiments of this sort over the past eight decades have verified that the speed of light received from a moving star is the *same* as that from a stationary star—670 million miles per hour—to within the impressive accuracy of ever more refined measuring devices. Moreover, a wealth of other detailed experiments has been carried out during the past century—experiments that directly measure the speed of light in various circumstances, as well as test many of the implications arising from this characteristic of light, as discussed shortly—and all have confirmed the constancy of the speed of light.

If you find this property of light hard to swallow, you are not alone. At the turn of the century physicists went to great length to refute it. They couldn't. Einstein, to the contrary, embraced the constancy of the speed of light, for here was the answer to the paradox that had troubled him since he was a teenager: No matter how hard you chase after a light beam, it still retreats from you at light speed. You can't make the apparent speed with which light departs one iota less than 670 million miles per hour, let alone slow it down to the point of appearing stationary. Case closed. But this triumph over conflict was no small victory. Einstein realized that the constancy of light's speed spelled the downfall of Newtonian physics.

Truth and Consequences

Speed is a measure of how far an object can travel in a given duration of time. If we are in a car going 65 miles per hour, this means of course that we will travel 65 miles if we persist in this state of motion for an hour. Phrased in this manner, speed is a rather mundane concept, and you may wonder about the fuss we have made regarding the speed of baseballs, snowballs, and photons. However, let's note that *distance* is a notion about space—in particular it is a measure of how much space there is between two points. Also note that *duration* is a notion about time—how much time elapses between two events. Speed, therefore, is intimately connected with our notions of space and time. When we phrase it this way, we see that any experimental fact that defies our common conception about speed, such as the constancy of the speed of light, has the potential to defy our common conceptions of space and time themselves. It is for this reason that the strange fact about the speed of light deserves de-

...ed scrutiny—scrutiny given to it by Einstein, leading him to remarkable conclusions.

The Effect on Time: Part I

With minimal effort, we can make use of the constancy of the speed of light to show that the familiar everyday conception of time is plain wrong. Imagine that the leaders of two warring nations, sitting at opposite ends of a long negotiating table, have just concluded an agreement for a cease-fire, but neither wants to sign the accord before the other. The secretary-general of the United Nations comes up with a brilliant resolution. A light bulb, initially turned off, will be placed midway between the two presidents. When it is turned on, the light it emits will reach each of the presidents simultaneously, since they are equidistant from the bulb. Each president agrees to sign a copy of the accord when he or she sees the light. The plan is carried out and the agreement is signed to the satisfaction of both sides.

Flushed with success, the secretary-general makes use of the same approach with two other embattled nations that have also reached a peace agreement. The only difference is that the presidents involved in this negotiation are sitting at opposite ends of a table inside a train traveling along at constant velocity. Fittingly, the president of Forwardland is facing in the direction of the train's motion while the president of Backwardland is facing in the opposite direction. Familiar with the fact that the laws of physics take precisely the same form regardless of one's state of motion so long as this motion is unchanging, the secretary-general takes no heed of this difference, and carries out the light bulb–initiated signing ceremony as before. Both presidents sign the agreement, and along with their entourage of advisers, celebrate the end of hostilities.

Just then, word arrives that fighting has broken out between people from each country who had been watching the signing ceremony from the platform outside the moving train. All those on the negotiation train are dismayed to hear that the reason for the renewed hostilities is the claim by people from Forwardland that they have been duped, as their president signed the agreement *before* the president of Backwardland. As everyone on the train—from both sides—agrees that the accord was signed simultaneously, how can it be that the outside observers watching the ceremony think otherwise?

Let's consider in more detail the perspective of an observer on the platform. Initially the bulb on the train is dark, and then at a particular moment it illuminates, sending beams of light speeding toward both presidents. From the perspective of a person on the platform, the president of Forwardland is heading toward the emitted light while the president of Backwardland is retreating. This means, to the platform observers, that the light beam does not have to travel as far to reach the president of Forwardland, who moves toward the approaching light, as it does to reach the president of Backwardland, who moves away from it. This is not a statement about the *speed* of the light as it travels toward the two presidents—we have already noted that regardless of the state of motion of the source or the observer, the speed of light is always the same. Instead, we are describing only how *far*, from the vantage point of the platform observers, the initial flash of light must travel to reach each of the presidents. Since this distance is less for the president of Forwardland than it is for the president of Backwardland, and since the speed of light toward each is the same, the light will reach the president of Forwardland first. This is why the citizens of Forwardland claim to have been duped.

When CNN broadcasts the eyewitness account, the secretary-general, the two presidents, and all of their advisers can't believe their ears. They all agree that the light bulb was secured firmly, exactly midway between the two presidents and that therefore, without further ado, the light it emitted traveled the *same* distance to reach each of them. Since the speed of the emitted light to the left and to the right is the same, they believe, and in fact observed, that the light clearly reached each president simultaneously.

Who is right, those on or off the train? The observations of each group and their supporting explanations are impeccable. The answer is that *both* are right. Like our two space inhabitants George and Gracie, each perspective has an equal claim on truth. The only subtlety here is that the respective truths seem to be contradictory. An important political issue is at stake: Did the presidents sign the agreement simultaneously? The observations and reasoning above ineluctably lead us to the conclusion that *according to those on the train they did* while *according to those on the platform they did not.* In other words, things that are simultaneous from the viewpoint of some observers will not be simultaneous from the viewpoint of others, if the two groups are in relative motion.

This is a startling conclusion. It is one of the deepest insights into the nature of reality ever discovered. Nevertheless, if long after you set down this book you remember nothing of this chapter except for the ill-fated attempt at détente, you will have retained the essence of Einstein's discovery. Without highbrow mathematics or a convoluted chain of logic, this completely unexpected feature of time follows directly from the constancy of the speed of light, as the scenario illustrates. Notice that if the speed of light were not constant but behaved according to our intuition based on slow-moving baseballs and snowballs, the platform observers would agree with those on the train. A platform observer would still claim that the photons have to travel farther to reach the president of Backwardland than they do to reach the president of Forwardland. However, usual intuition implies that the light approaching the president of Backwardland would be moving more quickly, having received a "kick" from the forward-moving train. Similarly, these observers would see that the light approaching the president of Forwardland would be moving more slowly, being "dragged" back by the train's motion. When these (erroneous) effects were considered, the observers on the platform would see that that the light beams reached each president simultaneously. However, in the real world light does not speed up or slow down, it cannot be kicked to a higher speed or dragged to a slower one. Platform observers will therefore justifiably claim that the light reached the president of Forwardland first.

The constancy of the speed of light requires that we give up the age-old notion that simultaneity is a universal concept that everyone, regardless of their state of motion, agrees upon. The universal clock previously envisioned to dispassionately tick off identical seconds here on earth and on Mars and on Jupiter and in the Andromeda galaxy and in each and every nook and cranny of the cosmos does not exist. On the contrary, observers in relative motion will not agree on which events occur at the same time. Once again, the reason that this conclusion—a bona fide characteristic of the world we inhabit—is so unfamiliar is that the effects are extremely small when the speeds involved are those commonly encountered in everyday experience. If the negotiating table were 100 feet long and the train were moving at 10 miles per hour, platform observers would "see" that the light reached the president of Forwardland about a millionth of a billionth of a second before it reached the president of Backwardland. Although this represents a genuine difference, it is so tiny that it cannot be detected directly by human senses. If the train were moving considerably faster, say at 600 million miles per hour, from the perspective of someone on the platform the light would take almost 20 times as long to reach the president of Backwardland compared with the time to reach the president of Forwardland. At high speeds, the startling effects of special relativity become increasingly pronounced.

The Effect on Time: Part II

It is difficult to give an abstract definition of time—attempts to do so often wind up invoking the word "time" itself, or else go through linguistic contortions simply to avoid doing so. Rather than proceeding down such a path, we can take a pragmatic viewpoint and define time to be that which is measured by clocks. Of course, this shifts the burden of definition to the word "clock"; here we can somewhat loosely think of a clock as a device that undergoes perfectly regular cycles of motion. We will measure time by counting the number of cycles our clock goes through. A familiar clock such as a wristwatch meets this definition; it has hands that move in regular cycles of motion and we do indeed measure elapsed time by counting the number of cycles (or fractions thereof) that the hands swing through between chosen events.

Of course, the meaning of "perfectly regular cycles of motion" implicitly involves a notion of time, since "regular" refers to equal time durations elapsing for each cycle. From a practical standpoint we address this by building clocks out of simple physical components that, on fundamental grounds, we expect to undergo repetitive cyclical evolutions that do not change in any manner from one cycle to the next. Grandfather clocks with pendulums that swing back and forth and atomic clocks based on repetitive atomic processes provide simple examples.

Our goal is to understand how motion affects the passage of time, and since we have defined time operationally in terms of clocks, we can translate our question into how motion affects the "ticking" of clocks. It is crucial to emphasize at the outset that our discussion is not concerned with how the mechanical elements of a particular clock happen to respond to shaking or jostling that might result from bumpy motion. In fact, we will consider only the simplest and most serene kind of motion—motion at absolutely constant velocity—and therefore there will not be any shaking or jostling at all. Rather, we are interested in the universal question of how motion affects the passage of time and therefore how it fundamentally affects the ticking of *any* and *all* clocks regardless of their particular design or construction.

For this purpose we introduce the world's conceptually simplest (yet most impractical) clock. It is known as a "light clock" and consists of two small mirrors mounted on a bracket facing one another, with a single photon of light bouncing back and forth between them (see Figure 2.1). If the mirrors are about six inches apart, it will take the photon about a billionth of a second to complete one round-trip journey. "Ticks" on the light clock may be thought of as occurring every time the photon completes a round-trip—a billion ticks means that one second has elapsed.

We can use the light clock like a stopwatch to measure the time elapsed between events: We simply count how many ticks occur during the period of interest and multiply by the time corresponding to one tick. For instance, if we are timing a horse race and count that between the start and finish the number of round-trip photon journeys is 55 billion, we can conclude that the race took 55 seconds.

The reason we use the light clock in our discussion is that its mechanical simplicity pares away extraneous details and therefore provides

Figure 2.1 A light clock consists of two parallel mirrors with a photon that bounces between them. The clock "ticks" each time the photon completes a round-trip journey.

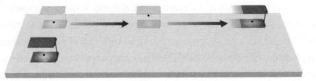

Figure 2.2 A stationary light clock in the foreground while a second light clock slides by at constant speed.

us with the clearest insight into how motion affects the passage of time. To see this, imagine that we are idly watching the passage of time by looking at a ticking light clock placed on a nearby table. Then, all of sudden, a second light clock slides by on the table, moving at constant velocity (see Figure 2.2) The question we ask is whether the moving light clock will tick at the same rate as the stationary light clock?

To answer the question, let's consider the path, from our perspective, that the photon in the sliding clock must take in order for it to result in a tick. The photon starts at the base of the sliding clock, as in Figure 2.2, and first travels to the upper mirror. Since, from our perspective, the clock is moving, the photon must travel at an angle, as shown in Figure 2.3. If the photon did not travel along this path, it would miss the upper mirror and fly off into space. As the sliding clock has every right to claim that it's stationary and everything else is moving, we know that the photon *will* hit the upper mirror and hence the path we have drawn is correct. The photon bounces off the upper mirror and again travels a diagonal path to hit the lower mirror, and the sliding clock ticks. The simple but essential point is that the double diagonal path that we see the photon traverse is *longer* than the straight up-and-down path taken by the photon in the stationary clock; in addition to traversing the up-and-down distance, the pho-

Figure 2.3 From our perspective, the photon in the sliding clock travels on a diagonal path.

ton in the sliding clock must also travel to the right, from our perspective. Moreover, the constancy of the speed of light tells us that the sliding clock's photon travels at exactly the same speed as the stationary clock's photon. But since it must travel farther to achieve one tick it will tick *less frequently*. This simple argument establishes that the moving light clock, from our perspective, ticks more slowly than the stationary light clock. And since we have agreed that the number of ticks directly reflects how much time has passed, we see that the passage of time has slowed down for the moving clock.

You might wonder whether this merely reflects some special feature of light clocks and would not apply to grandfather clocks or Rolex watches. Would time as measured by these more familiar clocks also slow down? The answer is a resounding yes, as can be seen by an application of the principle of relativity. Let's attach a Rolex watch to the top of each of our light clocks, and rerun the preceding experiment. As discussed, a stationary light clock and its attached Rolex measure identical time durations, with a billion ticks on the light clock occurring for every one second of elapsed time on the Rolex. But what about the moving light clock and its attached Rolex? Does the rate of ticking on the moving Rolex slow down so that it stays synchronized with the light clock to which it is attached? Well, to make the point most forcefully, imagine that the light clock–Rolex watch combination is moving because it is bolted to the floor of a windowless train compartment gliding along perfectly straight and smooth tracks at constant speed. By the principle of relativity, there is no way for an observer on this train to detect any influence of the train's motion. But if the light clock and Rolex were to fall out of synchronization, this would be a noticeable influence indeed. And so the moving light clock and its attached Rolex *must* still measure equal time durations; the Rolex *must* slow down in exactly the same way that the light clock does. Regardless of brand, type, or construction, clocks that are moving relative to one another record the passage of time at different rates.

The light clock discussion also makes clear that the precise time difference between stationary and moving clocks depends on how much farther the sliding clock's photon must travel to complete each round-trip journey. This in turn depends on how quickly the sliding clock is moving— from the viewpoint of a stationary observer, the faster the clock is sliding, the farther the photon must travel to the right. We conclude that in comparison to a stationary clock, the rate of ticking of the sliding clock becomes slower and slower as it moves faster and faster.[3]

To get a sense of scale, note that the photon traverses one round-trip in about a billionth of a second. For the clock to be able to travel an appreciable distance during the time for one tick it must therefore be traveling enormously quickly—that is, some significant fraction of the speed of light. If it is traveling at more commonplace speeds like 10 miles per hour, the distance it can move to the right before one tick is completed is minuscule—just about 15 billionths of a foot. The extra distance that the sliding photon must travel is tiny and it has a correspondingly tiny effect on the rate of ticking of the moving clock. And again, by the principle of relativity, this is true for all clocks—that is, for time itself. This is why beings such as ourselves who travel relative to one another at such slow speeds are generally unaware of the distortions in the passage of time. The effects, although present to be sure, are incredibly small. If, on the other hand, we were able to grab hold of the sliding clock and move with it at, say, three-quarters the speed of light, the equations of special relativity can be used to show that stationary observers would see our moving clock ticking at just about two-thirds the rate of their own. A significant effect, indeed.

Life on the Run

We have seen that the constancy of the speed of light implies that a moving light clock ticks more slowly than a stationary light clock. And by the principle of relativity, this must be true not only for light clocks but also for any clock—it must be true of time itself. Time elapses more slowly for an individual in motion than it does for a stationary individual. If the fairly simple reasoning that has led us to this conclusion is correct, then, for instance, shouldn't one be able to live longer by being in motion rather than staying stationary? After all, if time elapses more slowly for an individual in motion than for an individual at rest, then this disparity should apply not just to time as measured by watches but also to time as measured by heartbeats and the decay of body parts. This *is* the case, as has been directly confirmed—not with the life expectancy of humans, but with certain particles from the microworld: muons. There is one important catch, however, that prevents us from proclaiming a newfound fountain of youth.

When sitting at rest in the laboratory, muons disintegrate by a process closely akin to radioactive decay, in an average of about two millionths of a second. This disintegration is an experimental fact supported by an enormous amount of evidence. It's as if a muon lives its life with a gun to its head; when it reaches two millionths of a second in age, it pulls the trigger and explodes apart into electrons and neutrinos. But if these muons are not sitting at rest in the laboratory and instead are traveling through a piece of equipment known as a particle accelerator that boosts them to just shy of light-speed, their average life expectancy as measured by scientists in the laboratory increases dramatically. This *really* happens. At 667 million miles per hour (about 99.5 percent of light speed), the muon lifetime is seen to increase by a factor of about ten. The explanation, according to special relativity, is that "wristwatches" worn by the muons tick much more slowly than the clocks in the laboratory, so long after the laboratory clocks say that the muons should have pulled their triggers and exploded, the watches on the fast-moving muons have yet to reach doom time. This is a very direct and dramatic demonstration of the effect of motion on the passage of time. If people were to zip around as quickly as these muons, their life expectancy would also increase by the same factor. Rather than living seventy years, people would live 700 years.[4]

Now for the catch. Although laboratory observers see fast-moving muons living far longer than their stationary brethren, this is due to *time elapsing more slowly* for the muons in motion. This slowing of time applies not just to the watches worn by the muons but also to all activities they might undertake. For instance, if a stationary muon can read 100 books in its short lifetime, its fast-moving cousin will also be able to read the same 100 books, because although it appears to live longer than the stationary muon, its rate of reading—as well as everything else in its life—has slowed down as well. From the laboratory perspective, it's as if the moving muon is living its life in slow motion; from this viewpoint the moving muon will live longer than a stationary one, but the "amount of life" the muon will experience is precisely the same. The same conclusion, of course, holds true for the fast-moving people with a life expectancy of centuries. From *their* perspective, it's life as usual. From our perspective they are living life in hyper-slow motion and therefore one of their normal life cycles takes an enormous amount of *our* time.

Who Is Moving, Anyway?

The relativity of motion is both the key to understanding Einstein's theory and a potential source of confusion. You may have noticed that a reversal of perspective interchanges the roles of the "moving" muons, whose watches we have argued run slowly, and their "stationary" counterparts. Just as both George and Gracie had an equal right to declare that they were stationary and that the other was moving, the muons we have described as being in motion are fully justified in proclaiming that, from their perspective, they are motionless and that it is the "stationary" muons that are moving, in the opposite direction. The arguments presented can be applied equally well from this perspective, leading to the seemingly opposite conclusion that watches worn by the muons we christened as stationary are running slow compared with those worn by the muons we described as moving.

We have already met a situation, the signing ceremony with the light bulb, in which different viewpoints lead to results that seem to be completely at odds. In that case we were forced by the basic reasoning of special relativity to give up the ingrained idea that everyone, regardless of state of motion, agrees about which events happen at the same time. The present incongruity, though, appears to be worse. How can two observers each claim that the other's watch is running slower? More dramatically, the different but equally valid muon perspectives seem to lead us to the conclusion that each group will claim that it is the other group that dies first. We are learning that the world can have some unexpectedly strange features, but we would hope that it does not cross into the realm of logical absurdity. So what's going on?

As with all apparent paradoxes arising from special relativity, under close examination these logical dilemmas resolve to reveal new insights into the workings of the universe. To avoid ever more severe anthropomorphizing, let's switch from muons back to George and Gracie, who now, in addition to their flashing lights, have bright digital clocks on their space-suits. From George's perspective, he is stationary while Gracie with her flashing green light and large digital clock appears in the distance and then passes him in the blackness of empty space. He notices that Gracie's clock is running slow in comparison to his (with the rate of slowdown depending on how fast they pass one another). Were he a bit more astute, he would also note that in addition to the passage of time on her clock, everything about Gracie—the way she waves as she passes, the speed with which she blinks her eyes, and so on—is occurring in slow motion. From Gracie's perspective, exactly the same observations apply to George.

Although this seems paradoxical, let's try to pinpoint a precise experiment that would reveal a logical absurdity. The simplest possibility is to arrange things so that when George and Gracie pass one another they both set their clocks to read 12:00. As they travel apart, each claims that the other's clock is running slower. To confront this disagreement head on, George and Gracie must rejoin each other and directly compare the time elapsed on their clocks. But how can they do this? Well, George has a jet-pack that he can use, from his perspective, to catch up with Gracie. But if he does this, the symmetry of their two perspectives, which is the cause of the apparent paradox, is broken since George will have undergone *accelerated,* non-force-free motion. When they rejoin in this manner, less time will indeed have elapsed on George's clock as he can now definitively say that he was in motion, since he could feel it. No longer are George's and Gracie's perspectives on equal footing. By turning on the jet-pack, George relinquishes his claim to being at rest.

If George chases after Gracie in this manner, the time difference that their clocks will show depends on their relative velocity and the details of how George uses his jet-pack. As is by now familiar, if the speeds involved are small, the difference will be minuscule. But if substantial fractions of light speed are involved, the differences can be minutes, days, years, centuries, or more. As one concrete example, imagine that the relative speed of George and Gracie when they pass and are moving apart is 99.5 percent of light speed. Further, let's say that George waits 3 years, according to his clock, before firing up his jet-pack for a momentary blast that sends him closing in on Gracie at the same speed that they were previously moving apart, 99.5 percent of light speed. When he reaches Gracie, 6 years will have elapsed on his clock since it will take him 3 years to catch her. However, the mathematics of special relativity shows that 60 years will have elapsed on her clock. This is no sleight of hand: Gracie will have to search her distant memory, some 60 years before, to recall passing George in space. For George, on the other hand, it was a mere 6 years ago. In a real sense, George's motion has made him a time traveler, albeit in a very precise sense: He has traveled into Gracie's future.

Getting the two clocks back together for direct comparison might seem to be merely a logistical nuisance, but it is really at the heart of the matter. We can imagine a variety of tricks to circumvent this chink in the paradox armor, but all ultimately fail. For instance, rather than bringing the clocks back together, what if George and Gracie compare their clocks by cellular telephone communication? If such communication were instantaneous, we would be faced with an insurmountable inconsistency: reasoning from Gracie's perspective, George's clock is running slow and hence he must communicate less elapsed time; reasoning from George's perspective, Gracie's clock is running slow and hence she must communicate less elapsed time. They both can't be right, and we would be sunk. The key point of course is that cell phones, like all forms of communication, do not transmit their signals instantaneously. Cell phones operate with radio waves, a form of light, and the signal they transmit therefore travels at light speed. This means that it takes time for the signals to be received—just enough time delay, in fact, to make each perspective compatible with the other.

Let's see this, first, from George's perspective. Imagine that every hour, on the hour, George recites into his cell phone, "It's twelve o'clock and all is well," "It's one o'clock and all is well," and so forth. Since from his perspective Gracie's clock runs slow, at first blush he thinks that Gracie will receive these messages prior to her clock's reaching the appointed hour. In this way, he concludes, Gracie will have to agree that hers is the slow clock. But then he rethinks it: "Since Gracie is receding from me, the signal I send to her by cell phone must travel ever longer distances to reach her. Maybe this additional travel time compensates for the slowness of her clock." George's realization that there are competing effects—the slowness of Gracie's clock vs. the travel time of his signal—inspires him to sit down and quantitatively work out their combined effect. The result he finds is that the travel time effect *more than compensates* for the slowness of Gracie's clock. He comes to the surprising conclusion that Gracie will receive his signals proclaiming the passing of an hour on his clock *after* the appointed hour has passed on hers. In fact, since George is aware of Gracie's expertise in physics, he knows that she will take the signal's travel time into account when drawing conclusions about *his* clock based on his cell phone communications. A little more calculation quantitatively shows that even taking the travel time into account, Gracie's analysis of his signals will lead her to the conclusion that George's clock ticks more slowly than hers.

Exactly the same reasoning applies when we take Gracie's perspective, with her sending out hourly signals to George. At first the slowness of George's clock from her perspective leads her to think that he will receive her hourly messages prior to broadcasting his own. But when she takes into account the ever longer distances her signal must travel to catch George as he recedes into the darkness, she realizes that George will actually receive them *after* sending out his own. Once again, she realizes that even if George takes the travel time into account, he will conclude from Gracie's cell phone communications that her clock is running slower than his.

So long as neither George nor Gracie accelerates, their perspectives are on precisely equal footing. Even though it seems paradoxical, in this way they both realize that it is perfectly consistent for each to think the other's clock is running slow.

Motion's Effect on Space

The preceding discussion reveals that observers see moving clocks ticking more slowly than their own—that is, time is affected by motion. It is a short step to see that motion has an equally dramatic effect on space. Let's return to Slim and Jim on the drag strip. While in the showroom, as we mentioned, Slim had carefully measured the length of his new car with a tape measure. As Slim is speeding along the drag strip, Jim cannot apply this method to measure the length of the car, so he must proceed in an indirect manner. One such approach, as we indicated earlier, is this: Jim starts his stopwatch just when the front bumper of the car reaches him and stops it just as the rear bumper passes. By multiplying the elapsed time by the speed of the car, Jim can determine the car's length.

Using our newfound appreciation of the subtleties of time, we realize that from Slim's perspective he is stationary while Jim is moving, and hence Slim sees Jim's clock as running slow. As a result, Slim realizes that Jim's indirect measurement of the car's length will yield a *shorter* result than he measured in the showroom, since in Jim's calculation (length equals speed multiplied by elapsed time) Jim measures the elapsed time on a watch that is running slow. If it runs slow, the elapsed time he finds will be less and the result of his calculation will be a shorter length.

Thus Jim will perceive the length of Slim's car, when it is in motion, to be less than its length when measured at rest. This is an example of the general phenomenon that observers perceive a moving object as being shortened along the direction of its motion. For instance, the equations of special relativity show that if an object is moving at about 98 percent of light speed, then a stationary observer will view it as being 80 percent shorter than if it were at rest. This phenomenon is illustrated in Figure 2.4.[5]

Motion through Spacetime

The constancy of the speed of light has resulted in a replacement of the traditional view of space and time as rigid and objective structures with a new conception in which they depend intimately on the relative motion between observer and observed. We could end our discussion here, having realized that moving objects evolve in slow motion and are foreshortened. Special relativity, though, provides a more deeply unified perspective to encompass these phenomena.

To understand this perspective, let's imagine a rather impractical automobile that rapidly attains its cruising speed of 100 miles per hour and sticks to this speed, no more, no less, until it is shut off and rolls to a halt.

Figure 2.4 A moving object is shortened in the direction of its motion.

Let's also imagine that, due to his growing reputation as a skilled driver, Slim is asked to test-drive the vehicle on a long, straight, and wide track in the middle of a flat stretch of desert. As the distance between the start and finish lines is 10 miles, the car should cover this distance in one-tenth of an hour, or six minutes. Jim, who moonlights as an automobile engineer, inspects the data recorded from dozens of test-drives and is disturbed to see that although most were timed to be six minutes, the last few are a good deal longer: 6.5, 7, and even 7.5 minutes. At first he suspects a mechanical problem, since those times seem to indicate that the car was traveling slower than 100 miles per hour on the last three runs. Yet after examining the car extensively he convinces himself that it is in perfect condition. Unable to explain the anomalously long times, he consults Slim and asks him about the final few runs. Slim has a simple explanation. He tells Jim that, since the track runs from east to west, as it got later in the day, the sun was glaring into his view. During the last three runs it was so bad that he drove from one end of the track to the other at a slight angle. He draws a rough sketch of the path he took on the last three runs, and it is shown in Figure 2.5. The explanation for the three longer times is now perfectly clear: the path from start to finish is longer when traveling at an angle and therefore, at the same speed of 100 miles per hour, it will take more time to cover. Put another way, when traveling at an angle, part of the 100 miles per hour is expended on going from south to north, leaving a bit less to accomplish the trip from east to west. This implies that it will take a little longer to traverse the strip.

As stated, Slim's explanation is easy to understand; however, it is worth

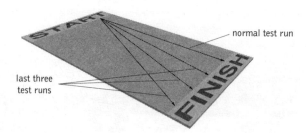

Figure 2.5 Due to the glaring late-afternoon sun, Slim drove at an increasingly greater angle on the last three runs.

rephrasing it slightly for the conceptual leap we are about to take. The north-south and east-west directions are two independent spatial dimensions in which a car can move. (It can also move vertically, when traversing a mountain pass, for example, but we will not need that ability here.) Slim's explanation illustrates that even though the car was traveling at 100 miles per hour on each and every run, during the last few runs it shared this speed between the two dimensions and hence appeared to be going slower than 100 miles per hour in the east-west direction. During the previous runs, all 100 miles per hour were devoted to purely east-west motion; during the last three, part of this speed was used for north-south motion as well.

Einstein found that precisely this idea—the sharing of motion between different dimensions—underlies all of the remarkable physics of special relativity, so long as we realize that not only can spatial dimensions share an object's motion, but the *time* dimension can share this motion as well. In fact, in the majority of circumstances, *most* of an object's motion is through time, not space. Let's see what this means.

Motion through space is a concept we learn about early in life. Although we often don't think of things in such terms, we also learn that we, our friends, our belongings, and so forth all *move through time*, as well. When we look at a clock or a wristwatch, even while we idly sit and watch TV, the reading on the watch is constantly changing, constantly "moving forward in time." We and everything around us are aging, inevitably passing from one moment in time to the next. In fact, the mathematician Hermann Minkowski, and ultimately Einstein as well, advocated thinking about time as another dimension of the universe—the fourth dimension—in some ways quite similar to the three spatial dimensions in which we find ourselves immersed. Although it sounds abstract, the notion of time as a dimension is actually concrete. When we want to meet someone, we tell them where "in space" we will expect to see them—for instance, the 9th floor of the building on the corner of 53rd Street and 7th Avenue. There are three pieces of information here (9th floor, 53rd Street, 7th Avenue) reflecting a particular location in the three spatial dimensions of the universe. Equally important, however, is our specification of *when* we expect to meet them—for instance, at 3 P.M. This piece of information tells us where "in time" our meeting will take place. Events are therefore specified by *four* pieces of information: three in space and one in time. Such data, it is said, specifies the location of the event in space and in time, or in *spacetime,* for short. In this sense, time is another dimension.

Since this view proclaims that space and time are simply different examples of dimensions, can we speak of an object's speed through time in a manner resembling the concept of its speed through space? We can.

A big clue for how to do this comes from a central piece of information we have already encountered. When an object moves through space relative to us, its clock runs slow compared to ours. That is, the speed of its *motion through time slows down.* Here's the leap: Einstein proclaimed that all objects in the universe are *always* traveling through spacetime at one fixed speed—that of light. This is a strange idea; we are used to the notion that objects travel at speeds considerably less than that of light. We have repeatedly emphasized this as the reason relativistic effects are so unfamiliar in the everyday world. All of this is true. We are presently talking about an object's combined speed through *all four* dimensions—three space and one time—and it is the object's speed in this generalized sense that is equal to that of light. To understand this more fully and to reveal its importance, we note that like the impractical single-speed car discussed above, this one fixed speed can be shared between the different dimensions—different space *and* time dimensions, that is. If an object is sitting still (relative to us) and consequently does not move through space at all, then in analogy to the first runs of the car, all of the object's motion is used to travel through one dimension—in this case, the time dimension. Moreover, all objects that are at rest relative to us and to each other move through time—they age—at exactly the same rate or speed. If an object does move through space, however, this means that some of the previous motion through time must be diverted. Like the car traveling at an angle, this sharing of motion implies that the object will travel more slowly through time than its stationary counterparts, since some of its motion is now being used to move through space. That is, its clock will tick more slowly if it moves through space. This is exactly what we found earlier. We now see that time slows down when an object moves relative to us because this diverts some of its motion through time into motion through space. The speed of an object through space is thus merely a reflection of how much of its motion through time is diverted.[6]

We also see that this framework immediately incorporates the fact that there is a limit to an object's spatial velocity: the maximum speed through space occurs if *all* of an object's motion through time is diverted to motion through space. This occurs when all of its previous light-speed motion through time is diverted to light-speed motion through space. But having used up all of its motion through time, this is the *fastest* speed through space that the object—any object—can possibly achieve. This is analogous to our car being test-driven directly in the north-south direction. Just as the car will have no speed left for motion in the east-west dimension, something traveling at light speed through space will have no speed left for motion through time. Thus light does not get old; a photon that emerged from the big bang is the same age today as it was then. There is no passage of time at light speed.

What about $E = mc^2$?

Although Einstein did not advocate calling his theory "relativity" (suggesting instead the name "invariance" theory to reflect the unchanging character of the speed of light, among other things), the meaning of the term is now clear. Einstein's work showed that concepts such as space and time, which had previously seemed to be separate and absolute, are actually interwoven and relative. Einstein went on to show that other physical properties of the world are unexpectedly interwoven as well. His most famous equation provides one of the most important examples. In it, Einstein asserted that the energy (E) of an object and its mass (m) are not independent concepts; we can determine the energy from knowledge of the mass (by multiplying the latter twice by the speed of light, c^2) or we can determine the mass from knowledge of the energy (by dividing the latter twice by the speed of light). In other words, energy and mass—like dollars and francs—are convertible currencies. Unlike money, however, the exchange rate given by two factors of the speed of light is always and forever fixed. Since this exchange-rate factor is so large (c^2 is a big number), a little mass goes an extremely long way in producing energy. The world grasped the devastating destructive power arising from the conversion of less than 1 percent of two pounds of uranium into energy at Hiroshima; one day, through fusion power plants, we may productively use Einstein's formula to meet the energy demands of the whole world with our endless supply of seawater.

From the viewpoint of the concepts we have emphasized in this chapter, Einstein's equation gives us the most concrete explanation for the central fact that nothing can travel faster than light speed. You may have wondered, for instance, why we can't take some object, a muon say, that an accelerator has boosted up to 667 million miles per hour—99.5 percent of light speed—and "push it a bit harder," getting it to 99.9 percent of light speed, and then "*really* push it harder" impelling it to cross the light-speed barrier. Einstein's formula explains why such efforts will never succeed. The faster something moves the more energy it has and from Einstein's formula we see that the more energy something has the more massive it becomes. Muons traveling at 99.9 percent of light speed, for example, weigh a lot more than their stationary cousins. In fact, they are about 22 times as heavy—literally. (The masses recorded in Table 1.1 are for particles at rest.) But the more massive an object is, the harder it is to increase its speed. Pushing a child on a bicycle is one thing, pushing a Mack truck is quite another. So, as a muon moves more quickly it gets ever more difficult to further increase its speed. At 99.999 percent of light speed the mass of a muon has increased by a factor of 224; at 99.99999999 percent of light speed it has increased by a factor of more than 70,000. Since the mass of the muon increases without limit as its speed approaches that of light, it would require a push with an *infinite* amount of energy to reach or to cross the light barrier. This, of course, is impossible and hence absolutely nothing can travel faster than the speed of light.

As we shall see in the next chapter, this conclusion plants the seeds for the second major conflict faced by physics during the past century and ultimately spells doom for another venerable and cherished theory—Newton's universal theory of gravity.

Chapter 3

Of Warps and Ripples

Through special relativity Einstein resolved the conflict between the "age-old intuition" about motion and the constancy of the speed of light. In short, the solution is that our intuition is wrong—it is informed by motion that typically is extremely slow compared to the speed of light, and such low speeds obscure the true character of space and time. Special relativity reveals their nature and shows them to differ radically from previous conceptions. Tinkering with our understanding of the foundations of space and time, though, was no small undertaking. Einstein soon realized that of the numerous reverberations following from the revelations of special relativity, one was especially profound: The dictum that nothing can outrun light proves to be incompatible with Newton's revered universal theory of gravity, proposed in the latter half of the seventeenth century. And so, while resolving one conflict, special relativity gave rise to another. After a decade of intense, sometimes tormented study, Einstein resolved the dilemma with his general theory of relativity. In this theory, Einstein once again revolutionized our understanding of space and time by showing that they warp and distort to communicate the force of gravity.

Newton's View of Gravity

Isaac Newton, born in 1642 in Lincolnshire, England, changed the face of scientific research by bringing the full force of mathematics to the service of physical inquiry. Newton's was such a monumental intellect that, for example, when he found that the mathematics required for some of his investigations did not exist, he invented it. Nearly three centuries would pass before the world would host a comparable scientific genius. Of Newton's numerous profound insights into the workings of the universe, the one that primarily concerns us here is his universal theory of gravity.

The force of gravity pervades everyday life. It keeps us and all of the objects around us fixed to the earth's surface; it keeps the air we breathe from escaping to outer space; it keeps the moon in orbit around the earth and it keeps the earth bound in orbit around the sun. Gravity dictates the rhythm of the cosmic dance that is tirelessly and meticulously executed by billions upon billions of cosmic inhabitants, from asteroids to planets to stars to galaxies. More than three centuries of Newton's influence causes us to take for granted that a single force—gravity—is responsible for this wealth of terrestrial and extraterrestrial happenings. But before Newton there was no understanding that an apple falling to earth from a tree bore witness to the same physical principle that keeps the planets revolving around the sun. With an audacious step in the service of scientific hegemony, Newton united the physics governing both heaven and earth and declared the force of gravity to be the invisible hand at work in each realm.

Newton's view of gravity might be called the great equalizer. He declared that absolutely everything exerts an attractive gravitational force on absolutely everything else. Regardless of physical composition, *everything* exerts as well as feels the force of gravity. Based on a close study of Johannes Kepler's analysis of planetary motion, Newton deduced that the strength of the gravitational attraction between two bodies depends on *precisely* two things: the amount of stuff composing each of the bodies and the distance between them. "Stuff" means matter—this comprises the total number of protons, neutrons, and electrons, which in turn determines the *mass* of the object. Newton's universal theory of gravity asserts that the strength of attraction between two objects is larger for larger-mass objects and smaller for smaller-mass objects; it also asserts that the strength of attraction is larger for smaller separations between the objects and smaller for larger separations.

Newton went much further than this qualitative description and wrote down equations that quantitatively describe the strength of the gravitational force between two objects. In words, these equations state that the gravitational force between two bodies is proportional to the product of their masses and inversely proportional to the square of the distance between them. This "law of gravity" can be used to predict the motion of planets and comets around the sun, the moon about the earth, and rockets heading off for planetary explorations, as well as more earthbound applications such as baseballs flying through the air and divers spiraling poolward from springboards. The agreement between the predictions and the actual observed motion of such objects is spectacular. This success gave Newton's theory unequivocal support until the early part of the twentieth century. Einstein's discovery of special relativity, however, raised what proved to be an insurmountable obstacle for Newton's theory.

The Incompatibility of Newtonian Gravity and Special Relativity

A central feature of special relativity is the absolute speed barrier set by light. It is important to realize that this limit applies not only to material objects but also to signals and influences of any kind. There is simply no way to communicate information or a disturbance from one place to another at faster than light speed. Of course, the world is full of ways for transmitting disturbances at *slower* than the speed of light. Your speech and all other sounds, for example, are carried by vibrations that travel at about 700 miles per hour through air, a feeble rate compared with light's 670 million miles per hour. This speed difference becomes obvious when you watch a baseball game, from seats that are far from home plate. When a batter hits the ball, the sound reaches you moments *after* you see the ball being hit. A similar thing happens in a thunderstorm. Although lightning and thunder are produced simultaneously, you see the lightning before hearing the thunder. Again, this reflects the substantial speed difference between light and sound. The success of special relativity informs us that the reverse situation, in which some signal reaches us *before* the light it emits, is just not possible. Nothing outruns photons.

Here's the rub. In Newton's theory of gravity, one body exerts a gravitational pull on another with a strength determined solely by the mass of the objects involved and the magnitude of their separation. The strength has nothing to do with how long the objects have been in each other's presence. This means that if their mass or their separation should change, the objects will, according to Newton, *immediately* feel a change in their mutual gravitational attraction. For instance, Newton's theory of gravity claims that if the sun were suddenly to explode, the earth—some 93 million miles away—would instantaneously suffer a departure from its usual elliptical orbit. Even though it would take light from the explosion eight minutes to travel from the sun to the earth, in Newton's theory knowledge that the sun had exploded would be instantaneously transmitted to the earth through the sudden change in the gravitational force governing its motion.

This conclusion is in direct conflict with special relativity, since the latter ensures that no information can be transmitted faster than the speed of light—instantaneous transmission violates this precept maximally.

In the early part of the twentieth century, therefore, Einstein realized that the tremendously successful Newtonian theory of gravity was in conflict with his special theory of relativity. Confident in the veracity of special relativity and notwithstanding the mountain of experimental support for Newton's theory, Einstein sought a new theory of gravity compatible with special relativity. This ultimately led him to the discovery of general relativity, in which the character of space and time again went through a remarkable transformation.

Einstein's Happiest Thought

Even before the discovery of special relativity, Newton's theory of gravity was lacking in one important respect. Although it can be used to make highly accurate predictions about how objects will move under the influence of gravity, it offers no insight into what gravity *is*. That is, how is it that two bodies that are physically separate from another, possibly hundreds of millions of miles apart if not more, nonetheless influence each other's motion? By what means does gravity execute its mission? This is a problem of which Newton himself was well aware. In his own words,

It is inconceivable, that inanimate brute matter, should, without the mediation of something else, which is not material, operate upon and affect other matter without mutual contact. That Gravity should be innate, inherent and essential to matter so that one body may act upon another at a distance thro' a vacuum without the mediation of anything else, by and through which their action and force may be conveyed, from one to another, is to me so great an absurdity that I believe no Man who has in philosophical matters a competent faculty of thinking can ever fall into it. Gravity must be caused by an agent acting constantly according to certain laws; but whether this agent be material or immaterial, I have left to the consideration of my readers.[1]

That is, Newton accepted the existence of gravity and went on to develop equations that accurately describe its effects, but he never offered any insight into how it actually works. He gave the world an "owner's manual" for gravity which delineated how to "use" it—instructions that physicists, astronomers, and engineers have exploited successfully to plot the course of rockets to the moon, Mars, and other planets in the solar system; to predict solar and lunar eclipses; to predict the motion of comets, and so on. But he left the inner workings—the contents of the "black box" of gravity—a complete mystery. When you use your CD player or your personal computer, you may find yourself in a similar state of ignorance regarding how it works internally. So long as you know how to operate the equipment neither you nor anyone else needs to know *how* it accomplishes the tasks you set for it. But if your CD player or personal computer breaks, its repair relies crucially on knowledge of its internal workings. Similarly, Einstein realized that hundreds of years of experimental confirmation notwithstanding, special relativity implied that in some subtle way Newton's theory was "broken" and that its repair required coming to grips with the question of the true and full nature of gravity.

In 1907, while pondering these issues at his desk in the patent office in Bern, Switzerland, Einstein had the central insight that, through fits and starts, would eventually lead him to a radically new theory of gravity—an approach that would not merely fill in the gap in Newton's theory, but, rather, would completely reformulate thinking about gravity and, of utmost importance, would do so in a manner fully consistent with special relativity.

The insight Einstein had is relevant for a question that may have troubled you in Chapter 2. There we emphasized that we were interested in understanding how the world appears to individuals undergoing constant-velocity relative motion. By carefully comparing the observations of such individuals, we found some dramatic implications for the nature of space and time. But what about individuals who are experiencing *accelerated* motion? The observations of such individuals will be more complicated to analyze than those of constant-velocity observers, whose motion is more serene, but nevertheless we can ask whether there is some way of taming this complexity and bringing accelerated motion squarely into our new-found understanding of space and time.

Einstein's "happiest thought" showed how to do so. To understand his insight, imagine the year is 2050, you are the FBI's chief explosives expert, and you have just received a frantic call to investigate what appears to be a sophisticated bomb planted in the heart of Washington, D.C. After rushing to the scene and examining the device, your worst nightmare is confirmed: The bomb is nuclear and of such powerful design that even if it were buried deeply in the earth's crust or submerged in an ocean's depth, the damage from its blast would be devastating. After gingerly studying the bomb's detonation mechanism you realize that there is no hope to disarm it and, furthermore, you see that it has a novel booby-trap feature. The bomb is mounted on a scale. Should the reading on the scale deviate from its present value by more than 50 percent, the bomb will detonate. According to the timing mechanism, you see that you have but one week and counting. The fate of millions of people rests on your shoulders—what do you do?

Well, having determined that there is no safe place anywhere on or in the earth to detonate the device, you appear to have only one option: You must launch the device into the depths of outer space where its explosion will cause no damage. You present this idea to a meeting of your team at the FBI and almost immediately your plan is dashed by a young assistant. "There is a serious problem with your plan," your assistant Isaac begins. "As the device gets farther from the earth, its weight will decrease, since its gravitational attraction with the earth will diminish. This means that the reading on the scale inside the device will decrease, causing detonation well before reaching the safety of deep space." Before you have time to fully contemplate this criticism, another young assistant pipes up: "In fact, come to think of it, there is even another problem," your assistant Albert says. "This problem is as important as Isaac's objection but somewhat more subtle, so bear with me as I explain it." Wanting a moment to think through Isaac's objection, you try to hush Albert, but as usual, once he begins there is no stopping him.

"In order to launch the device into outer space we will have to mount it on a rocket. As the rocket *accelerates* upward in order to penetrate outer space, the reading on the scale will *increase,* again causing the device to detonate prematurely. You see, the base of the bomb—which rests on the scale—will push harder on the scale than when the device is at rest in the same way that your body is squeezed back into the seat of an accelerating car. The bomb will 'squeeze' the scale just as your back squeezes the cushion in the car seat. When a scale is squeezed, of course, its reading increases—and this will cause the bomb to detonate if the resulting increase is more than 50 percent."

You thank Albert for his comment but, having tuned out his explanation to mentally confirm Isaac's remark, you dejectedly proclaim that it takes only one fatal blow to kill an idea, and Isaac's obviously correct observation has definitively done that. Feeling somewhat hopeless you ask for new suggestions. At that moment, Albert has a stunning revelation: "On second thought," he continues, "I do not think that your idea is dead at all. Isaac's observation that gravity diminishes as the device is lifted into space means that the reading on the scale will go *down.* My observation that the upward acceleration of the rocket will cause the device to push harder against the scale means that the reading will go *up.* Taken together, this means that if we carefully adjust the precise moment-to-moment acceleration of the rocket as it moves upward, these two effects can *cancel each other out!* Specifically, in the early stages of liftoff, when the rocket still feels the full force of the earth's gravity, it can accelerate, just not too severely, so that we stay within the 50 percent window. As the rocket gets farther and farther from the earth—and therefore feels the earth's gravity less and less—we need to increase its upward acceleration to compensate. The increase in the reading from upward acceleration can exactly equal the decrease in the reading from the diminishing gravitational attraction, so, in fact, we can keep the actual reading on the scale from changing at all!"

Albert's suggestion slowly begins to make sense. "In other words," you respond, "an upward acceleration can provide a stand-in or a substitute for gravity. We can imitate the effect of gravity through suitably accelerated motion."

"Exactly," responds Albert.

"So," you continue, "we *can* launch the bomb into space and by judiciously adjusting the acceleration of the rocket we can ensure that the reading on the scale does not change, thus avoiding detonation until it is a safe distance from earth." And so by playing off gravity and accelerated motion—using the precision of twenty-first-century rocket science—you are able to stave off disaster.

The recognition that gravity and accelerated motion are profoundly interwoven is the key insight that Einstein had one happy day in the Bern patent office. Although the bomb experience highlights the essence of his idea, it is worth rephrasing it in a framework closer to that of Chapter 2. For this purpose, recall that if you are put into a sealed, windowless compartment that is *not* accelerating, there is no way for you to determine your speed. The compartment looks the same and any experiments you do yield identical results regardless of how fast you are moving. More fundamentally, without outside benchmarks for comparison there is no way that a velocity can even be assigned to your state of motion. On the other hand, if you are accelerating, then even with your perceptions limited to the confines of your sealed compartment, you will *feel* a force on your body. For instance, if your forward-facing chair is bolted to the floor and your compartment is being accelerated forward, you will feel the force of your seat on your back just as with the car described by Albert. Similarly, if your compartment is being accelerated upward you will feel the force of the floor on your feet. Einstein's realization was that within the confines of your tiny compartment, you will not be able to distinguish these accelerated situations from ones *without acceleration* but *with gravity:* When their magnitudes are judiciously adjusted, the force you feel from a gravitational field or from accelerated motion are indistinguishable. If your compartment is placidly sitting upright on the earth's surface, you will feel the familiar force of the floor on your feet, just as in the scenario of upward acceleration; this is exactly the same equivalence Albert exploited in his solution for launching the terrorist bomb into space. If your compartment is resting on its back end you will feel the force of your seat on your back (preventing you from falling), just as when you were accelerating horizontally. Einstein called the indistinguishability between accelerated motion and gravity the *equivalence principle*.[2] It plays a central role in general relativity.

This description shows that general relativity finishes a job initiated by special relativity. Through its principle of relativity, the special theory of relativity declares a democracy of observational vantage points: the laws of physics appear identical to all observers undergoing constant-velocity motion. But this is limited democracy indeed, for it excludes an enormous number of other viewpoints—those of individuals who are accelerating. Einstein's 1907 insight now shows us how to embrace *all* points of view—constant velocity and accelerating—within one egalitarian framework. Since there is no difference between an accelerated vantage point *without* a gravitational field and a nonaccelerated vantage point *with* a gravitational field, we can invoke the latter perspective and declare that *all observers, regardless of their state of motion, may proclaim that they are stationary and "the rest of the world is moving by them," so long as they include a suitable gravitational field in the description of their own surroundings*. In this sense, through the inclusion of gravity, general relativity ensures that all possible observational vantage points are on equal footing. (As we shall see later, this means that distinctions between observers in Chapter 2 that relied on accelerated motion—as when George chased after Gracie by turning on his jet-pack and aged less than she—admit an equivalent description without acceleration, but with gravity.)

This deep connection between gravity and accelerated motion is certainly a remarkable realization, but why did it make Einstein so happy? The reason, simply put, is that gravity is mysterious. It is a grand force permeating the life of the cosmos, but it is elusive and ethereal. On the other hand, accelerated motion, although somewhat more complicated than constant-velocity motion, is concrete and tangible. By finding a fundamental link between the two, Einstein realized that he could use his understanding of motion as a powerful tool toward gaining a similar understanding of gravity. Putting this strategy into practice was no small task, even for the genius of Einstein, but ultimately this approach bore the fruit of general relativity. Achieving this end required that Einstein forge a second link in the chain uniting gravity and accelerated motion: the *curvature* of space and time, to which we now turn.

Acceleration and the Warping of Space and Time

Einstein worked on the problem of understanding gravity with extreme, almost obsessive, intensity. About five years after his happy revelation in the Bern patent office, he wrote to the physicist Arnold Sommerfeld, "I am now working exclusively on the gravity problem. . . . [O]ne thing is certain—that never in my life have I tormented myself anything like this. . . . Compared to this problem the original [i.e., special] relativity theory is child's play."[3]

He appears to have made the next key breakthrough, a simple yet subtle consequence of applying special relativity to the link between gravity and accelerated motion, in 1912. To understand this step in Einstein's reasoning it is easiest to focus, as apparently he did, on a particular example of accelerated motion.[4] Recall that an object is accelerating if either the speed or the direction of its motion changes. For simplicity we will focus on accelerated motion in which *only* the direction of our object's motion changes while its speed stays fixed. Specifically, we consider motion in a circle such as what one experiences on the Tornado ride in an amusement park. In case you have never tested the stability of your constitution on this ride, you stand with your back against the inside of a circular Plexiglas structure that spins at a high speed. Like all accelerated motion, you can feel this motion—you feel your body being pulled radially away from the ride's center and you feel the circular wall of Plexiglas pressing on your back, keeping you moving in a circle. (In fact, although not relevant for the present discussion, the spinning motion "pins" your body to the Plexiglas with such a force that when the ledge on which you are standing drops away you do not slip downward.) If the ride is extremely smooth and you close your eyes, the pressure of the ride on your back—like the support of a bed—can almost make you feel that you are lying down. The "almost" comes from the fact that you still feel ordinary "vertical" gravity, so your brain cannot be fully fooled. But if you were to ride the Tornado in outer space, and if it were to spin at just the right rate, it would feel just like lying in a stationary bed on earth. Moreover, were you to "get up" and walk along the interior of the spinning Plexiglas, your feet would press against it just as they do against an earthbound floor. In fact, space stations are designed to spin in this manner to create an artificial feeling of gravity in outer space.

Having used the accelerated motion of the spinning Tornado to imitate gravity, we can now follow Einstein and set out to see how space and time appear to someone on the ride. His reasoning, adapted to this situation, went as follows. We stationary observers can easily measure the circumference and the radius of the spinning ride. For instance, to measure the circumference we can carefully lay out a ruler—head to tail—alongside the ride's spinning girth; for its radius we can similarly use the head-to-tail method working our way from the central axle of the ride to its outer rim. As we anticipate from high-school geometry, we find that their ratio is two times the number pi—about 6.28—just as it is for any circle drawn on a flat sheet of paper. But what do things look like from the perspective of someone on the ride itself?

To find out, we ask Slim and Jim, who are currently enjoying a spin on the Tornado, to take a few measurements for us. We toss one of our rulers to Slim, who sets out to measure the circumference of the ride, and another to Jim, who sets out to measure the radius. To get the clearest perspective, let's take a bird's-eye view of the ride, as in Figure 3.1. We have adorned this snapshot of the ride with an arrow that indicates the momentary direction of motion at each point. As Slim begins to measure the circumference, we immediately see from our bird's-eye perspective that he is going to get a different answer than we did. As he lays the ruler out along the circumference, we notice that the ruler's *length is shortened.* This is nothing but the Lorentz contraction discussed in Chapter 2, in which the length of an object appears shortened along the direction of its motion. A shorter ruler means that he will have to lay it out—head to tail—*more* times to traverse the whole circumference. Since he still considers the ruler to be one foot long (since there is no relative motion between Slim and his ruler, he perceives it as having its usual length of one foot), this means that Slim will measure a *longer* circumference than did we.

What about the radius? Well, Jim also uses the head-to-tail method to find the length of a radial strut, and from our bird's-eye view we see that he is going to find the same answer as we did. The reason is that the ruler

Figure 3.1 Slim's ruler is contracted, since it lies along the direction of the ride's motion. But Jim's ruler lies along a radial strut, perpendicular to the direction of the ride's motion, and therefore its length is not contracted.

is not pointing along the instantaneous direction of the motion of the ride (as it is when measuring the circumference). Instead, it is pointed at a ninety-degree angle to the motion, and therefore it is *not* contracted along its length. Jim will therefore find exactly the same radial length as we did.

But now, when Slim and Jim calculate the ratio of the circumference of the ride to its radius they will get a number that is larger than our answer of two times pi, since the circumference is longer but the radius is the same. This is weird. How in the world can something in the shape of a circle violate the ancient Greek realization that for any circle this ratio is *exactly* two times pi?

Here is Einstein's explanation. The ancient Greek result holds true for circles drawn on a flat surface. But just as the warped or curved mirrors in an amusement park fun-house distort the normal spatial relationships of your reflection, if a circle is drawn on a warped or curved surface, its usual spatial relationships will also be distorted: the ratio of its circumference to its radius will generally *not* be two times pi.

For instance, Figure 3.2 compares three circles whose radii are identical. Notice, however, that their circumferences are *not* the same. The circumference of the circle in (b), drawn on the curved surface of a sphere,

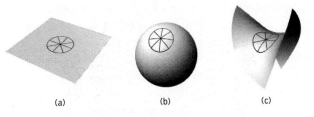

Figure 3.2 A circle drawn on a sphere (b) has a shorter circumference than one drawn on a flat sheet of paper (a), while a circle drawn on the surface of a saddle (c) has a longer circumference, even though they all have the same radius.

is less than the circumference of the circle drawn on the flat surface in (a), even though they have the same radius. The curved nature of the sphere's surface causes the radial lines of the circle to converge toward each other slightly, resulting in a small decrease in the circle's circumference. The circumference of the circle in (c), again drawn on a curved surface—a saddle shape—is *greater* than that drawn on a flat surface; the curved nature of the saddle's surface causes the radial lines of the circle to splay outward from each other slightly, resulting in a small increase in the circle's circumference. These observations imply that the ratio of the circumference to the radius of the circle in (b) will be less than two times pi, while the same ratio in (c) will be greater than two times pi. But this deviation from two times pi, especially the larger value found in (c), is just what we found for the spinning Tornado ride. This led Einstein to propose an idea—the curving of space—as an explanation for the violation of "ordinary," Euclidean geometry. The flat geometry of the Greeks, taught to schoolchildren for thousands of years, simply does not apply to someone on the spinning ride. Rather, its curved space generalization as schematically drawn in part (c) of Figure 3.2 takes its place.[5]

And so Einstein realized that the familiar geometrical spatial relationships codified by the Greeks, relationships that pertain to "flat" space figures like a circle on a flat table, *do not hold* from the perspective of an accelerated observer. Of course, we have discussed only one particular kind of accelerated motion, but Einstein showed that a similar result—the warping of space—holds in all instances of accelerated motion.

In fact, accelerated motion not only results in a warping of space, it also results in an analogous warping of time. (Historically, Einstein first focused on the warping of time and subsequently realized the importance of the warping of space.[6]) On one level, it should not be too surprising that time is also affected, since we have already seen in Chapter 2 that special relativity articulates a union between space and time. This merger was summarized by the poetic words of Minkowski, who during a lecture on special relativity in 1908 said, "Henceforward space on its own and time on its own will decline into mere shadows, and only a kind of union between the two will preserve its independence."[7] In more down-to-earth but similarly imprecise language, by knitting space and time together into the unified structure of spacetime, special relativity declares, "What's true for space is true for time." But this raises a question: Whereas we can picture warped space by its having a curved shape, what do we really mean by warped time?

To get a feel for the answer, let's once again impose upon Slim and Jim on the Tornado ride and ask them to carry out the following experiment. Slim will stand with his back against the ride, at the far end of one of the ride's radial struts, while Jim will slowly crawl toward him along the strut, starting from the ride's center. Every few feet, Jim will stop his crawling and the two brothers are to compare the readings on their watches. What will they find? From our stationary, bird's-eye perspective, we can again predict the answer: Their watches will not agree. We come to this conclusion because we realize that Slim and Jim are travelling at different speeds—on the Tornado ride, the farther out along a radial strut you are, the farther you must travel to complete one rotation, and therefore the faster you must go. But from special relativity, the faster you go, the slower your watch ticks, and hence we realize that Slim's watch will tick more slowly than Jim's. Furthermore, Slim and Jim will find that, as Jim gets closer to Slim, the ticking rate of Jim's watch will slow down, approaching that of Slim's. This reflects the fact that as Jim gets farther out along the strut, his circular speed increases toward that of Slim.

We conclude that to observers on the spinning ride, such as Slim and Jim, the rate of passage of time depends upon their precise position—in this case, their distance from the center of the ride. This is an illustration of what we mean by warped time: Time is warped if its rate of passage differs from one location to another. And of particular importance to our present discussion, Jim will also notice something else as he crawls out along the strut. He will feel an increasingly strong outward pull because not only does speed increase, but his acceleration increases as well, the farther he is from the spinning ride's center. On the Tornado ride, then, we see that greater acceleration is tied up with slower clocks—that is, greater acceleration results in a more significant warping of time.

These observations took Einstein to the final leap. Since he had already shown gravity and accelerated motion to be effectively indistinguishable, and since he now had shown that accelerated motion is associated with the warping of space and time, he made the following proposal for the innards of the "black box" of gravity—the mechanism by which gravity operates. Gravity, according to Einstein, *is* the warping of space and time. Let's see what this means.

The Basics of General Relativity

To get a feel for this new view of gravity, let's consider the prototypical situation of a planet, such as the earth, revolving around a star, such as the sun. In Newtonian gravity the sun keeps the earth in orbit with an unidentified gravitational "tether" that somehow instantaneously reaches out across vast distances of space and grabs hold of the earth (and, similarly, the earth reaches out and grabs hold of the sun). Einstein provided a new conception of what actually happens. It will aid in our discussion of Einstein's approach to have a concrete visual model of spacetime that we can conveniently manipulate. To do so, we will simplify things in two ways. First, for the moment, we will ignore time and focus solely on a visual model of space. We will reincorporate time in our discussion shortly. Second, in order to allow us to draw and manipulate visual images on the pages of this book, we will often refer to a *two*-dimensional analog of three-dimensional space. Most of the insight we gain from thinking in terms of this lower-dimensional model is directly applicable to the physical three-dimensional setting, so the simpler model provides a powerful pedagogical device.

In Figure 3.3, we make use of these simplifications and draw a two-dimensional model of a spatial region of our universe. The grid-like structure indicates a convenient means of specifying positions just as a street grid gives a means of specifying locations in a city. In a city, of course, one gives an address by specifying a location on the two-dimensional street grid

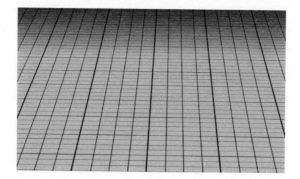

Figure 3.3 A schematic representation of flat space.

and also giving a location in the vertical direction, such as a floor number. It is the latter information, location in the third spatial dimension, that our two-dimensional analogy suppresses for visual clarity.

In the absence of any matter or energy, Einstein envisioned that space would be *flat*. In our two-dimensional model, this means that the "shape" of space should be like the surface of a smooth table, as drawn in Figure 3.3. This is the image of our spatial universe commonly held for thousands of years. But what happens to space if a massive object like the sun is present? Before Einstein the answer was *nothing*; space (and time) were thought to provide an inert theater, merely setting the stage on which the events of the universe play themselves out. The chain of Einstein's reasoning that we have been following, however, leads to a different conclusion.

A massive body like the sun, and indeed any body, exerts a gravitational force on other objects. In the example of the terrorist bomb, we learned that gravitational forces are indistinguishable from accelerated motion. In the example of the Tornado ride, we learned that a mathematical description of accelerated motion *requires* the relations of curved space. These links between gravity, accelerated motion, and curved space led Einstein to the remarkable suggestion that the presence of mass, such as the sun, causes the fabric of space around it to *warp*, as shown in Figure 3.4. A useful, and oft-quoted, analogy is that much like a rubber membrane on which a bowling ball has been placed, the fabric of space becomes distorted due to the presence of a massive object like the sun.

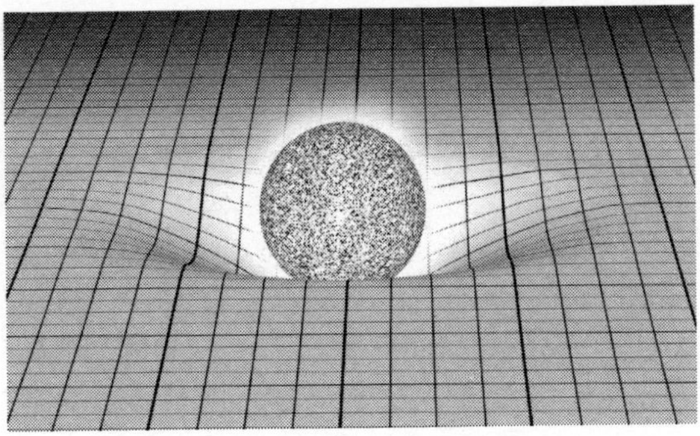

Figure 3.4 A massive body like the sun causes the fabric of space to warp, somewhat like the effect of a bowling ball placed on a rubber sheet.

According to this radical proposal, space is not merely a passive forum providing the arena for the events of the universe; rather, the shape of space *responds* to objects in the environment.

This warping, in turn, affects other objects moving in the vicinity of the sun, as they now must traverse the distorted spatial fabric. Using the rubber membrane–bowling ball analogy, if we place a small ball-bearing on the membrane and set it off with some initial velocity, the path it will follow depends on whether or not the bowling ball is sitting in the center. If the bowling ball is absent, the rubber membrane will be flat and the ball bearing will travel along a straight line. If the bowling ball is present and thereby warps the membrane, the ball bearing will travel along a curved path. In fact, ignoring friction, if we set the ball bearing moving with just the right speed in just the right direction, it will continue to move in a recurring curved path around the bowling ball—in effect, it will "go into orbit." Our language presages the application of this analogy to gravity.

The sun, like the bowling ball, warps the fabric of space surrounding it, and the earth's motion, like that of the ball bearing, is determined by the shape of the warp. The earth, like the ball bearing, will move in orbit around the sun if its speed and orientation have suitable values. This effect on the motion of the earth is what we normally would refer to as the gravitational influence of the sun, and is illustrated in Figure 3.5. The difference, now, is that unlike Newton, Einstein has specified the *mechanism* by which gravity is transmitted: the warping of space. In Einstein's

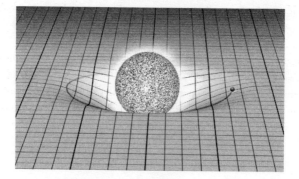

Figure 3.5 The earth is kept in orbit around the sun because it rolls along a valley in the warped spatial fabric. In more precise language, it follows a "path of least resistance" in the distorted region around the sun.

view, the gravitational tether holding the earth in orbit is not some mysterious instantaneous action of the sun; rather, it is the warping of the spatial fabric caused by the sun's presence.

This picture allows us to understand the two essential features of gravity in a new way. First, the more massive the bowling ball, the greater the distortion it causes in the rubber membrane; similarly, in Einstein's description of gravity the more massive an object is, the greater the distortion it causes in the surrounding space. This implies that the more massive an object, the greater the gravitational influence it can exert on other bodies, precisely in accord with our experiences. Second, just as the distortion of the rubber membrane due to the bowling ball gets smaller as one gets farther from it, the amount of spatial warping due to a massive body such as the sun decreases as one's distance from it increases. This, again, jibes with our understanding of gravity, whose influence becomes weaker as the distance between objects becomes larger.

An important point to note is that the ball bearing itself warps the rubber membrane, although only slightly. Similarly, the earth, being a massive body in its own right, also warps the fabric of space, although far less than the sun. This is how, in the language of general relativity, the earth keeps the moon in orbit, and it is also how the earth keeps each of us bound to its surface. As a skydiver plunges earthward, he or she is sliding down a depression in the spatial fabric caused by the earth's mass. Moreover, each of us—like any massive object—also warps the spatial fabric in close proximity to our bodies, although the comparatively small mass of a human body makes this a minuscule indentation.

In summary then, Einstein fully agreed with Newton's statement that "Gravity must be caused by an agent" and rose to Newton's challenge in which the identity of the agent was left "to the consideration of my readers." The agent of gravity, according to Einstein, is the fabric of the cosmos.

A Few Caveats

The rubber membrane–bowling ball analogy is valuable because it gives us a visual image with which we can grasp tangibly what we mean by a warp in the spatial fabric of the universe. Physicists often use this and similar analogies to guide their own intuition regarding gravitation and curvature. However, its usefulness notwithstanding, the rubber membrane–bowling ball analogy is not perfect and for clarity we call attention to a few of its shortcomings.

First, when the sun causes the fabric of space around it to warp this is not due to its "being pulled downward" by gravity as in the case of the bowling ball, which warps the rubber membrane because it is pulled earthward by gravity. In the case of the sun, there is no other object to "do the pulling." Instead, Einstein has taught us that the warping of space *is* gravity. The mere presence of an object with mass causes space to respond by warping. Similarly, the earth is not kept in orbit because the gravitational pull of some other external object guides it along the valleys in the warped spatial environment, as occurs for a ball bearing on the warped rubber membrane. Instead, Einstein showed that objects move through space (spacetime, more precisely) along the shortest possible paths—the "easiest possible paths" or the "paths of least resistance." If the space is warped, such paths will be curved. And so, although the rubber membrane–bowling ball model provides a good visual analogy of how an object such as the sun warps the space around it and thereby influences the motion of other bodies, the physical mechanism by which these distortions occur is totally different. The former appeals to our intuition about gravity in the traditional Newtonian framework, whereas the latter expresses a reformulation of gravity in terms of curved space.

A second shortcoming of the analogy stems from the rubber membrane's being two-dimensional. In reality, although harder to visualize, the sun (and all other massive objects) actually warps the three-dimensional space surrounding it. Figure 3.6 is a rough attempt to depict this; *all* of the space surrounding the sun—"below," "on the sides," on "top"—suffers the same kind of distortion, and Figure 3.6 schematically shows a partial sampling. A body, like the earth, travels *through* the three-dimensional warped spatial environment caused by the sun's presence. You may find this figure troubling—why doesn't the earth slam into the "vertical part" of curved space in the image? Bear in mind, though, that space, unlike the rubber membrane, is not a solid barrier. Instead, the warped grids in the image are but a couple of thin slices through the full three-dimensional warped space in which you, the earth, and everything else are immersed fully and move freely. Perhaps you find that this only makes the problem seem worse: Why don't we *feel* space if we are immersed within its fabric? But we do. We feel gravity, and space is the medium by which the gravitational force is communicated. As the eminent physicist John Wheeler has often said in describing gravity, "mass grips space by telling it how to curve, space grips mass by telling it how to move."[8]

A third, related shortcoming of the analogy is that we have suppressed the time dimension. We have done this for visual clarity because, notwithstanding the declaration of special relativity that we should think of the

Figure 3.6 A sampling of the warped three-dimensional space surrounding the sun.

time dimension on par with the three familiar spatial dimensions, it is significantly harder to "see" time. But, as illustrated by the example of the Tornado ride, acceleration—and hence gravity—warps *both space and time*. (In fact, the mathematics of general relativity shows that in the case of a relatively slow-moving body like the earth revolving around a typical star like the sun, the warping of time actually has a far more significant impact on the earth's motion than does the warping of space.) We will return to a discussion of the warping of time after the next section.

Important as these three caveats are, so long as you hold them in the back of your mind, it is perfectly acceptable to invoke the warped-space image provided by the bowling ball on the rubber membrane as an intuitive summary of Einstein's new view of gravity.

Conflict Resolution

By introducing space and time as dynamic players, Einstein provided a clear conceptual image of how gravity works. The central question, though, is whether this reformulation of the gravitational force resolves the conflict with special relativity that afflicts Newton's theory of gravity. It does. Again, the rubber membrane analogy gives the essential idea. Imagine that we have a ball bearing rolling in a straight line along the flat membrane in the absence of the bowling ball. As we place the bowling ball on the membrane the motion of the ball bearing will be affected, but *not instantaneously*. If we were to film this sequence of events and view it in slow motion we would see that the disturbance caused by the introduction of the bowling ball spreads like ripples in a pond and eventually reaches the position of the ball bearing. After a short time, transitory oscillations along the rubber surface would settle down, leaving us with a static warped membrane.

The same is true for the fabric of space. When no mass is present, space is flat, and a small object will blissfully be at rest or will travel at a constant velocity. If a large mass comes on the scene, space will warp—but as in the case of the membrane, the distortion will not be instantaneous. Rather, it will spread outward from the massive body, ultimately settling down into a warped shape that communicates the gravitational pull of the new body. In our analogy, disturbances to the rubber membrane travel along its extent at a speed dictated by its particular material composition. In the real setting of general relativity, Einstein was able to calculate how fast disturbances to the fabric of the universe travel and he found that they travel at *precisely the speed of light*. This means, for instance, that in the hypothetical example discussed earlier in which the demise of the sun affects the earth by virtue of changes in their mutual gravitational attraction, the influence will not be instantaneously communicated. Rather, as an object changes its position or even blows apart, it causes a change in the distortion of the spacetime fabric that spreads outward at light speed, precisely in keeping with the cosmic speed limit of special relativity. Thus, we on earth would visually learn of the sun's destruction at the same moment that we would feel the gravitational consequences—about eight minutes after it explodes. Einstein's formulation thereby resolves the conflict; gravitational disturbances keep pace with, but do not outrun, photons.

The Warping of Time, Revisited

Illustrations such as those of Figures 3.2, 3.4, and 3.6 capture the essence of what "warped space" means. A warp distorts the shape of space. Physicists have invented analogous images to try to convey the meaning of "warped time," but they are significantly more difficult to decipher, so we will not introduce them here. Instead, let's follow up the example of Slim and Jim on the Tornado ride, and try to get a sense of the experience of gravitationally induced warped time.

To do so, we revisit George and Gracie, no longer in the deep darkness of empty space, but floating near the outskirts of the solar system. They are still each wearing large digital clocks on their space suits that are initially synchronized. To keep things simple, we ignore the effects of the planets and consider only the gravitational field of the sun. Let's further imagine that a spaceship hovering near George and Gracie has reeled out a long cable extending all the way down to the vicinity of the sun's surface. George uses this cable to slowly lower himself toward the sun. As he does so, he periodically stops so that he and Gracie can compare the rate at which time is elapsing on their clocks. The warping of time predicted by Einstein's general relativity implies that George's clock will run slower and slower compared with Gracie's as the gravitational field he experiences gets stronger and stronger. That is, the closer he gets to the sun the slower his clock will run. It is in this sense that gravity distorts time as well as space.

You should note that unlike the case in Chapter 2 in which George and Gracie were in empty space moving relative to each other with constant velocity, in the present setting there is no symmetry between them. George, unlike Gracie, *feels* the force of gravity getting stronger and stronger—he has to hold the cable tighter and tighter as he gets closer to the sun to avoid being pulled in. Each of them agrees that George's clock is running slow. There is no "equally valid perspective" that exchanges their roles and reverses this conclusion. This is, in fact, what we found in Chapter 2 when George experienced an acceleration by turning on his jet-pack to catch up with Gracie. The acceleration George felt resulted in his clock definitively running slow relative to Gracie's. Since we now know that feeling accelerated motion is the same as feeling a gravitational force, the present situation of George on the cable involves the same principle, and once again we see that George's clock, and everything else in his life, runs in slow motion compared with Gracie's.

In a gravitational field such as that at the surface of an ordinary star like the sun, the slowing of clocks is quite small. If Gracie stays put at a billion miles from the sun, then when George has climbed to within a few miles of its surface, the rate of ticking of his clock will be about 99.9998 percent of Gracie's. Slower, but not by much.[9] If, however, George lowered himself on a cable so that he hovered just above the surface of a neutron star whose mass, roughly equal to that of the sun, is crushed to a density some million billion times that of solar density, the larger gravitational field would cause his clock to tick at about 76 percent of the rate of Gracie's. Stronger gravitational fields, such as those just outside a black hole (as discussed below), cause the flow of time to slow even further; stronger gravitational fields cause a more severe warping of time.

Experimental Verification of General Relativity

Most people who study general relativity are captivated by its aesthetic elegance. By replacing the cold, mechanistic Newtonian view of space, time, and gravity with a dynamic and geometric description involving curved spacetime, Einstein wove gravity into the basic fabric of the universe. Rather than being imposed as an additional structure, gravity becomes part and parcel of the universe at its most fundamental level. Breathing life into space and time by allowing them to curve, warp, and ripple results in what we commonly refer to as gravity.

Aesthetics aside, the ultimate test of a physical theory is its ability to explain and predict physical phenomena accurately. Since its inception in the late 1600s until the beginning of this century, Newton's theory of gravity passed this test with flying colors. Whether applied to balls thrown up in the air, objects dropped from leaning towers, comets whirling around the sun, or planets going about their solar orbits, Newton's theory provides extremely accurate explanations of all observations as well as predictions that have been verified innumerable times in a wealth of situations. The motivation for questioning this experimentally successful theory, as we have emphasized, was its property of instantaneous transmission of the gravitational force, in conflict with special relativity.

The effects of special relativity, although central to a fundamental understanding of space, time, and motion, are extremely small in the slow-velocity world we typically inhabit. Similarly, the deviations between Einstein's general relativity—a theory of gravity compatible with special relativity—and Newton's theory of gravity are also extremely small in most common situations. This is both good and bad. It is good because any theory purporting to supplant Newton's theory of gravity had better closely agree with it when applied in those arenas in which Newton's theory has been experimentally verified. It is bad because it makes it difficult to adjudicate between the two theories experimentally. Distinguishing between Newton's and Einstein's theories requires extremely precise measurements applied to experiments that are very sensitive to the ways in which the two theories differ. If you throw a baseball, Newtonian and Einsteinian gravity can be used to predict where it will land, and the answers will be different, but the differences will be so slight that they are generally beyond our capacity to detect experimentally. A more clever experiment is called for, and Einstein suggested one.[10]

We see stars at night, but of course they are also there during the day. We usually don't see them because their distant, pinpoint light is overwhelmed by the light emitted by the sun. During a solar eclipse, however, the moon temporarily blocks the light of the sun and distant stars become visible. Nevertheless, the presence of the sun still has an effect. Light from some of the distant stars must pass close to the sun on the way to earth. Einstein's general relativity predicts that the sun will cause the surrounding space and time to warp and such distortion *will influence the path taken by the starlight.* After all, the photons of distant origin travel along the fabric of the universe; if the fabric is warped, the motion of the photons will be affected much as for a material body. The bending of the path of light is greatest for those light signals that just graze the sun on their way to earth. A solar eclipse makes it possible to see such sun-grazing starlight without its being completely obscured by sunlight itself.

The angle through which the light path is bent can be measured in a simple way. The bending of the starlight's path results in a shift in the *apparent* position of the star. The shift can be accurately measured by comparing this apparent position with the star's *actual* location known from observations of the star at night (in the absence of the sun's warping influence), carried out when the earth is at an appropriate position, some six months earlier or later. In November of 1915, Einstein used his new understanding of gravity to calculate the angle through which starlight signals that just graze the sun would be bent and found the answer to be about .00049 of a degree (1.75 arcseconds, where an arcsecond is $\frac{1}{3600}$ of a degree). This tiny angle is equal to that subtended by a quarter placed upright and viewed from nearly two miles away. The detection of such a small angle was, however, within reach of the technology of the day. At the urging of Sir Frank Dyson, director of the Greenwich observatory, Sir Arthur Eddington, a well-known astronomer and secretary of the Royal Astronomical Society in England, organized an expedition to the island of Principe off the coast of West Africa to test Einstein's prediction during the solar eclipse of May 29, 1919.

On November 6, 1919, after some five months of analysis of the photographs taken during the eclipse at Principe (and of other photographs of the eclipse taken by a second British team led by Charles Davidson and Andrew Crommelin in Sobral, Brazil), it was announced at a joint meeting of the Royal Society and the Royal Astronomical Society that Einstein's prediction based on general relativity had been confirmed. It took little time for word of this success—a complete overturning of previous conceptions of space and time—to spread well beyond the confines of the physics community, making Einstein a celebrated figure worldwide. On November 7, 1919, the headline in the London *Times* read "REVOLUTION IN SCIENCE—NEW THEORY OF THE UNIVERSE—NEWTONIAN IDEAS OVERTHROWN."[11] This was Einstein's moment of glory.

In the years following this experiment, Eddington's confirmation of general relativity came under some critical scrutiny. Numerous difficult and subtle aspects of the measurement made it hard to reproduce and raised some questions regarding the trustworthiness of the original experiment. Nevertheless, in the last 40 years a variety of experiments making use of technological advancements have tested numerous aspects of general relativity with great precision. The predictions of general relativity have been uniformly confirmed. There is no longer any doubt that Einstein's description of gravity is not only compatible with special relativity, but yields predictions closer to experimental results than those of Newton's theory.

Black Holes, the Big Bang, and the Expansion of Space

Whereas special relativity is most manifest when things are moving fast, general relativity comes into its own when things are very massive and the warps in space and time are correspondingly severe. Let's describe two examples.

The first is a discovery made by the German astronomer Karl Schwarzschild while studying Einstein's revelations on gravity in between his own calculations of artillery trajectories at the Russian front during World War I in 1916. Remarkably, just months after Einstein had put the finishing touches on general relativity, Schwarzschild was able to use the theory to gain a complete and exact understanding of the way space and time warp in the vicinity of a perfectly spherical star. Schwarzschild sent his results from the Russian front to Einstein, who presented them on Schwarzschild's behalf to the Prussian Academy.

Beyond confirming and making mathematically precise the warping that was schematically illustrated in Figure 3.5, Schwarzschild's work—which has now come to be known as "Schwarzschild's solution"—revealed a stunning implication of general relativity. He showed that if the mass of a star is concentrated in a small enough spherical region, so that its mass divided by its radius exceeds a particular critical value, the resulting space-time warp is so radical that *anything,* including light, that gets too close to the star will be unable to escape its gravitational grip. Since not even light can escape such "compressed stars," they were initially called *dark* or *frozen stars.* A more catchy name was coined years later by John Wheeler, who called them *black holes*—black because they cannot emit light, holes because anything getting too close falls into them, never to return. The name stuck.

We illustrate Schwarzschild's solution in Figure 3.7. Although black holes have a reputation for rapacity, objects that pass by them at a "safe" distance are deflected in much the same way that they would be by an ordinary star, and can proceed on their merry way. But objects of any composition whatsoever that get too close—closer than what has been termed the black hole's *event horizon*—are doomed: they will be drawn inexorably toward the center of the black hole and subject to an ever increasing and ultimately destructive gravitational strain. For example, if you dropped feet first through the event horizon, as you approached the black hole's center you would find yourself getting increasingly uncomfortable. The gravitational force of the black hole would increase so dramatically that its

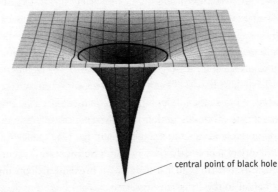

central point of black hole

Figure 3.7 A black hole warps the surrounding spacetime fabric so severely that anything that comes within its "event horizon"—illustrated by the dark circle—can't escape from its gravitational grip. No one knows exactly what happens at the deepest interior point of a black hole.

pull on your feet would be much stronger than its pull on your head (since in a feet-first fall your feet are always a bit closer than your head to the black hole's center); so much stronger, in fact, that you would be stretched with a force that would quickly tear your body to shreds.

If, on the contrary, you were more prudent in your wanderings near a black hole and took great care not to trespass beyond the event horizon, you could make use of the black hole for a rather amazing feat. Imagine, for example, that you were to discover a black hole whose mass was about 1,000 times the mass of the sun, and that you were to lower yourself on a cable, much as George did near the sun, to about an inch above the black hole's event horizon. As we have discussed, gravitational fields cause a warping of time, and this means that your passage through time would slow down. In fact, since black holes have such strong gravitational fields, your passage through time would slow *way down*. Your watch would tick about ten thousand times more slowly than those of your friends back on earth. If you were to hover just above the black hole's event horizon in this manner for a year, and then climb up the cable to your waiting starship for a short, yet leisurely, journey home, upon arrival at earth you would find that more than ten thousand years had passed since your initial departure. You would have successfully used the black hole as a kind of time machine, allowing you to travel to earth's distant future.

To get a sense of the extreme scales involved, a star with the mass of the sun would be a black hole if its radius were not its actual value (about 450,000 miles), but, instead, just under 2 miles. Imagine: The whole of the sun squeezed to fit comfortably within upper Manhattan. A teaspoonful of such a compressed sun would weigh about as much as Mount Everest. To make a black hole out of the earth we would need to crush it into a sphere whose radius is less than half an inch. For a long time physicists were skeptical about whether such extreme configurations of matter could ever actually occur, and many thought that black holes were merely a reflection of an overworked theoretician's imagination.

Nevertheless, during the last decade, an increasingly convincing body of experimental evidence for the existence of black holes has accumulated. Of course, since they are black, they cannot be observed directly by scanning the sky with telescopes. Instead, astronomers search for black holes by seeking anomalous behavior of other more ordinary light-emitting stars that may be positioned just outside a black hole's event horizon. For instance, as dust and gas from the outer layers of nearby ordinary stars fall toward the event horizon of a black hole, they are accelerated to nearly the speed of light. At such speeds, friction within the maelstrom of downward-swirling material generates an enormous amount of heat, causing the dust-gas mixture to "glow," giving off both ordinary visible light and X rays. Since this radiation is produced just outside the event horizon, it can escape the black hole and travel through space to be observed and studied directly. General relativity makes detailed predictions about properties that such X ray emissions will have; observation of these predicted properties gives strong, albeit indirect, evidence for the existence of black holes. For example, mounting evidence indicates that there is a very massive black hole, some two and a half million times as massive as the sun, sitting in the center of our own Milky Way galaxy. And even this seemingly gargantuan black hole pales in comparison to what astronomers believe to reside in the core of the astonishingly luminous quasars that are scattered throughout the cosmos: black holes whose masses may well be *billions* of times that of the sun.

Schwarzschild died only a few months after finding his solution, from a skin disease he contracted at the Russian front. He was 42. His tragically brief encounter with Einstein's theory of gravity uncovered one of the most striking and mysterious facets of the natural world.

The second example in which general relativity flexes its muscle concerns the origin and evolution of the whole universe. As we have seen, Einstein showed that space and time respond to the presence of mass and energy. This distortion of spacetime affects the motion of other cosmic bodies moving in the vicinity of the resulting warps. In turn, the precise way in which these bodies move, by virtue of their own mass and energy, has a further effect on the warping of spacetime, which further affects the motion of the bodies, and on and on the interconnected cosmic dance goes. Through the equations of general relativity, equations rooted in geometrical insights into curved space spearheaded by the great nineteenth-century mathematician Georg Bernhard Riemann (more about Riemann later), Einstein was able to describe the mutual evolution of space, time, and matter quantitatively. To his great surprise, when the equations are applied beyond an isolated context within the universe, such as a planet or a comet orbiting a star, to the universe as a whole, a remarkable conclusion is reached: *the overall size of the spatial universe must be changing in time.* That is, either the fabric of the universe is stretching or it is shrinking, but it is not simply staying put. The equations of general relativity show this explicitly.

This conclusion was too much even for Einstein. He had overturned the collective intuition regarding the nature of space and time built up through everyday experiences over thousands of years, but the notion of an always existing, never changing universe was too ingrained for even this radical thinker to abandon. For this reason, Einstein revisited his equations and modified them by introducing something known as a *cosmological constant,* an additional term that allowed him to avoid this prediction and once again bask in the comfort of a static universe. However, 12 years later, through detailed measurements of distant galaxies, the American astronomer Edwin Hubble experimentally established that the universe *is expanding.* In a now-famous story in the annals of science, Einstein then returned to the original form of his equations, citing his temporary modification of them as the biggest blunder of his life.[12] His initial unwillingness to accept the conclusion notwithstanding, Einstein's theory predicted the expansion of the universe. In fact, in the early 1920s—years before Hubble's measurements—the Russian meteorologist Alexander Friedmann had used Einstein's original equations to show, in some detail, that all galaxies would be carried along on the substrate of stretching spatial fabric, thereby speedily moving away from all others. Hubble's observations and numerous subsequent ones have thoroughly verified this astonishing conclusion of general relativity. By offering the explanation for the expansion of the universe, Einstein achieved one of the greatest intellectual feats of all time.

If the fabric of space is stretching, thereby increasing the distance between galaxies that are carried along on the cosmic flow, we can imagine running the evolution backward in time to learn about the origin of the universe. In reverse, the fabric of space shrinks, bringing all galaxies closer and closer to each other. Like the contents of a pressure cooker, as the shrinking universe compresses the galaxies together, the temperature dramatically increases, stars disintegrate and a hot plasma of matter's elementary constituents is formed. As the fabric continues to shrink, the temperature rises unabated, as does the density of the primordial plasma. As we imagine running the clock backward from the age of the presently observed universe, about 15 billion years, the universe as we know it is crushed to an ever smaller size. The matter making up *everything*—every car, house, building, mountain on earth; the earth itself; the moon; Saturn, Jupiter, and every other planet; the sun and every other star in the Milky Way; the Andromeda galaxy with its 100 billion stars and each and every other of the more than 100 billion galaxies—is squeezed by a cosmic vise to astounding density. And as the clock is turned back to ever earlier times, the whole of the cosmos is compressed to the size of an orange, a lemon, a pea, a grain of sand, and to yet tinier size still. Extrapolating all the way back to "the beginning," the universe would appear to have begun as a *point*—an image we will critically re-examine in later chapters—in which all matter and energy is squeezed together to unimaginable density and temperature. It is believed that a cosmic fireball, the big bang, erupted from this volatile mixture spewing forth the seeds from which the universe as we know it evolved.

The image of the big bang as a cosmic explosion ejecting the material contents of the universe like shrapnel from an exploding bomb is a useful one to bear in mind, but it is a little misleading. When a bomb explodes, it does so at a particular location *in space* and at a particular moment *in time*. Its contents are ejected into the surrounding space. In the big bang, there is no surrounding space. As we devolve the universe backward toward the beginning, the squeezing together of all material content occurs because *all of space* is shrinking. The orange-size, the pea-size, the grain of sand–size devolution describes the *whole* of the universe—not something within the universe. Carrying on to the beginning, there is simply no space outside the primordial pinpoint grenade. Instead, the big bang is the eruption of compressed space whose unfurling, like a tidal wave, carries along matter and energy even to this day.

Is General Relativity Right?

No deviations from the predictions of general relativity have been found in experiments performed with our present level of technology. Only time will tell if greater experimental precision will ultimately uncover some, thereby showing this theory, too, to be only an approximate description of how nature actually works. The systematic testing of theories to greater and greater levels of accuracy is, certainly, one of the ways science progresses, but it is not the only way. In fact, we have already seen this: The search for a new theory of gravity was initiated, not by an experimental refutation of Newton's theory, but rather by the conflict of Newtonian gravity with another *theory*—special relativity. It was only after the discovery of general relativity as a competing theory of gravity that experimental flaws in Newton's theory were identified by seeking out tiny but measurable ways in which the two theories differ. Thus, internal theoretical inconsistencies can play as pivotal a role in driving progress as do experimental data.

For the last half century, physics has been faced with still another theoretical conflict whose severity is on par with that between special relativity and Newtonian gravity. General relativity appears to be fundamentally incompatible with another extremely well-tested theory: *quantum mechanics*. Regarding the material covered in this chapter, the conflict prevents physicists from understanding what really happens to space, time, and matter when crushed together fully at the moment of the big bang or at the central point of a black hole. But more generally, the conflict alerts us to a fundamental deficiency in our conception of nature. The resolution of this conflict has eluded attempts by some of the greatest theoretical physicists, giving it a well-deserved reputation as *the* central problem of modern theoretical physics. Understanding the conflict requires familiarity with some basic features of quantum theory, to which we now turn.

Chapter 4

Microscopic Weirdness

A bit worn out from their trans-solar-system expedition, George and Gracie return to earth and head over to the H-Bar for some post-space-sojourning refreshments. George orders the usual—papaya juice on the rocks for himself and a vodka tonic for Gracie—and kicks back in his chair, hands clasped behind his head, to enjoy a freshly lit cigar. Just as he prepares to inhale, though, he is stunned to find that the cigar has vanished from between his teeth. Thinking that the cigar must somehow have slipped from his mouth, George sits forward expecting to find it burning a hole in his shirt or trousers. But it is not there. The cigar is not to be found. Gracie, roused by George's frantic movement, glances over and spots the cigar lying on the counter directly *behind* George's chair. "Strange," George says, "how in the heck could it have fallen over there? It's as if it went right through my head—but my tongue isn't burned and I don't seem to have any new holes." Gracie examines George and reluctantly confirms that his tongue and head appear to be perfectly normal. As the drinks have just arrived, George and Gracie shrug their shoulders and chalk up the fallen cigar to one of life's little mysteries. But the weirdness at the H-Bar continues.

George looks into his papaya juice and notices that the ice cubes are incessantly rattling around—bouncing off of each other and the sides of the glass like overcharged automobiles in a bumper-car arena. And this time he is not alone. Gracie holds up her glass, which is about half the size of George's, and both of them see that her ice cubes are bouncing around even more frantically. They can hardly make out the individual cubes as they all blur together into an icy mass. But none of this compares to what happens next. As George and Gracie stare at her rattling drink with wide-eyed wonderment, they see a single ice cube *pass through* the side of her glass and drop down to the bar. They grab the glass and see that it is fully intact; somehow the ice cube went right through the solid glass without causing any damage. "Must be post-space-walk hallucinations," says George. They each fight off the frenzy of careening ice cubes to down their drinks in one go, and head home to recover. Little do George and Gracie realize that in their haste to leave, they mistook a decorative door painted on a wall of the bar for the real thing. The patrons of the H-Bar, though, are well accustomed to people passing through walls and hardly take note of George and Gracie's abrupt departure.

A century ago, while Conrad and Freud were illuminating the heart and the soul of darkness, the German physicist Max Planck shed the first ray of light on quantum mechanics, a conceptual framework that proclaims, among other things, that the H-Bar experiences of George and Gracie—when scaled down to the microscopic realm—need not be attributed to clouded faculties. Such unfamiliar and bizarre happenings are typical of how our universe, on extremely small scales, actually behaves.

The Quantum Framework

Quantum mechanics is a conceptual framework for understanding the microscopic properties of the universe. And just as special relativity and general relativity require dramatic changes in our worldview when things are moving very quickly or when they are very massive, quantum mechanics reveals that the universe has equally if not more startling properties when examined on atomic and subatomic distance scales. In 1965, Richard Feynman, one of the greatest practitioners of quantum mechanics, wrote,

There was a time when the newspapers said that only twelve men understood the theory of relativity. I do not believe there ever was such a time. There might have been a time when only one man did because he was the only guy who caught on, before he wrote his paper. But after people read the paper a lot of people understood the theory of relativity in one way or other, certainly more than twelve. On the other hand I think I can safely say that nobody understands quantum mechanics.[1]

Although Feynman expressed this view more than three decades ago, it applies equally well today. What he meant is that although the special and general theories of relativity require a drastic revision of previous ways of seeing the world, when one fully accepts the basic principles underlying them, the new and unfamiliar implications for space and time follow directly from careful logical reasoning. If you ponder the descriptions of Einstein's work in the preceding two chapters with adequate intensity, you will—if even for just a moment—recognize the inevitability of the conclusions we have drawn. Quantum mechanics is different. By 1928 or so, many of the mathematical formulas and rules of quantum mechanics had been put in place and, ever since, it has been used to make *the* most precise and successful numerical predictions in the history of science. But in a real sense those who use quantum mechanics find themselves following rules and formulas laid down by the "founding fathers" of the theory—calculational procedures that are straightforward to carry out—without really understanding *why* the procedures work or *what* they really mean. Unlike relativity, few if any people ever grasp quantum mechanics at a "soulful" level.

What are we to make of this? Does it mean that on a microscopic level the universe operates in ways so obscure and unfamiliar that the human mind, evolved over eons to cope with phenomena on familiar everyday scales, is unable to fully grasp "what really goes on"? Or, might it be that through historical accident physicists have constructed an extremely awkward formulation of quantum mechanics that, although quantitatively successful, obfuscates the true nature of reality? No one knows. Maybe some time in the future some clever person will see clear to a new formulation that will fully reveal the "whys" and the "whats" of quantum mechanics. And then again, maybe not. The only thing we know with certainty is that quantum mechanics absolutely and unequivocally shows us that a number of basic concepts essential to our understanding of the familiar everyday world *fail to have any meaning* when our focus narrows to the microscopic realm. As a result, we must significantly modify both our language and our reasoning when attempting to understand and explain the universe on atomic and subatomic scales.

In the following sections we will develop the basics of this language and describe a number of the remarkable surprises it entails. If along the way quantum mechanics seems to you to be altogether bizarre or even ludicrous, you should bear in mind two things. First, beyond the fact that it is a mathematically coherent theory, the only reason we believe in quantum mechanics is because it yields predictions that have been verified to astounding accuracy. If someone can tell you volumes of intimate details of your childhood in excruciating detail, it's hard not to believe their claim of being your long-lost sibling. Second, you are not alone in having this reaction to quantum mechanics. It is a view held to a greater or lesser extent by some of the most revered physicists of all time. Einstein refused to accept quantum mechanics fully. And even Niels Bohr, one of the central pioneers of quantum theory and one of its strongest proponents, once remarked that if you do not get dizzy sometimes when you think about quantum mechanics, then you have not really understood it.

It's Too Hot in the Kitchen

The road to quantum mechanics began with a puzzling problem. Imagine that your oven at home is perfectly insulated, that you set it to some temperature, say 400 degrees Fahrenheit, and you give it enough time to heat up. Even if you had sucked all the air from the oven before turning it on, by heating its walls you generate waves of radiation in its interior. This is the same kind of radiation—heat and light in the form of electromagnetic waves—that is emitted by the surface of the sun, or a glowing-hot iron poker.

Here's the problem. Electromagnetic waves carry energy—life on earth, for example, relies crucially on solar energy transmitted from the sun to the earth by electromagnetic waves. At the beginning of the twentieth century, physicists calculated the total energy carried by all of the electromagnetic radiation inside an oven at a chosen temperature. Using well-established calculational procedures they came up with a ridiculous answer: For any chosen temperature, the total energy in the oven is *infinite*.

It was clear to everyone that this was nonsense—a hot oven can embody significant energy but surely not an infinite amount. To understand the resolution proposed by Planck it is worth understanding the problem in a bit more detail. It turns out that when Maxwell's electromagnetic theory is applied to the radiation in an oven it shows that the waves generated by the hot walls must have a *whole* number of peaks and troughs that fit perfectly between opposite surfaces. Some examples are shown in Figure 4.1. Physicists use three terms to describe these waves: wavelength, frequency, and amplitude. The *wavelength* is the distance between successive peaks or successive troughs of the waves, as illustrated in Figure 4.2. More peaks and troughs mean a shorter wavelength, as they must all be crammed in between the fixed walls of the oven. The *frequency* refers to the number of up-and-down cycles of oscillation that a wave completes every second. It turns out that the frequency is determined by the wavelength and vice versa: longer wavelengths imply lower frequency; shorter wavelengths imply higher frequency. To see why, think of what happens when you produce waves by shaking a long rope that is tied down at one end. To generate a long wavelength, you leisurely shake your end up and down. The frequency of the waves matches the number of cycles per second your arm goes through and is consequently fairly low. But to generate short wavelengths you shake your end more frantically—more frequently, so to speak—and this yields a higher-frequency wave. Finally, physicists use the term *amplitude* to describe the maximum height or depth of a wave, as also illustrated in Figure 4.2.

In case you find electromagnetic waves a bit abstract, another good

Figure 4.1 Maxwell's theory tells us that the radiation waves in an oven have a whole number of crests and troughs—they fill out complete wave-cycles.

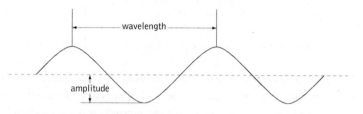

Figure 4.2 The wavelength is the distance between successive peaks or troughs of a wave. The amplitude is the maximal height or depth of the wave.

analogy to keep in mind are the waves that are produced by plucking a violin string. Different wave frequencies correspond to different musical notes: the higher the frequency, the higher the note. The amplitude of a wave on a violin string is determined by how hard you pluck it. A harder pluck means that you put more energy into the wave disturbance; more energy therefore corresponds to a larger amplitude. You can hear this, as the resulting tone is louder. Similarly, less energy corresponds to a smaller amplitude and a lower volume of sound.

By making use of nineteenth-century thermodynamics, physicists were able to determine how much energy the hot walls of the oven would pump into electromagnetic waves of each allowed wavelength—how hard the walls would, in effect, "pluck" each wave. The result they found is simple to state: Each of the allowed waves—*regardless of its wavelength*—carries the same amount of energy (with the precise amount determined by the temperature of the oven). In other words, all of the possible wave patterns within the oven are on completely equal footing when it comes to the amount of energy they embody.

At first this seems like an interesting, albeit innocuous, result. It isn't. It spells the downfall of what has come to be known as classical physics. The reason is this: Even though requiring that all waves have a whole number of peaks and troughs rules out an enormous variety of conceivable wave patterns in the oven, there are still an infinite number that are possible—those with ever more peaks and troughs. Since each wave pattern carries the same amount of energy, an infinite number of them translates into an infinite amount of energy. At the turn of the century, there was a gargantuan fly in the theoretical ointment.

Making Lumps at the Turn of the Century

In 1900 Planck made an inspired guess that allowed a way out of this puzzle and would earn him the 1918 Nobel Prize in physics.[2] To get a feel for his resolution, imagine that you and a huge crowd of people—"infinite" in number—are crammed into a large, cold warehouse run by a miserly landlord. There is a fancy digital thermostat on the wall that controls the temperature but you are shocked when you discover the charges that the landlord levies for heat. If the thermostat is set to 50 degrees Fahrenheit everyone must give the landlord $50. If it is set to 55 degrees everyone must pay $55, and so on. You realize that since you are sharing the warehouse with an infinite number of companions, the landlord will earn an infinite amount of money if you turn on the heat at all.

But on closer reading of the landlord's rules of payment you see a loophole. Because the landlord is a very busy man he does not want to give change, especially not to an infinite number of individual tenants. So he works on an honor system. Those who can pay exactly what they owe, do so. Otherwise, they pay only as much as they can without requiring change. And so, wanting to involve everyone but wanting to avoid the exorbitant charges for heat, you compel your comrades to organize the wealth of the group in the following manner: One person carries all of the pennies, one person carries all of the nickels, one carries all of the dimes, one carries all of the quarters, and so on through dollar bills, five-dollar bills, ten-dollar bills, twenties, fifties, hundreds, thousands, and ever larger (and unfamiliar) denominations. You brazenly set the thermostat to 80 degrees and await the landlord's arrival. When he does come, the person carrying pennies goes to pay first and turns over 8,000. The person carrying nickels then turns over 1,600 of them, the person carrying dimes turns over 800, the person with quarters turns over 320, the person with dollars gives the landlord 80, the person with five-dollar bills turns over 16, the person with ten-dollar bills gives him 8, the person with twenties gives him 4, and the person with fifties hands over one (since 2 fifty-dollar bills would exceed the necessary payment, thereby requiring change). But everyone else carries only a denomination—a minimal "lump" of money— that exceeds the required payment. Therefore they cannot pay the landlord and hence rather than getting the infinite amount of money he expected, the landlord leaves with the paltry sum of $690.

Planck made use of a very similar strategy to reduce the ridiculous result of infinite energy in an oven to one that is finite. Here's how. Planck boldly guessed that the energy carried by an electromagnetic wave in the oven, like money, comes in lumps. The energy can be one times some fundamental "energy denomination," or two times it, or three times it, and so forth—but that's it. Just as you can't have one-third of a penny or two and a half quarters, Planck declared that when it comes to energy, no fractions are allowed. Now, our monetary denominations are determined by the United States Treasury. Seeking a more fundamental explanation, Planck suggested that the energy denomination of a wave—the minimal lump of energy that it can have—is determined by its frequency. Specifically, he posited that the *minimum* energy a wave can have is *proportional to its frequency:* larger frequency (shorter wavelength) implies larger minimum energy; smaller frequency (longer wavelength) implies smaller minimum energy. Roughly speaking, just as gentle ocean waves are long and luxurious while harsh ones are short and choppy, long-wavelength radiation is intrinsically less energetic than short-wavelength radiation.

Here's the punch line: Planck's calculations showed that this lumpiness of the allowed energy in each wave cured the previous ridiculous result of infinite total energy. It's not hard to see why. When an oven is heated to some chosen temperature, the calculations based on nineteenth-century thermodynamics predicted the common energy that each and every wave would supposedly contribute to the total. But like those comrades who cannot contribute the common amount of money they each owe the landlord because the monetary denomination they carry is too large, if the minimum energy a particular wave can carry exceeds the energy it is supposed to contribute, it can't contribute and instead lies dormant. Since, according to Planck, the minimum energy a wave can carry is proportional to its frequency, as we examine waves in the oven of ever larger frequency (shorter wavelength), sooner or later the minimum energy they can carry *is* bigger than the expected energy contribution. Like the comrades in the warehouse entrusted with denominations larger than fifty-dollar bills, these waves with ever-larger frequencies cannot contribute the amount of energy demanded by nineteenth-century physics. And so, just as only a finite number of comrades are able to contribute to the total heat payment—leading to a finite amount of total money—only a finite number of waves are able to contribute to the oven's total energy—again leading to a finite amount of total energy. Be it energy or money, the lumpiness of the fundamental units—and the ever increasing size of these lumps as we go to higher frequencies or to larger monetary denominations—changes an infinite answer to one that is finite.[3]

By eliminating the manifest nonsense of an infinite result, Planck had taken an important step. But what really made people believe that his guess had validity is that the finite answer that his new approach gave for the energy in an oven agreed spectacularly with experimental measurements. Specifically, Planck found that by adjusting *one* parameter that entered into his new calculations, he could predict accurately the measured energy of an oven for any selected temperature. This one parameter is the proportionality factor between the frequency of a wave and the minimal lump of energy it can have. Planck found that this proportionality factor—now known as *Planck's constant* and denoted $\hbar$ (pronounced "h-bar")—is about a billionth of a billionth of a billionth in everyday units.[4] The tiny value of Planck's constant means that the size of the energy lumps are typically very small. This is why, for example, it *seems* to us that we can cause the energy of a wave on a violin string—and hence the volume of sound it produces—to change continuously. In reality, though, the energy of the wave passes through discrete steps, à la Planck, but the size of the steps is so small that the discrete jumps from one volume to another appear to be smooth. According to Planck's assertion, the size of these jumps in energy grows as the frequency of the waves gets higher and higher (while wavelengths get shorter and shorter). This is the crucial ingredient that resolves the infinite-energy paradox.

As we shall see, Planck's quantum hypothesis does far more than allow us to understand the energy content of an oven. It overturns much about the world that we hold to be self-evident. The smallness of $\hbar$ confines most of these radical departures from life-as-usual to the microscopic realm, but if $\hbar$ happened to be much larger than it is, the strange happenings at the H-Bar would actually be commonplace. As we shall see, their microscopic counterparts certainly are.

What Are the Lumps?

Planck had no justification for his pivotal introduction of lumpy energy. Beyond the fact that it worked, neither he nor anyone else could give a compelling reason for why it should be true. As the physicist George Gamow once said, it was as if nature allowed one to drink a whole pint of beer or no beer at all, but nothing in between.[5] In 1905, Einstein found an explanation and for this insight he was awarded the 1921 Nobel Prize in physics.

Einstein came up with his explanation by puzzling over something known as the photoelectric effect. The German physicist Heinrich Hertz in 1887 was the first to find that when electromagnetic radiation—light— shines on certain metals, they emit electrons. By itself this is not particularly remarkable. Metals have the property that some of their electrons are only loosely bound within atoms (which is why they are such good conductors of electricity). When light strikes the metallic surface it relinquishes its energy, much as it does when it strikes the surface of your skin, causing you to feel warmer. This transfered energy can agitate electrons in the metal, and some of the loosely bound ones can be knocked clear off the surface.

But the strange features of the photoelectric effect become apparent when one studies more detailed properties of the ejected electrons. At first sight you would think that as the intensity of the light—its brightness— is increased, the speed of the ejected electrons will also increase, since the impinging electromagnetic wave has more energy. But this does *not* happen. Rather, the *number* of ejected electrons increases, but their speed stays fixed. On the other hand, it has been experimentally observed that the speed of the ejected electrons *does* increase if the *frequency* of the impinging light is increased, and, equivalently, their speed decreases if the frequency of the light is decreased. (For electromagnetic waves in the visible part of the spectrum, an increase in frequency corresponds to a change in color from red to orange to yellow to green to blue to indigo and finally to violet. Frequencies higher than that of violet are not visible and correspond to ultraviolet and, subsequently, X rays; frequencies lower than that of red are also not visible, and correspond to infrared radiation.) In fact, as the frequency of the light used is decreased, there comes a point when the speed of the emitted electrons drops to zero and they stop being ejected from the surface, *regardless of the possibly blinding intensity of the light source.* For some unknown reason, the *color* of the impinging light beam—not its total energy—controls whether or not electrons are ejected, and if they are, the energy they have.

To understand how Einstein explained these puzzling facts, let's go back to the warehouse, which has now heated up to a balmy 80 degrees. Imagine that the landlord, who hates children, requires everyone under the age of fifteen to live in the sunken basement of the warehouse, which the adults can view from a huge wraparound balcony. Moreover, the only way any of the enormous number of basement-bound children can leave the warehouse is if they can pay the guard an 85-cent exit fee. (This landlord is *such* an ogre.) The adults, who at your urging have arranged the collective wealth by denomination as described above, can give money to the children only by throwing it down to them from the balcony. Let's see what happens.

The person carrying pennies begins by tossing a few down, but this is far too meagre a sum for any of the children to be able to afford the departure fee. And because there is an essentially "infinite" sea of children all ferociously fighting in a turbulent tumult for the falling money, even if the penny-entrusted adult throws enormous numbers down, no individual child will come anywhere near collecting the 85 he or she needs to pay the guard. The same is true for the adults carrying nickels, dimes, and quarters. Although each tosses down a staggeringly large total amount of money, any single child is lucky if he or she gets even one coin (most get nothing at all) and certainly no child collects the 85 cents necessary to leave. But then, when the adult carrying dollars starts throwing them down—even comparatively tiny sums, dollar by single dollar—those lucky children who catch a single bill are able to leave immediately. Notice, though, that even as this adult loosens up and throws down barrels of dollar bills, the number of children who are able to leave increases enormously, but each has exactly 15 cents left after paying the guard. This is true regardless of the total number of dollars tossed.

Here is what all this has to do with the photoelectric effect. Based on the experimental data reviewed above, Einstein suggested incorporating Planck's lumpy picture of wave energy into a new description of light. A light beam, according to Einstein, should actually be thought of as a *stream of tiny packets*—tiny particles of light—which were ultimately christened *photons* by the chemist Gilbert Lewis (an idea we made use of in our example of the light clock of Chapter 2). To get a sense of scale, according to this particle view of light, a typical one-hundred-watt bulb emits about a hundred billion billion (10^{20}) photons per second. Einstein used this new conception to suggest a microscopic mechanism underlying the photoelectric effect: An electron is knocked off a metallic surface, he proposed, if it gets hit by a sufficiently energetic photon. And what determines the energy of an individual photon? To explain the experimental data, Einstein followed Planck's lead and proposed that the energy of *each* photon is proportional to the frequency of the light wave (with the proportionality factor being Planck's constant).

Now, like the children's minimum departure fee, the electrons in a metal must be jostled by a photon posessing a certain minimum energy in order to be kicked off the surface. (As with the children fighting for money, it is extremely unlikely that any one electron gets hit by more than one photon—most don't get hit at all.) But if the impinging light beam's frequency is too low, its individual photons will lack the punch necessary to eject electrons. Just as no children can afford to leave regardless of the huge total number of coins the adults shower upon them, no electrons are jostled free regardless of the huge total energy embodied in the impinging light beam, if its frequency (and thus the energy of its individual photons) is too low.

But just as children are able to leave the warehouse as soon as the monetary denomination showered upon them gets large enough, electrons will be knocked off the surface as soon as the frequency of the light shone on them—its energy denomination—gets high enough. Moreover, just as the dollar-entrusted adult increases the total money thrown down by increasing the number of individual bills tossed, the total intensity of a light beam of a chosen frequency is increased by increasing the number of photons it contains. And just as more dollars result in more children being able to leave, more photons result in more electrons being hit and knocked clear off the surface. But notice that the leftover energy that each of these electrons has after ripping free of the surface depends solely on the energy of the photon that hits it—and this is determined by the frequency of the light beam, not its total intensity. Just as children leave the basement with

15 cents no matter how many dollar bills are thrown down, each electron leaves the surface with the same energy—and hence the same speed—regardless of the total intensity of the impinging light. More total money simply means more children can leave; more total energy in the light beam simply means more electrons are knocked free. If we want children to leave the basement with more money, we must increase the monetary denomination tossed down; if we want electrons to leave the surface with greater speed, we must increase the frequency of the impinging light beam—that is, we must increase the energy denomination of the photons we shine on the metallic surface.

This is precisely in accord with the experimental data. The frequency of the light (its color) determines the speed of the ejected electrons; the total intensity of the light determines the number of ejected electrons. And so Einstein showed that Planck's guess of lumpy energy actually reflects a fundamental feature of electromagnetic waves: They are composed of particles—photons—that are little bundles, or *quanta,* of light. The lumpiness of the energy embodied by such waves is due to their being composed of lumps.

Einstein's insight represented great progress. But, as we shall now see, the story is not as tidy as it might appear.

Is It a Wave or Is It a Particle?

Everyone knows that water—and hence water waves—are composed of a huge number of water molecules. So is it really surprising that light waves are also composed of a huge number of particles, namely photons? It is. But the surprise is in the details. You see, more than three hundred years ago Newton proclaimed that light consisted of a stream of particles, so the idea is not exactly new. However, some of Newton's colleagues, most notably the Dutch physicist Christian Huygens, disagreed with him and argued that light is a wave. The debate raged but ultimately experiments carried out by the English physicist Thomas Young in the early 1800s showed that Newton was wrong.

A version of Young's experimental setup—known as the double-slit experiment—is schematically illustrated in Figure 4.3. Feynman was fond of saying that all of quantum mechanics can be gleaned from carefully

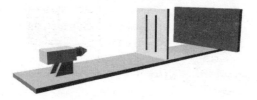

Figure 4.3 In the double-slit experiment, a beam of light is shone on a barrier in which two slits have been cut. The light that passes through the barrier is then recorded on a photographic plate, when either or both of the slits are open.

thinking through the implications of this single experiment, so it's well worth discussing. As we see from Figure 4.3, light is shone on a thin solid barrier in which two slits are cut. A photographic plate records the light that gets through the slits—brighter areas of the photograph indicate more incident light. The experiment consists of comparing the images on photographic plates that result when either or both of the slits in the barrier are kept open and the light source is turned on.

If the left slit is covered and the right slit is open, the photograph looks like that shown in Figure 4.4. This makes good sense, since the light that hits the photographic plate must pass through the only open slit and will therefore be concentrated around the right part of the photograph. Similarly, if the right slit is covered and the left slit open, the photograph will look like that in Figure 4.5. If *both* slits are open, Newton's particle picture of light leads to the prediction that the photographic plate will look like that in Figure 4.6, an amalgam of Figures 4.4 and 4.5. In essence, if

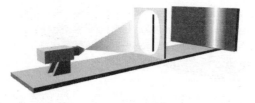

Figure 4.4 The right slit is open in this experiment, leading to an image on the photographic plate as shown.

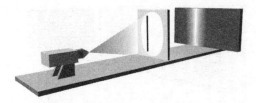

Figure 4.5 As in Figure 4.4, except now only the left slit is open.

you think of Newton's corpuscles of light as if they were little pellets you fire at the wall, the ones that get through will be concentrated in the two areas that line up with the two slits. The wave picture of light, on the contrary, leads to a very different prediction for what happens when both slits are open. Let's see this.

Imagine for a moment that rather than dealing with light waves we use water waves. The result we will find is the same, but water is easier to think about. When water waves strike the barrier, outgoing circular water waves emerge from each slit, much like those created by throwing a pebble into a pond, as illustrated in Figure 4.7. (It is simple to try this using a cardboard barrier with two slits in a pan of water.) As the waves emerging from each slit overlap with each other, something quite interesting happens. If two wave peaks overlap, the height of the water wave at that point increases: It's the sum of the heights of the two individual peaks. If two wave troughs overlap, the depth of the water depression at that point is similarly increased. And finally, if a wave peak emerging from one slit overlaps with a wave trough emerging from the other, they *cancel each*

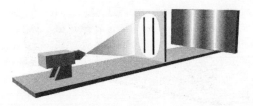

Figure 4.6 Newton's particle view of light predicts that when both slits are open, the photographic plate will be a merger of the images in Figures 4.4 and 4.5.

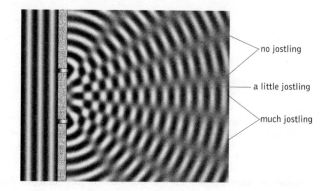

Figure 4.7 Circular water waves that emerge from each slit overlap with each other, causing the total wave to be increased at some locations and decreased at others.

other out. (In fact, this is the idea behind fancy noise-eliminating head-phones—they measure the shape of the incoming sound wave and then produce another whose shape is exactly "opposite," leading to a cancella-tion of the undesired noise.) In between these extreme overlaps—peaks with peaks, troughs with troughs, and peaks with troughs—are a host of partial height augmentations and cancellations. If you and a slew of com-panions form a line of little boats parallel to the barrier and you each de-clare how severely you are jostled by the resulting water wave as it passes, the result will look something like that shown on the far right of Figure 4.7. Locations of significant jostling are where wave peaks (or troughs) from each slit coincide. Regions of minimal or no jostling are where peaks from one slit coincide with troughs from the other, resulting in a cancellation.

Since the photographic plate records how much it is "jostled" by the in-coming light, exactly the same reasoning applied to the wave picture of a light beam tells us that when both slits are open the photograph will look like that in Figure 4.8. The brightest areas in Figure 4.8 are where light-wave peaks (or troughs) from each slit coincide. Dark areas are where wave peaks from one slit coincide with wave troughs from the other, re-sulting in a cancellation. The sequence of light and dark bands is known as an *interference pattern*. This photograph is significantly different from that shown in Figure 4.6, and hence there is a concrete experiment to dis-

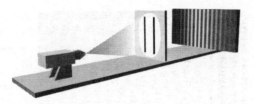

Figure 4.8 If light is a wave, then when both slits are open there will be interference between the portions of the wave emerging from each slit.

tinguish between the particle and the wave pictures of light. Young carried out a version of this experiment and his results matched Figure 4.8, thereby confirming the wave picture. Newton's corpuscular view was defeated (although it took quite some time before physicists accepted this). The prevailing wave view of light was subsequently put on a mathematically firm foundation by Maxwell.

But Einstein, the man who brought down Newton's revered theory of gravity, seems now to have resurrected Newton's particle model of light by his introduction of photons. Of course, we still face the same question: How can a particle perspective account for the interference pattern shown in Figure 4.8? At first blush you might make the following suggestion. Water is composed of H_2O molecules—the "particles" of water. Nevertheless, when a lot of these molecules stream along with one another they can produce water waves, with the attendant interference properties illustrated in Figure 4.7. And so, it might seem reasonable to guess that wave properties, such as interference patterns, can arise from a particle picture of light provided a huge number of photons, the particles of light, are involved.

In reality, though, the microscopic world is far more subtle. Even if the intensity of the light source in Figure 4.8 is turned down and down, finally to the point where *individual* photons are being fired *one by one* at the barrier—say at the rate of one every ten seconds—the resulting photographic plate will *still* look like that in Figure 4.8: So long as we wait long enough for a huge number of these separate bundles of light to make it through the slits and to each be recorded by a single dot where they hit the photographic plate, these dots will build up to form the image of an interference pattern, the image in Figure 4.8. This is astounding. How can *individual* photon particles that sequentially pass through the screen and separately hit the photographic plate conspire to produce the bright and dark bands of interfering waves? Conventional reasoning tells us that each and every photon passes through either the left slit or the right slit and we would therefore expect to find the pattern shown in Figure 4.6. But we don't.

If you are not bowled over by this fact of nature, it means that either you have seen it before and have become blasé or the description so far has not been sufficiently vivid. So, in case it's the latter, let's describe it again, but in a slightly different way. You close off the left slit and fire the photons one by one at the barrier. Some get through, some don't. The ones that do create an image on the photographic plate, dot by single dot, which looks like that in Figure 4.4. You then run the experiment again with a new photographic plate, but this time you open both slits. Naturally enough, you think that this will only increase the number of photons that pass through the slits in the barrier and hit the photographic plate, thereby exposing the film to more total light than in your first run of the experiment. But when you later examine the image produced, you find that not only are there places on the photographic plate that were dark in the first experiment and are now bright, as expected, there are also places on the photographic plate that were bright in your first experiment but are now dark, as in Figure 4.8. By *increasing* the number of individual photons that hit the photographic plate you have *decreased* the brightness in certain areas. Somehow, temporally separated, individual particulate photons are able to cancel each other out. Think about how crazy this is: Photons that would have passed through the right slit and hit the film in one of the dark bands in Figure 4.8 fail to do so when the *left* slit is opened (which is why the band is now dark). But how in the world can a tiny bundle of light that passes through one slit be at all affected by whether or not the *other* slit is open? As Feynman noted, it's as strange as if you fire a machine gun at the screen, and when both slits are open, independent, separately fired bullets somehow cancel one another out, leaving a pattern of unscathed positions on the target—positions that *are* hit when only one slit in the barrier is open.

Such experiments show that Einstein's particles of light are quite different from Newton's. Somehow photons—although they are particles—embody wave-like features of light as well. The fact that the energy of these particles is determined by a wave-like feature—frequency—is the first clue that a strange union is occurring. But the photoelectric effect and the double-slit experiment really bring the lesson home. The photoelectric effect shows that light has particle properties. The double-slit experiment shows that light manifests the interference properties of waves. Together they show that light has *both wave-like and particle-like properties.* The microscopic world demands that we shed our intuition that something is either a wave or a particle and embrace the possibility that it is *both.* It is here that Feynman's pronouncement that "nobody understands quantum mechanics" comes to the fore. We can utter words such as "wave-particle duality." We can translate these words into a mathematical formalism that describes real-world experiments with amazing accuracy. But it is extremely hard to understand at a deep, intuitive level this dazzling feature of the microscopic world.

Matter Particles Are Also Waves

In the first few decades of the twentieth century, many of the greatest theoretical physicists grappled tirelessly to develop a mathematically sound and physically sensible understanding of these hitherto hidden microscopic features of reality. Under the leadership of Niels Bohr in Copenhagen, for example, substantial progress was made in explaining the properties of light emitted by glowing-hot hydrogen atoms. But this and other work prior to the mid-1920s was more a makeshift union of nineteenth-century ideas with newfound quantum concepts than a coherent framework for understanding the physical universe. Compared with the clear, logical framework of Newton's laws of motion or Maxwell's electromagnetic theory, the partially developed quantum theory was in a chaotic state.

In 1923, the young French nobleman Prince Louis de Broglie added a new element to the quantum fray, one that would shortly help to usher in the mathematical framework of modern quantum mechanics and that earned him the 1929 Nobel Prize in physics. Inspired by a chain of reasoning rooted in Einstein's special relativity, de Broglie suggested that the wave-particle duality applied not only to light but to matter as well. He reasoned, roughly speaking, that Einstein's $E = mc^2$ relates mass to energy, that Planck and Einstein had related energy to the frequency of waves, and therefore, by combining the two, mass should have a wave-like incarnation as well. After carefully working through this line of thought, he suggested that just as light is a wave phenomenon that quantum theory shows to have an equally valid particle description, an electron—which we normally think of as being a particle—might have an equally valid description in terms of waves. Einstein immediately took to de Broglie's idea, as it was a natural outgrowth of his own contributions of relativity and of photons. Even so, nothing is a substitute for experimental proof. Such proof was soon to come from the work of Clinton Davisson and Lester Germer.

In the mid-1920s, Davisson and Germer, experimental physicists at the Bell telephone company, were studying how a beam of electrons bounces off of a chunk of nickel. The only detail that matters for us is that the nickel crystals in such an experiment act very much like the two slits in the experiment illustrated by the figures of the last section—in fact, it's perfectly okay to think of this experiment as being the same one illustrated there, except that a beam of electrons is used in place of a beam of light. We will adopt this point of view. When Davisson and Germer examined electrons making it through the two slits in the barrier by allowing them to hit a phosphorescent screen that recorded the location of impact of each electron by a bright dot—essentially what happens inside a television—they found something remarkable. A pattern very much akin to that of Figure 4.8 emerged. Their experiment therefore showed that electrons exhibit interference phenomena, the telltale sign of *waves*. At dark spots on the phosphorescent screen, electrons were somehow "canceling each other out" just like the overlapping peak and trough of water waves. Even if the beam of fired electrons was "thinned" so that, for instance, only one electron was emitted every ten seconds, the individual electrons still built up the bright and dark bands—one spot at a time. Somehow, as with photons, individual electrons "interfere" with themselves in the sense that individual electrons, over time, reconstruct the interference pattern associated with waves. We are inescapably forced to conclude that each electron embodies a wave-like character in conjunction with its more familiar depiction as a particle.

Although we have described this in the case of electrons, similar experiments lead to the conclusion that *all* matter has a wave-like character. But how does this jibe with our real-world experience of matter as being solid and sturdy, and in no way wave-like? Well, de Broglie set down a formula for the wavelength of matter waves, and it shows that the wavelength is proportional to Planck's constant $\hbar$. (More precisely, the wavelength is given by $\hbar$ divided by the material body's momentum.) Since $\hbar$ is so small, the resulting wavelengths are similarly minuscule compared with everyday scales. This is why the wave-like character of matter becomes directly apparent only upon careful microscopic investigation. Just as the large value of c, the speed of light, obscures much of the true nature of space and time, the smallness of $\hbar$ obscures the wave-like aspects of matter in the day-to-day world.

Waves of What?

The interference phenomenon found by Davisson and Germer made the wave-like nature of electrons tangibly evident. But waves of *what?* One early suggestion made by Austrian physicist Erwin Schrödinger was that the waves were "smeared-out" electrons. This captured some of the "feeling" of an electron wave, but it was too rough. When you smear something out, part of it is here and part of it is there. However, one never encounters half of an electron or a third of an electron or any other fraction, for that matter. This makes it hard to grasp what a smeared electron actually is. As an alternative, in 1926 German physicist Max Born sharply refined Schrödinger's interpretation of an electron wave, and it is his interpretation—amplified by Bohr and his colleagues—that is still with us today. Born's suggestion is one of the strangest features of quantum theory, but is supported nonetheless by an enormous amount of experimental data. He asserted that an electron wave must be interpreted from the standpoint of *probability*. Places where the magnitude (a bit more correctly, the square of magnitude) of the wave is *large* are places where the electron is more *likely* to be found; places where the magnitude is *small* are places where the electron is *less likely* to be found. An example is illustrated in Figure 4.9.

This is truly a peculiar idea. What business does probability have in the formulation of fundamental physics? We are accustomed to probability showing up in horse races, in coin tosses, and at the roulette table, but in those cases it merely reflects our *incomplete* knowledge. If we knew *pre-*

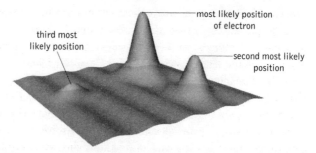

Figure 4.9 The wave associated with an electron is largest where the electron is most likely to be found, and progressively smaller at locations where it is less likely to be found.

cisely the speed of the roulette wheel, the weight and hardness of the white marble, the location and speed of the marble when it drops to the wheel, the exact specifications of the material constituting the cubicles and so on, and if we made use of sufficiently powerful computers to carry out our calculations we would, according to classical physics, be able to predict with certainty where the marble would settle. Gambling casinos rely on your inability to ascertain all of this information and to do the necessary calculations prior to placing your bet. But we see that probability as encountered at the roulette table does not reflect anything particularly fundamental about how the world works. Quantum mechanics, on the contrary, injects the concept of probability into the universe at a far deeper level. According to Born and more than half a century of subsequent experiments, the wave nature of matter implies that matter itself must be described fundamentally in a probabilistic manner. For macroscopic objects like a coffee cup or the roulette wheel, de Broglie's rule shows that the wave-like character is virtually unnoticeable and for most ordinary purposes the associated quantum-mechanical probability can be completely ignored. But at a microscopic level we learn that the best we can ever do is say that an electron has a particular probability of being found at any given location.

The probabilistic interpretation has the virtue that if an electron wave does what other waves can do—for instance, slam into some obstacle and develop all sorts of distinct ripples—it does not mean that the electron itself has shattered into separate pieces. Rather, it means that there are now a number of locations where the electron *might* be found with a non-negligible probability. In practice this means that if a particular experiment involving an electron is repeated over and over again in an absolutely identical manner, the same answer for, say, the measured position of an electron will *not* be found over and over again. Rather, the subsequent repeats of the experiment will yield a variety of different results with the property that the number of times the electron is found at any given location is governed by the shape of the electron's probability wave. If the probability wave (more precisely, the square of the probability wave) is twice as large at location A than at location B, then the theory predicts that in a sequence of many repeats of the experiment the electron will be found at location A twice as often as at location B. Exact outcomes of experiments cannot be predicted; the best we can do is predict the probability that any given outcome *may* occur.

Even so, as long as we can determine mathematically the precise form of probability waves, their probabilistic predictions *can* be tested by repeating a given experiment numerous times, thereby experimentally measuring the likelihood of getting one particular result or another. Just a few months after de Broglie's suggestion, Schrödinger took the decisive step toward this end by determining an equation that governs the shape and the evolution of probability waves, or as they came to be known, *wave functions*. It was not long before Schrödinger's equation and the probabilistic interpretation were being used to make wonderfully accurate predictions. By 1927, therefore, classical innocence had been lost. Gone were the days of a clockwork universe whose individual constituents were set in motion at some moment in the past and obediently fulfilled their inescapable, uniquely determined destiny. According to quantum mechanics, the universe evolves according to a rigorous and precise mathematical formalism, but this framework determines only the probability that any particular future will happen—not which future actually ensues.

Many find this conclusion troubling or even downright unacceptable. Einstein was one. In one of physics' most time-honored utterances, Einstein admonished the quantum stalwarts that "God does not play dice with the Universe." He felt that probability was turning up in fundamental physics because of a subtle version of the reason it turns up at the roulette wheel: some basic incompleteness in our understanding. The universe, in Einstein's view, had no room for a future whose precise form involves an element of chance. Physics should predict *how* the universe evolves, not merely the likelihood that any particular evolution might occur. But experiment after experiment—some of the most convincing ones being carried out after his death—convincingly confirm that Einstein was wrong. As the British theoretical physicist Stephen Hawking has said, on this point "Einstein was confused, not the quantum theory."[6]

Nevertheless, the debate about what quantum mechanics really means continues unabated. Everyone agrees on how to use the equations of quantum theory to make accurate predictions. But there is no consensus on what it really means to have probability waves, nor on how a particle "chooses" which of its many possible futures to follow, nor even on whether it really does choose or instead splits off like a branching tributary to live out all possible futures in an ever-expanding arena of parallel universes. These interpretational issues are worthy of a book-length discussion in their own right, and, in fact, there are many excellent books that espouse one or another way of thinking about quantum theory. But what appears certain is that no matter how you interpret quantum mechanics, it undeniably shows that the universe is founded on principles that, from the standpoint of our day-to-day experiences, are bizarre.

The meta-lesson of both relativity and quantum mechanics is that when we deeply probe the fundamental workings of the universe we may come upon aspects that are vastly different from our expectations. The boldness of asking deep questions may require unforeseen flexibility if we are to accept the answers.

Feynman's Perspective

Richard Feynman was one of the greatest theoretical physicists since Einstein. He fully accepted the probabilistic core of quantum mechanics, but in the years following World War II he offered a powerful new way of thinking about the theory. From the standpoint of numerical predictions, Feynman's perspective *agrees exactly* with all that went before. But its formulation is quite different. Let's describe it in the context of the electron two-slit experiment.

The troubling thing about Figure 4.8 is that we envision each electron as passing through either the left slit or the right slit and therefore we expect the union of Figures 4.4 and 4.5, as in Figure 4.6, to represent the resulting data accurately. An electron that passes through the right slit should not care that there also happens to be a left slit, and vice versa. But somehow it does. The interference pattern generated requires an overlapping and an intermingling between *something* sensitive to both slits, even if we fire electrons one by one. Schrödinger, de Broglie, and Born explained this phenomenon by associating a probability wave to each electron. Like the water waves in Figure 4.7, the electron's probability wave "sees" both slits and is subject to the same kind of interference from intermingling. Places where the probability wave is augmented by the intermingling, like the places of significant jostling in Figure 4.7, are locations where the electron is likely to be found; places where the probability wave is diminished by the intermingling, like the places of minimal or no jostling in Figure 4.7, are locations where the electron is unlikely or never to be found. Electrons hit the phosphorescent screen one by one, distributed according to this probability profile, and thereby build up an interference pattern like that in Figure 4.8.

Feynman took a different tack. He challenged the basic classical assumption that each electron either goes through the left slit or the right slit. You might think this to be such a basic property of how things work that challenging it is fatuous. After all, can't you *look* in the region between the slits and the phosphorescent screen to determine through which slit each electron passes? You can. But now you have *changed* the experiment. To *see* the electron you must *do* something to it—for instance, you can shine light on it, that is, bounce photons off it. Now, on everyday scales photons act as negligible little probes that bounce off trees, paintings, and people with essentially no effect on the state of motion of these comparatively large material bodies. But electrons are little wisps of matter. Regardless of how gingerly you carry out your determination of the slit through which it passed, photons that bounce off the electron necessarily affect its subsequent motion. And this change in motion changes the results of our experiment. If you disturb the experiment just enough to determine the slit through which each electron passes, experiments show that the results change from that of Figure 4.8 and become like that of Figure 4.6! The quantum world ensures that once it has been established that each electron has gone through either the left slit or the right slit, the interference between the two slits disappears.

And so Feynman was justified in leveling his challenge since—although our experience in the world seems to require that each electron pass through one or the other of the slits—by the late 1920s physicists realized that any attempt to verify this seemingly basic quality of reality ruins the experiment.

Feynman proclaimed that each electron that makes it through to the phosphorescent screen actually goes through *both* slits. It sounds crazy, but hang on: Things get even more wild. Feynman argued that in traveling from the source to a given point on the phosphorescent screen each individual electron actually traverses *every possible trajectory simultaneously;* a few of the trajectories are illustrated in Figure 4.10. It goes in a nice orderly way through the left slit. It simultaneously also goes in a nice orderly way through the right slit. It heads toward the left slit, but suddenly changes course and heads through the right. It meanders back and forth, finally passing through the left slit. It goes on a long journey to the Andromeda galaxy before turning back and passing through the left slit on its way to the screen. And on and on it goes—the electron, according to Feynman, simultaneously "sniffs" out *every* possible path connecting its starting location with its final destination.

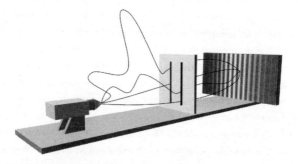

Figure 4.10 According to Feynman's formulation of quantum mechanics, particles must be viewed as travelling from one location to another along every possible path. Here, a few of the infinity of trajectories for a single electron travelling from the source to the phosphorescent screen are shown. Notice that this one electron actually goes through both slits.

Feynman showed that he could assign a number to each of these paths in such a way that their combined average yields exactly the same result for the probability calculated using the wave-function approach. And so from Feynman's perspective no probability wave needs to be associated with the electron. Instead, we have to imagine something equally if not more bizarre. The probability that the electron—always viewed as a particle through and through—arrives at any chosen point on the screen is built up from the combined effect of every possible way of getting there. This is known as Feynman's "sum-over-paths" approach to quantum mechanics.[7]

At this point your classical upbringing is balking: How can one electron *simultaneously* take different paths—and no less than an infinite number of them? This seems like a defensible objection, but quantum mechanics—the physics of our world—requires that you hold such pedestrian complaints in abeyance. The result of calculations using Feynman's approach agree with those of the wave function method, which agree with experiments. You must allow nature to dictate what is and what is not sensible. As Feynman once wrote, "[Quantum mechanics] describes nature as absurd from the point of view of common sense. And it fully agrees with experiment. So I hope you can accept nature as She is—absurd."[8]

But no matter how absurd nature is when examined on microscopic scales, things must conspire so that we recover the familiar prosaic happenings of the world experienced on everyday scales. To this end, Feynman showed that if you examine the motion of large objects—like baseballs, airplanes, or planets, all large in comparison with subatomic particles—his rule for assigning numbers to each path ensures that *all paths but one cancel each other out* when their contributions are combined. In effect, only one of the infinity of paths matters as far as the motion of the object is concerned. And this trajectory is precisely the one emerging from Newton's laws of motion. This is why in the everyday world it *seems* to us that objects—like a ball tossed in the air—follow a single, unique, and predictable trajectory from their origin to their destination. But for microscopic objects, Feynman's rule for assigning numbers to paths shows that many different paths can and often do contribute to an object's motion. In the double-slit experiment, for example, some of these paths pass through different slits, giving rise to the interference pattern observed. In the microscopic realm we therefore cannot assert that an electron passes through only one slit or the other. The interference pattern and Feynman's alternative formulation of quantum mechanics emphatically attest to the contrary.

Just as we may find that varying interpretations of a book or a film can be more or less helpful in aiding our understanding of different aspects of the work, the same is true of the different approaches to quantum mechanics. Although their predictions always agree completely, the wave function approach and Feynman's sum-over-paths approach give us different ways of thinking about what's going on. As we shall see later on, for some applications, one or the other approach can provide an invaluable explanatory framework.

Quantum Weirdness

By now you should have some sense of the dramatically new way that the universe works according to quantum mechanics. If you have not as yet fallen victim to Bohr's dizziness dictum, the quantum weirdness we now discuss should at least make you feel a bit lightheaded.

Even more so than with the theories of relativity, it is hard to embrace quantum mechanics viscerally—to think like a miniature person born and raised in the microscopic realm. There is, though, one aspect of the theory that can act as a guidepost for your intuition, as it is the hallmark feature that fundamentally differentiates quantum from classical reasoning. It is the *uncertainty principle,* discovered by the German physicist Werner Heisenberg in 1927.

This principle grows out of an objection that may have occurred to you earlier. We noted that the act of determining the slit through which each electron passes (its position) necessarily disturbs its subsequent motion (its velocity). But just as we can assure ourselves of someone's presence either by gently touching them or by giving them an overzealous slap on the back, why can't we determine the electron's position with an "ever gentler" light source in order to have an ever decreasing impact on its motion? From the standpoint of nineteenth-century physics we can. By using an ever dimmer lamp (and an ever more sensitive light detector) we can have a vanishingly small impact on the electron's motion. But quantum mechanics itself illuminates a flaw in this reasoning. As we turn down the intensity of the light source we now know that we are decreasing the number of photons it emits. Once we get down to emitting individual photons we cannot dim the light any further without actually turning it off. There is a fundamental quantum-mechanical limit to the "gentleness" of our probe. And hence, there is always a minimal disruption that we cause to the electron's velocity through our measurement of its position.

Well, that's almost correct. Planck's law tells us that the energy of a single photon is proportional to its frequency (inversely proportional to its wavelength). By using light of lower and lower frequency (larger and larger wavelength) we can therefore produce ever gentler individual photons. But here's the catch. When we bounce a wave off of an object, the information we receive is only enough to determine the object's position to within a *margin of error equal to the wave's wavelength*. To get an intuitive feel for this important fact, imagine trying to pinpoint the location of a large, slightly submerged rock by the way it affects passing ocean waves. As the waves approach the rock, they form a nice orderly train of one up-and-down wave cycle followed by another. After passing by the rock, the individual wave cycles are distorted—the telltale sign of the submerged rock's presence. But like the finest set of tick marks on a ruler, the individual up-and-down wave cycles are the finest units making up the wave-train, and therefore by examining solely how they are disrupted we can determine the rock's location only to within a margin of error equal to the length of the wave cycles, that is, the wave's wavelength. In the case of light, the constituent photons are, roughly speaking, the individual wave cycles (with the height of the wave cycles being determined by the number of photons); a photon, therefore, can be used to pinpoint an object's location only to within a precision of one wavelength.

And so we are faced with a quantum-mechanical balancing act. If we use high-frequency (short wavelength) light we can locate an electron with greater precision. But high-frequency photons are very energetic and therefore sharply disturb the electron's velocity. If we use low-frequency (long wavelength) light we minimize the impact on the electron's motion, since the constituent photons have comparatively low energy, but we sacrifice precision in determining the electron's position. Heisenberg quantified this competition and found a mathematical relationship between the precision with which one measures the electron's position and the precision with which one measures its velocity. He found—in line with our discussion—that each is inversely proportional to the other: Greater precision in a position measurement necessarily entails greater imprecision in a velocity measurement, and vice versa. And of utmost importance, although we have tied our discussion to one particular means for determining the electron's whereabouts, Heisenberg showed that the trade-off between the precision of position and velocity measurements is a fundamental fact that holds true regardless of the equipment used or the procedure employed. Unlike the framework of Newton or even of Einstein, in which the motion of a particle is described by giving its location and its velocity, quantum mechanics shows that at a microscopic level *you cannot possibly know both of these features with total precision.* Moreover, the more precisely you know one, the less precisely you know the other. And although we have described this for electrons, the ideas directly apply to *all* constituents of nature.

Einstein tried to minimize this departure from classical physics by arguing that although quantum reasoning certainly does appear to limit one's *knowledge* of the position and velocity, the electron still *has* a definite position and velocity exactly as we have always thought. But during the last couple of decades theoretical progress spearheaded by the late Irish physicist John Bell and the experimental results of Alain Aspect and his collaborators have shown convincingly that Einstein was wrong. Electrons—and everything else for that matter—cannot be described as simultaneously being at such-and-such location *and* having such-and-such speed. Quantum mechanics shows that not only could such a statement never be experimentally verified—as explained above—but it directly contradicts other, more recently established experimental results.

In fact, if you were to capture a single electron in a big, solid box and then slowly crush the sides to pinpoint its position with ever greater precision, you would find the electron getting more and more frantic. Almost as if it were overcome with claustrophobia, the electron will go increasingly haywire—bouncing off of the walls of the box with increasingly frenetic and unpredictable speed. Nature does not allow its constituents to be cornered. In the H-Bar, where we imagine $\hbar$ to be *much* larger than in the real world, thereby making everyday objects directly subject to quantum effects, the ice cubes in George's and Gracie's drinks frantically rattle around as they too suffer from quantum claustrophobia. Although the H-Bar is a fantasy land—in reality, $\hbar$ is terribly small—precisely this kind of quantum claustrophobia is a pervasive feature of the microscopic realm. The motion of microscopic particles becomes increasingly wild when they are examined and confined to ever smaller regions of space.

The uncertainty principle also gives rise to a striking effect known as *quantum tunneling*. If you fire a plastic pellet against a ten-foot-thick concrete wall, classical physics confirms what your instincts tell you will happen: The pellet will bounce back at you. The reason is that the pellet simply does not have enough energy to penetrate such a formidable obstacle. But at the level of fundamental particles, quantum mechanics shows unequivocally that the wave functions—that is, the probability waves—of the particles making up the pellet all have a tiny piece that *spills out* through the wall. This means that there is a small—but not zero—chance that the pellet actually *can* penetrate the wall and emerge on the other side. How can this be? The reason comes down, once again, to Heisenberg's uncertainty principle.

To see this, imagine that you are completely destitute and suddenly learn that a distant relative has passed on in a far-off land, leaving you a tremendous fortune to claim. The only problem is that you don't have the money to buy a plane ticket to get there. You explain the situation to your friends: if only they will allow you to surmount the barrier between you and your new fortune by temporarily lending you the money for a ticket, you can pay them back handsomely after your return. But no one has the money to lend. You remember, though, that an old friend of yours works for an airline and you implore him with the same request. Again, he cannot afford to lend you the money but he does offer a solution. The accounting system of the airline is such that if you wire the ticket payment within 24 hours of arrival at your destination, no one will ever know that it was not paid for prior to departure. In this way you are able to claim your inheritance.

The accounting procedures of quantum mechanics are quite similar. Just as Heisenberg showed that there is a trade-off between the precision of measurements of position and velocity, he also showed that there is a similar trade-off in the precision of *energy* measurements and *how long* one takes to do the measurement. Quantum mechanics asserts that you can't say that a particle has precisely such-and-such energy at precisely such-and-such moment in time. Ever increasing precision of energy measurements require ever longer durations to carry them out. Roughly speaking, this means that the energy a particle has can wildly fluctuate so long as this fluctuation is over a short enough time scale. So, just as the accounting system of the airline "allows" you to "borrow" the money for a plane ticket provided you pay it back quickly enough, quantum mechanics allows a particle to "borrow" energy so long as it can relinquish it within a time frame determined by Heisenberg's uncertainty principle.

The mathematics of quantum mechanics shows that the greater the energy barrier, the lower the probability that this creative microscopic accounting will actually occur. But for microscopic particles facing a concrete slab, they can and sometimes do borrow enough energy to do what is impossible from the standpoint of classical physics—momentarily penetrate and tunnel through a region that they do not initially have enough energy to enter. As the objects we study become increasingly complicated, consisting of more and more particle constituents, such quantum tunneling can still occur, but it becomes very unlikely since *all* of the individual particles must be lucky enough to tunnel together. But the shocking episodes of George's disappearing cigar, of an ice cube passing right through the wall of a glass, and of George and Gracie's passing right through a wall of the bar, *can* happen. In a fantasy land such as the H-Bar, in which we imagine that $\hbar$ is large, such quantum tunneling is commonplace. But the probability rules of quantum mechanics—and, in particular, the actual smallness of $\hbar$ in the real world—show that if you walked into a solid wall every second, you would have to wait longer than the current age of the universe to have a good chance of passing through it on one of your attempts. With eternal patience (and longevity), though, you could—sooner or later—emerge on the other side.

The uncertainty principle captures the heart of quantum mechanics. Features that we normally think of as being so basic as to be beyond question—that objects have definite positions and speeds and that they have definite energies at definite moments—are now seen as mere artifacts of Planck's constant being so tiny on the scales of the everyday world. Of prime importance is that when this quantum realization is applied to the fabric of spacetime, it shows fatal imperfections in the "stitches of gravity" and leads us to the third and primary conflict physics has faced during the past century.

Chapter 5

The Need for a New Theory: General Relativity vs. Quantum Mechanics

Our understanding of the physical universe has deepened profoundly during the past century. The theoretical tools of quantum mechanics and general relativity allow us to understand and make testable predictions about physical happenings from the atomic and subatomic realms all the way through phenomena occurring on the scales of galaxies, clusters of galaxies, and beyond to the structure of the whole universe itself. This is a monumental achievement. It is truly inspiring that beings confined to one planet orbiting a run-of-the-mill star in the far edges of a fairly ordinary galaxy have been able, through thought and experiment, to ascertain and comprehend some of the most mysterious characteristics of the physical universe. Nevertheless, physicists by their nature will not be satisfied until they feel that the deepest and most fundamental understanding of the universe has been unveiled. This is what Stephen Hawking has alluded to as a first step toward knowing "the mind of God."[1]

There is ample evidence that quantum mechanics and general relativity do not provide this deepest level of understanding. Since their usual domains of applicability are so different, most situations require the use of quantum mechanics *or* general relativity, but not both. Under certain extreme conditions, however, where things are very massive *and* very small—near the central point of black holes or the whole universe at the moment of the big bang, to name two examples—we require both general relativity and quantum mechanics for proper understanding. But like the mixing of fire and gunpowder, when we try to combine quantum mechanics and general relativity, their union brings violent catastrophe. Well-formulated physical problems elicit nonsensical answers when the equations of both these theories are commingled. The nonsense often takes the form of a prediction that the quantum-mechanical probability for some process is not 20 percent or 73 percent or 91 percent but *infinity*. What in the world does a probability greater than one mean, let alone one that is infinite? We are forced to conclude that there is something seriously wrong. By closely examining the basic properties of general relativity and quantum mechanics, we can identify what that something is.

The Heart of Quantum Mechanics

When Heisenberg discovered the uncertainty principle, physics turned a sharp corner, never to retrace its steps. Probabilities, wave functions, interference, and quanta all involve radically new ways of seeing reality. Nevertheless, a die-hard "classical" physicist might still have hung on to a thread of hope that when all was said and done these departures would add up to a framework not too distant from old ways of thinking. But the uncertainty principle cleanly and definitively undercut any attempt to cling to the past.

The uncertainty principle tells us that the universe is a frenetic place when examined on smaller and smaller distances and shorter and shorter time scales. We saw some evidence of this in our attempt, described in the preceding chapter, to pinpoint the location of elementary particles such as electrons: By shining light of ever higher frequency on electrons, we measure their position with ever greater precision, but at a cost, since our observations become ever more disruptive. High-frequency photons have a lot of energy and therefore give the electrons a sharp "kick," significantly changing their velocities. Like the frenzy in a room full of children all of whose momentary positions you know with great accuracy but over whose velocities—the speeds and directions in which they are moving—you have almost no control, this inability to know both the positions and velocities of elementary particles implies that the microscopic realm is intrinsically turbulent.

Although this example conveys the basic relationship between uncertainty and frenzy, it actually reveals only part of the story. It might lead you to think, for instance, that uncertainty arises only when we clumsy observers of nature stumble onto the scene. This is *not* true. The example of an electron violently reacting to being confined in a small box by rattling around at high speed takes us a bit closer to the truth. Even without "direct hits" from an experimenter's disruptive photon, the electron's velocity severely and unpredictably changes from one moment to the next. But even this example does not fully reveal the stunning microscopic features of nature entailed by Heisenberg's discovery. Even in the most quiescent setting imaginable, such as an empty region of space, the uncertainty principle tells us that from a microscopic vantage point there is a tremendous amount of activity. And this activity gets increasingly agitated on ever smaller distance and time scales.

Quantum accounting is essential to understand this. We saw in the preceding chapter that just as you might temporarily borrow money to overcome an important financial obstacle, a particle such as an electron can temporarily borrow energy to overcome a literal physical barrier. This is true. But quantum mechanics forces us to take the analogy one important step further. Imagine someone who is a compulsive borrower and goes from friend to friend asking for money. The shorter the time for which a friend can lend him money, the larger the loan he seeks. Borrow and return, borrow and return—over and over again with unflagging intensity he takes in money only to give it back in short order. Like stock prices on a wild, roller-coaster day on Wall Street, the amount of money the compulsive borrower possesses at any given moment goes through extreme fluctuations, but when all is said and done, an accounting of his finances shows that he is no better off than when he began.

Heisenberg's uncertainty principle asserts that a similar frantic shifting back and forth of energy and momentum is occurring perpetually in the universe on microscopic distance and time intervals. Even in an empty region of space—inside an empty box, for example—the uncertainty principle says that the energy and momentum are *uncertain*: They fluctuate between extremes that get larger as the size of the box and the time scale over which it is examined get smaller and smaller. It's as if the region of space inside the box is a compulsive "borrower" of energy and momentum, constantly extracting "loans" from the universe and subsequently "paying" them back. But what participates in these exchanges in, for instance, a quiet *empty* region of space? Everything. Literally. Energy (and momentum as well) is the ultimate convertible currency. $E = mc^2$ tells us that energy can be turned into matter and vice versa. Thus if an energy fluctuation is big enough it can momentarily cause, for instance, an electron and its antimatter companion the positron to erupt into existence, even if the region was initially empty! Since this energy must be quickly repaid, these particles will annihilate one another after an instant, relinquishing the energy borrowed in their creation. And the same is true for all of the other forms that energy and momentum can take—other particle eruptions and annihilations, wild electromagnetic-field oscillations, weak and strong force-field fluctuations—quantum-mechanical uncertainty tells us the universe is a teeming, chaotic, frenzied arena on microscopic scales. As Feynman once jested, "Created and annihilated, created and annihilated—what a waste of time."[2] Since the borrowing and repaying on average cancel each other out, an empty region of space looks calm and placid when examined with all but microscopic precision. The uncertainty principle, however, reveals that macroscopic averaging obscures a wealth of microscopic activity.[3] As we will see shortly, this frenzy is *the* obstacle to merging general relativity and quantum mechanics.

Quantum Field Theory

Over the course of the 1930s and 1940s theoretical physicists, led by the likes of Paul Dirac, Wolfgang Pauli, Julian Schwinger, Freeman Dyson, Sin-Itiro Tomonaga, and Feynman, to name a few, struggled relentlessly to find a mathematical formalism capable of dealing with this microscopic obstreperousness. They found that Schrödinger's quantum wave equation (mentioned in Chapter 4) was actually only an approximate description of microscopic physics—an approximation that works extremely well when one does not probe too deeply into the microscopic frenzy (either experimentally or theoretically), but that certainly fails if one does.

The central piece of physics that Schrödinger ignored in his formulation of quantum mechanics is special relativity. In fact, Schrödinger *did* try to incorporate special relativity initially, but the quantum equation to which this led him made predictions that proved to be at odds with experimental measurements of hydrogen. This inspired Schrödinger to adopt the time-honored tradition in physics of divide and conquer: Rather than trying, through one leap, to incorporate all we know about the physical universe in developing a new theory, it is often far more profitable to take many small steps that sequentially include the newest discoveries from the forefront of research. Schrödinger sought and found a mathematical framework encompassing the experimentally discovered wave-particle duality, but he did not, at that early stage of understanding, incorporate special relativity.[4]

But physicists soon realized that special relativity was central to a proper quantum-mechanical framework. This is because the microscopic frenzy requires that we recognize that energy can manifest itself in a huge variety of ways—a notion that comes from the special relativistic declaration $E = mc^2$. By ignoring special relativity, Schrödinger's approach ignored the malleability of matter, energy, and motion.

Physicists focused their initial pathbreaking efforts to merge special relativity with quantum concepts on the electromagnetic force and its interactions with matter. Through a series of inspirational developments, they created *quantum electrodynamics*. This is an example of what has come to be called a *relativistic quantum field theory*, or a *quantum field theory*, for short. It's quantum because all of the probabilistic and uncertainty issues are incorporated from the outset; it's a field theory because it merges the quantum principles into the previous classical notion of a force field—in this case, Maxwell's electromagnetic field. And finally, it's relativistic because special relativity is also incorporated from the outset.(If you'd like a visual metaphor for a quantum field, you can pretty much invoke the image of a classical field—say, as an ocean of invisible field lines permeating space—but you should refine this image in two ways. First, you should envision a quantum field as composed of particulate ingredients, such as photons for the electromagnetic field. Second, you should imagine energy, in the form of particles' masses and their motion, endlessly shifting back and forth from one quantum field to another as they continually vibrate through space and time.)

Quantum electrodynamics is arguably the most precise theory of natural phenomena ever advanced. An illustration of its precision can be found in the work of Toichiro Kinoshita, a particle physicist from Cornell University, who has, over the last 30 years, painstakingly used quantum electrodynamics to calculate certain detailed properties of electrons. Kinoshita's calculations fill thousands of pages and have ultimately required the most powerful computers in the world to complete. But the effort has been well worth it: the calculations yield predictions about electrons that have been experimentally verified to an accuracy of better than one part in a billion. This is an absolutely astonishing agreement between abstract theoretical calculation and the real world. Through quantum electrodynamics, physicists have been able to solidify the role of photons as the "smallest possible bundles of light" and to reveal their interactions with electrically charged particles such as electrons, in a mathematically complete, predictive, and convincing framework.

The success of quantum electrodynamics inspired other physicists in the 1960s and 1970s to try an analogous approach for developing a quantum-mechanical understanding of the weak, the strong, and the gravitational forces. For the weak and the strong forces, this proved to be an immensely fruitful line of attack. In analogy with quantum electrodynamics, physicists were able to construct quantum field theories for the strong and the weak forces, called *quantum chromodynamics* and *quantum electroweak theory.* "Quantum chromodynamics" is a more colorful name than the more logical "quantum strong dynamics," but it is just a name without any deeper meaning; on the other hand, the name "electroweak" does summarize an important milestone in our understanding of the forces of nature.

Through their Nobel Prize–winning work, Sheldon Glashow, Abdus Salam, and Steven Weinberg showed that the weak and electromagnetic forces are naturally *united* by their quantum field–theoretic description even though their manifestations seem to be utterly distinct in the world around us. After all, weak force fields diminish to almost vanishing strength on all but subatomic distance scales, whereas electromagnetic fields—visible light, radio and TV signals, X-rays—have an indisputable macroscopic presence. Nevertheless, Glashow, Salam, and Weinberg showed, in essence, that at high enough energy and temperature—such as occurred a mere fraction of a second after the big bang—electromagnetic and weak force fields *dissolve* into one another, take on indistinguishable characteristics, and are more accurately called *electroweak* fields. When the temperature drops, as it has done steadily since the big bang, the electromagnetic and weak forces *crystallize* out in a different manner from their common high-temperature form—through a process known as *sym-*

metry breaking that we will describe later—and therefore appear to be distinct in the cold universe we currently inhabit.

And so, if you are keeping score, by the 1970s physicists had developed a sensible and successful quantum-mechanical description of three of the four forces (strong, weak, electromagnetic) and had shown that two of the three (weak and electromagnetic) actually share a common origin (the electroweak force). During the past two decades, physicists have subjected this quantum-mechanical treatment of the three nongravitational forces—as they act among themselves and the matter particles introduced in Chapter 1—to an enormous amount of experimental scrutiny. The theory has met all such challenges with aplomb. Once experimentalists measure some 19 parameters (the masses of the particles in Table 1.1, their force charges as recorded in the table in endnote 1 to Chapter 1, the strengths of the three nongravitational forces in Table 1.2, as well as a few other numbers we need not discuss), and theorists input these numbers into the quantum field theories of the matter particles and the strong, weak, and electromagnetic forces, the subsequent predictions of the theory regarding the microcosmos agree spectacularly with experimental results. This is true up to the energies capable of pulverizing matter into bits as small as a billionth of a billionth of a meter, the current technological limit. For this reason, physicists call the theory of the three nongravitational forces and the three families of matter particles the standard theory, or (more often) the *standard model* of particle physics.

Messenger Particles

According to the standard model, just as the photon is the smallest constituent of an electromagnetic field, the strong and the weak force fields have smallest constituents as well. As we discussed briefly in Chapter 1, the smallest bundles of the strong force are known as *gluons,* and those of the weak force are known as *weak gauge bosons* (or more precisely, the *W* and *Z* bosons). The standard model instructs us to think of these force particles as having no internal structure—in this framework they are every bit as elementary as the particles in the three families of matter.

The photons, gluons, and weak gauge bosons provide the microscopic mechanism for transmitting the forces they constitute. For example, when one electrically charged particle repels another of like electric charge, you can think of it roughly in terms of each particle being surrounded by an electric field—a "cloud" or "mist" of "electric-essence"—and the force each particle feels arises from the repulsion between their respective force fields. The more precise microscopic description of how they repel each other, though, is somewhat different. An electromagnetic field is composed of a swarm of photons; the interaction between two charged particles actually arises from their "shooting" photons back and forth between themselves. In rough analogy to the way in which you can affect a fellow ice-skater's motion and your own by hurling a barrage of bowling balls at him or her, two electrically charged particles influence each other by exchanging these smallest bundles of light.

An important failing of the ice-skater analogy is that the exchange of bowling balls is always "repulsive"—it always drives the skaters apart. On the contrary, two oppositely charged particles also interact through the exchange of photons, although the resulting electromagnetic force is attractive. It's as if the photon is not so much the transmitter of the force per se, but rather the transmitter of a *message* of how the recipient must respond to the force in question. For like-charged particles, the photon carries the message "move apart," while for oppositely charged particles it carries the message "come together." For this reason the photon is sometimes referred to as the *messenger particle* for the electromagnetic force. Similarly, the gluons and weak gauge bosons are the messenger particles for the strong and weak nuclear forces. The strong force, which keeps quarks locked up inside of protons and neutrons, arises from individual quarks exchanging gluons. The gluons, so to speak, provide the "glue" that keeps these subatomic particles stuck together. The weak force, which is responsible for certain kinds of particle transmutations involved in radioactive decay, is mediated by the weak gauge bosons.

Gauge Symmetry

You may have realized that the odd man out in our discussion of the quantum theory of the forces of nature is gravity. Given the successful approach physicists have used with the other three forces, you might suggest that physicists seek a quantum field theory of the gravitational force—a theory in which the smallest bundle of a gravitational force field, the *graviton,* would be its messenger particle. At first sight, as we now note, this suggestion would appear to be particularly apt because the quantum field theory of the three nongravitational forces reveals that there is a tantalizing similarity between them and an aspect of the gravitational force we encountered in Chapter 3.

Recall that the gravitational force allows us to declare that all observers—regardless of their state of motion—are on absolutely equal footing. Even those whom we would normally think of as accelerating may claim to be at rest, since they can attribute the force they feel to their being immersed in a gravitational field. In this sense, gravity enforces the symmetry: it ensures the equal validity of all possible observational points of view, all possible frames of reference. The similarity with the strong, weak, and electromagnetic forces is that they too are all connected with enforcing symmetries, albeit ones that are significantly more abstract than the one associated with gravity.

To get a rough feel for these rather subtle symmetry principles, let's consider one important example. As we recorded in the table in endnote 1 of Chapter 1, each quark comes in three "colors" (fancifully called red, green, and blue, although these are merely labels and have no relation to color in the usual visual sense), which determine how it responds to the strong force in much the same way that its electric charge determines how it responds to the electromagnetic force. All the data that have been collected establish that there is a symmetry among the quarks in the sense that the interactions between any two like-colored quarks (red with red, green with green, or blue with blue) are all identical, and similarly, the interactions between any two unlike-colored quarks (red with green, green with blue, or blue with red) are also identical. In fact, the data support something even more striking. If the three colors—the three different strong charges—that a quark can carry were all shifted in a particular manner (roughly speaking, in our fanciful chromatic language, if red, green, and blue were shifted, for instance, to yellow, indigo, and violet), and even if the details of this shift were to change from moment to moment or from place to place, the interactions between the quarks would be, again, completely unchanged. For this reason, just as we say that a sphere exemplifies rotational symmetry because it looks the same regardless of how we rotate it around in our hands or how we shift the angle from which we view it, we say that the universe exemplifies *strong force symmetry*: Physics is unchanged by—it is completely insensitive to—these force-charge shifts. For historical reasons, physicists also say that the strong force symmetry is an example of a *gauge symmetry*.[5]

Here is the essential point. Just as the symmetry between all possible observational vantage points in general relativity requires the existence of the gravitational force, developments relying on work of Hermann Weyl in the 1920s and Chen-Ning Yang and Robert Mills in the 1950s showed that gauge symmetries require the existence of yet other forces. Much like a sensitive environmental-control system that keeps temperature, air pressure, and humidity in an area completely constant by compensating perfectly for any exterior influences, certain kinds of force fields, according to Yang and Mills, will provide perfect compensation for shifts in force charges, thereby keeping the physical interactions between the particles completely unchanged. For the case of the gauge symmetry associated with shifting quark-color charges, the required force is none other than the strong force itself. That is, without the strong force, physics *would* change under the kinds of shifts of color charges indicated above. This realization shows that, although the gravitational force and the strong force have vastly different properties (recall, for example, that gravity is far feebler than the strong force and operates over enormously larger distances), they do have a somewhat similar heritage: they are each required in order that the universe embody particular symmetries. Moreover, a similar discussion applies to the weak and electromagnetic forces, showing that their existence, too, is bound up with yet other gauge symmetries—the so-called weak and electromagnetic gauge symmetries. And hence, all four forces are directly associated with principles of symmetry.

This common feature of the four forces would seem to bode well for the suggestion made at the beginning of this section. Namely, in our effort to incorporate quantum mechanics into general relativity we should seek a quantum field theory of the gravitational force, much as physicists have discovered successful quantum field theories of the other three forces. Over the years, such reasoning has inspired a prodigious and distinguished group of physicists to follow this path vigorously, but the terrain has proven to be fraught with danger, and no one has succeeded in traversing it completely. Let's see why.

General Relativity vs. Quantum Mechanics

The usual realm of applicability of general relativity is that of large, astronomical distance scales. On such distances Einstein's theory implies that the absence of mass means that space is flat, as illustrated in Figure 3.3. In seeking to merge general relativity with quantum mechanics we must now change our focus sharply and examine the *microscopic* properties of space. We illustrate this in Figure 5.1 by zooming in and sequentially magnifying ever smaller regions of the spatial fabric. At first, as we zoom in, not much happens; as we see in the first three levels of magnification in Figure 5.1, the structure of space retains the same basic form. Reasoning from a purely classical standpoint, we would expect this placid and flat image of space to persist all the way to arbitrarily small length scales. But quantum mechanics changes this conclusion radically. *Everything* is subject to the quantum fluctuations inherent in the uncertainty principle—even the gravitational field. Although classical reasoning implies that empty space has zero gravitational field, quantum mechanics shows that on average it is zero, but that its actual value undulates up and down due to quantum fluctuations. Moreover, the uncertainty principle tells us that the size of the undulations of the gravitational field gets larger as we focus our attention on smaller regions of space. Quantum mechanics shows that nothing likes to be cornered; narrowing the spatial focus leads to ever larger undulations.

As gravitational fields are reflected by curvature, these quantum fluctuations manifest themselves as increasingly violent distortions of the surrounding space. We see the glimmers of such distortions emerging in the fourth level of magnification in Figure 5.1. By probing to even smaller distance scales, as we do in the fifth level of Figure 5.1, we see that the random quantum mechanical undulations in the gravitational field correspond to such severe warpings of space that it no longer resembles a gently curving geometrical object such as the rubber-membrane analogy used in our discussion in Chapter 3. Rather, it takes on the frothing, turbulent, twisted form illustrated in the uppermost part of the figure. John Wheeler coined the term *quantum foam* to describe the frenzy revealed by such an

Figure 5.1 By sequentially magnifying a region of space, its ultramicroscopic properties can be probed. Attempts to merge general relativity and quantum mechanics run up against the violent quantum foam emerging at the highest level of magnification.

ultramicroscopic examination of space (and time)—it describes an unfamiliar arena of the universe in which the conventional notions of left and right, back and forth, up and down (and even of before and after) lose their meaning. It is on such short distance scales that we encounter the fundamental incompatibility between general relativity and quantum mechanics. *The notion of a smooth spatial geometry, the central principle of general relativity, is destroyed by the violent fluctuations of the quantum world on short distance scales.* On ultramicroscopic scales, the central feature of quantum mechanics—the uncertainty principle—is in direct conflict with the central feature of general relativity—the smooth geometrical model of space (and of spacetime).

In practice, this conflict rears its head in a very concrete manner. Calculations that merge the equations of general relativity and those of quantum mechanics typically yield one and the same ridiculous answer: infinity. Like a sharp rap on the wrist from an old-time schoolteacher, an infinite answer is nature's way of telling us that we are doing something that is quite wrong.[6] The equations of general relativity cannot handle the roiling frenzy of quantum foam.

Notice, however, that as we recede to more ordinary distances (following the sequence of drawings in Figure 5.1 in reverse), the random, violent small-scale undulations cancel each other out—in much the same way that, on average, our compulsive borrower's bank account shows no evidence of his compulsion—and the concept of a smooth geometry for the fabric of the universe once again becomes accurate. It's like what you experience when you look at a dot-matrix picture: From far away the dots that compose the picture blend together and create the impression of a smooth image whose variations in lightness seamlessly and gently change from one area to another. When you inspect the picture on finer distance scales you realize, however, that it markedly differs from its smooth, long-distance appearance. It is nothing but a collection of discrete dots, each quite separate from the others. But note that you become aware of the discrete nature of the picture only when you examine it on the smallest of scales; from far away it looks smooth. Similarly, the fabric of spacetime appears to be smooth except when examined with ultramicroscopic precision. This is why general relativity works on large enough distance (and time) scales—the scales relevant for many typical astronomical applications—but is rendered inconsistent on short distance (and time) scales. The central tenet of a smooth and gently curving geometry is justified in the large but breaks down due to quantum fluctuations when pushed to the small.

The basic principles of general relativity and quantum mechanics allow us to calculate the approximate distance scales below which one would have to shrink in order for the pernicious phenomenon of Figure 5.1 to become apparent. The smallness of Planck's constant—which governs the strength of quantum effects—and the intrinsic weakness of the gravitational force team up to yield a result called the *Planck length,* which is small almost beyond imagination: a millionth of a billionth of a billionth of a billionth of a centimeter (10^{-33} centimeter).[7] The fifth level in Figure 5.1 thus schematically depicts the ultramicroscopic, sub–Planck length landscape of the universe. To get a sense of scale, if we were to magnify an atom to the size of the known universe, the Planck length would barely expand to the height of an average tree.

And so we see that the incompatability between general relativity and quantum mechanics becomes apparent only in a rather esoteric realm of the universe. For this reason you might well ask whether it's worth worrying about. In fact, the physics community does not speak with a unified voice when addressing this issue. There are those physicists who are willing to note the problem, but happily go about using quantum mechanics and general relativity for problems whose typical lengths far exceed the Planck length, as their research requires. There are other physicists, however, who are deeply unsettled by the fact that the two foundational pillars of physics as we know it are at their core fundamentally incompatible, regardless of the ultramicroscopic distances that must be probed to expose the problem. The incompatibility, they argue, points to an essential flaw in our understanding of the physical universe. This opinion rests on an unprovable but profoundly felt view that the universe, if understood at its deepest and most elementary level, can be described by a logically sound theory whose parts are harmoniously united. And surely, regardless of how central this incompatibility is to their own research, most physicists find it hard to believe that, at rock bottom, our deepest theoretical understanding of the universe will be composed of a mathematically inconsistent patchwork of two powerful yet conflicting explanatory frameworks.

Physicists have made numerous attempts at modifying either general relativity or quantum mechanics in some manner so as to avoid the conflict, but the attempts, although often bold and ingenious, have met with failure after failure.

That is, until the discovery of superstring theory.[8]

Part III

The Cosmic

Symphony

Chapter 6

Nothing but Music: The Essentials of Superstring Theory

Music has long since provided the metaphors of choice for those puzzling over questions of cosmic concern. From the ancient Pythagorean "music of the spheres" to the "harmonies of nature" that have guided inquiry through the ages, we have collectively sought the song of nature in the gentle wanderings of celestial bodies and the riotous fulminations of subatomic particles. With the discovery of superstring theory, musical metaphors take on a startling reality, for the theory suggests that the microscopic landscape is suffused with tiny strings whose vibrational patterns orchestrate the evolution of the cosmos. The winds of change, according to superstring theory, gust through an aeolian universe.

By contrast, the standard model views the elementary constituents of the universe as pointlike ingredients with no internal structure. As powerful as this approach is (as we have mentioned, essentially every prediction about the microworld made by the standard model has been verified down to about a billionth of a billionth of a meter, the present-day technological limit), the standard model cannot be a complete or final theory because it does not include gravity. Moreover, attempts to incorporate gravity into its quantum-mechanical framework have failed due to the violent fluctuations in the spatial fabric that appear at ultramicroscopic distances—that is, distances shorter than the Planck length. The unresolved conflict has impelled a search for an even deeper understanding of nature. In 1984, the physicists Michael Green, then of Queen Mary

College, and John Schwarz of the California Institute of Technology provided the first piece of convincing evidence that *superstring theory* (or string theory, for short) might well provide this understanding.

String theory offers a novel and profound modification to our theoretical description of the ultramicroscopic properties of the universe—a modification that, physicists slowly realized, alters Einstein's general relativity in just the right way to make it fully compatible with the laws of quantum mechanics. According to string theory, the elementary ingredients of the universe are *not* point particles. Rather, they are tiny, one-dimensional filaments somewhat like infinitely thin rubber bands, vibrating to and fro. But don't let the name fool you: Unlike an ordinary piece of string, which is itself composed of molecules and atoms, the strings of string theory are purported to lie deep within the heart of matter. The theory proposes that *they* are ultramicroscopic ingredients making up the particles out of which atoms themselves are made. The strings of string theory are so small—on average they are about as long as the Planck length—that they *appear* pointlike even when examined with our most powerful equipment.

Yet the simple replacement of point particles with strands of string as the fundamental ingredients of everything has far-reaching consequences. First and foremost, string theory appears to resolve the conflict between general relativity and quantum mechanics. As we shall see, the spatially extended nature of a string is the crucial new element allowing for a single harmonious framework incorporating both theories. Second, string theory provides a truly unified theory, since all matter and all forces are proposed to arise from one basic ingredient: oscillating strings. Finally, as discussed more fully in subsequent chapters, beyond these remarkable achievements, string theory once again radically changes our understanding of spacetime.[1]

A Brief History of String Theory

In 1968, a young theoretical physicist named Gabriele Veneziano was struggling to make sense of various experimentally observed properties of the strong nuclear force. Veneziano, then a research fellow at CERN, the European accelerator laboratory in Geneva, Switzerland, had worked on aspects of this problem for a number of years, until one day he came upon a striking revelation. Much to his surprise, he realized that an esoteric formula concocted for purely mathematical pursuits by the renowned Swiss mathematician Leonhard Euler some two hundred years earlier—the so-called Euler beta-function—seemed to describe numerous properties of strongly interacting particles in one fell swoop. Veneziano's observation provided a powerful mathematical encapsulation of many features of the strong force and it launched an intense flurry of research aimed at using Euler's beta-function, and various generalizations, to describe the surfeit of data being collected at various atom smashers around the world. Nevertheless, there was a sense in which Veneziano's observation was incomplete. Like memorized formulae used by a student who does not understand their meaning or justification, Euler's beta-function seemed to work, but no one knew why. It was a formula in search of an explanation. This changed in 1970 when the works of Yoichiro Nambu of the University of Chicago, Holger Nielsen of the Niels Bohr Institute, and Leonard Susskind of Stanford University revealed the hitherto-unknown physics lurking behind Euler's formula. These physicists showed that if one modeled elementary particles as little, vibrating, one-dimensional strings, their nuclear interactions could be described exactly by Euler's function. If the pieces of string were small enough, they reasoned, they would still look like point particles, and hence could be consistent with experimental observations.

Although this provided an intuitively simple and pleasing theory, it was not long before the string description of the strong force was shown to fail. During the early 1970s, high-energy experiments capable of probing the subatomic world more deeply showed that the string model made a number of predictions that were in direct conflict with observations. At the same time, the point-particle quantum field theory of quantum chromodynamics was being developed, and its overwhelming success in describing the strong force led to the dismissal of string theory.

Most particle physicists thought that string theory had been relegated to the dustbin of science, but a few dedicated researchers kept at it. Schwarz, for instance, felt that "the mathematical structure of string theory was so beautiful and had so many miraculous properties that it had to be pointing toward something deep."[2] One of the problems physicists found with string theory was that it seemed to have a true embarrassment of riches. The theory contained configurations of vibrating string that had properties akin to those of gluons, substantiating its early claim of being a theory of the strong force. But beyond these it contained *additional* messenger-like particles that did not appear to have any relevance to experimental observations of the strong force. In 1974, Schwarz and Joël Scherk of the Ecole Normale Supérieure made a bold leap that transformed this apparent vice into a virtue. After studying the puzzling messenger-like patterns of string vibration, they realized that their properties matched perfectly those of the hypothesized messenger particle of the gravitational force—the graviton. Although these "smallest bundles" of the gravitational force have, as yet, never been seen, theorists can confidently predict certain basic features that they must possess, and Scherk and Schwarz found these properties to be realized exactly by certain vibrational patterns. Based on this, Scherk and Schwarz suggested that string theory had failed in its initial attempt because physicists had unduly constrained its scope. String theory is *not* just a theory of the strong force, they proclaimed; it is a quantum theory that *includes gravity* as well.[3]

The physics community did not receive this suggestion with unbridled enthusiasm. In fact, Schwarz recounts that "our work was universally ignored."[4] The path of progress was already littered with numerous failed attempts to unite gravity and quantum mechanics. String theory had been shown wrong in its initial effort to describe the strong force, and it seemed to many that it was senseless to try to use the theory to pursue an even grander goal. Even more devastating, subsequent studies during the late 1970s and early 1980s showed that string theory and quantum mechanics suffered from their own subtle conflicts. It appeared that the gravitational force had, once again, resisted incorporation into the microscopic description of the universe.

Such was the case until 1984. In a landmark paper culminating more than a dozen years of intense research that had been largely ignored and often outright dismissed by most physicists, Green and Schwarz established that the subtle quantum conflict afflicting string theory could be resolved. Moreover, they showed that the resulting theory had sufficient breadth to encompass all of the four forces and all of matter as well. As word of this result spread throughout the worldwide physics community, particle physicists by the hundreds dropped their research projects to launch a full-scale assault on what appeared to be the last theoretical battleground in the ancient quest to understand the deepest workings of the universe.

I began graduate school at Oxford University in October 1984. Although I was excited to be learning about the likes of quantum field theory, gauge theory, and general relativity, there was a pervasive feeling among the older graduate students that there was little or no future for particle physics. The standard model was in place and its remarkable success at predicting experimental outcomes indicated that its verification was merely a matter of time and details. Going beyond its limits to include gravity and possibly to *explain* the experimental input on which it relies— the 19 numbers summarizing the elementary particle masses, their force charges, and the relative strengths of the forces, numbers that are known from experiment but are not understood theoretically—was so daunting a task that all but the most courageous physicists recoiled at the challenge. But six months later the mood had swung completely around. The success of Green and Schwarz finally trickled down even to first-year graduate students, and an electrifying sense of being on the inside of a profound moment in the history of physics displaced the previous ennui. A number of us consistently worked deep into the night to try to master the vast areas of theoretical physics and abstract mathematics that are required to understand string theory.

The period from 1984 to 1986 has come to be known as the "first superstring revolution." During those three years more than a thousand research papers on string theory were written by physicists from around the world. These works showed conclusively that numerous features of the standard model—features that had been painstakingly discovered over the course of decades of research—*emerged naturally* and simply from the grand structure of string theory. As Michael Green has said, "The moment you encounter string theory and realize that almost all of the major developments in physics over the last hundred years emerge—and emerge with such elegance—from such a simple starting point, you realize that this incredibly compelling theory is in a class of its own."[5] Moreover, for many of these features, as we shall discuss, string theory offers a far fuller and more satisfying explanation than is found in the standard model. These developments convinced many physicists that string theory was well on its way to fulfilling its promise of being the ultimate unified theory.

Nonetheless, over and over again string theorists encountered a significant stumbling block. In theoretical physics research, one is frequently confronted with equations that are just too hard to understand or to analyze. Typically, physicists don't give up, but try to solve the equations approximately. The situation in string theory is even more difficult. Even determining the *equations themselves* has proved to be so difficult that only approximate versions of them have so far been deduced. String theorists have thereby been limited to finding approximate solutions to approximate equations. After the few years of dramatic progress during the first superstring revolution, physicists found that the approximations being used were inadequate to answer a number of essential questions hindering further developments. With no concrete proposals for going beyond the approximate methods, many physicists working on string theory grew frustrated and returned to their previous lines of research. For those who remained, the late 1980s and early 1990s were trying times. Like a golden treasure securely locked in a safe and visible only through a tiny, tantalizing peephole, the beauty and promise of string theory beckoned, but no one had the key to unlock its power. Long dry spells were periodically punctuated by important discoveries, but it was clear to everyone in the field that new methods with the power to go beyond the previous approximations were required.

Then, in a breathtaking lecture at the Strings 1995 conference held at the University of Southern California—a lecture that stunned a packed audience of the world's top physicists—Edward Witten announced a plan for taking the next step, thereby igniting the "second superstring revolution." String theorists, as of this writing, are working vigorously to sharpen a set of new methods that promise to overcome the theoretical obstacles previously encountered. The difficulties that lie ahead will severely test the technical might of the world's superstring theorists, but the light at the end of the tunnel, although still distant, may finally be becoming visible.

In this chapter and a number that follow, we shall describe the understanding of string theory that emerged from the first superstring revolution and subsequent work prior to the second superstring revolution. From time to time we will indicate new insights stemming from the latter; our discussion of these most recent advances will come in Chapters 12 and 13.

The Greeks' Atoms, Again?

As we mentioned at the outset of this chapter and as illustrated in Figure 1.1, string theory claims that if the presumed point-particles of the standard model could be examined with a precision significantly beyond our present capacity, each would be seen to be made of a single, tiny, oscillating loop of string.

For reasons that will become clear, the length of a typical string loop is about the Planck length, about a hundred billion billion (10^{20}) times smaller than an atomic nucleus. It is no wonder that our present-day experiments are unable to resolve the microscopic stringy nature of matter: strings are minute even on the scales set by subatomic particles. We would need an accelerator to slam matter together with energies some million billion times more powerful than any previously constructed in order to reveal directly that a string is not a point-particle.

We will describe shortly the stunning implications that follow from replacing point-particles by strings, but let's first address a more basic question: What are strings made of?

There are two possible answers to this question. First, strings are truly fundamental—they are "atoms," *uncuttable constituents,* in the truest sense of the ancient Greeks. As the absolute smallest constituents of anything and everything, they represent the end of the line—the last of the Russian *matrioshka* dolls—in the numerous layers of substructure in the microscopic world. From this perspective, even though strings have spatial extent, the question of their composition is without any content. Were strings to be made of something smaller they would not be fundamental. Instead, whatever strings were composed of would immediately displace them and lay claim to being an even more basic constituent of the universe. Using our linguistic analogy, paragraphs are made of sentences, sentences are made of words, and words are made of letters. What makes up a letter? From a linguistic standpoint, that's the end of the line. Letters are letters—they are the fundamental building blocks of written language; there is no further substructure. Questioning their composition has no meaning. Similarly, a string is simply a string—as there is nothing more fundamental, it can't be described as being composed of any other substance.

That's the first answer. The second answer is based on the simple fact that as yet we do not know if string theory is a correct or final theory of nature. If string theory is truly off the mark, then, well, we can forget strings and the irrelevant question of their composition. Although this is a possibility, research since the mid-1980s overwhelmingly points toward its being extremely unlikely. But history surely has taught us that every time our understanding of the universe deepens, we find yet smaller microscopic ingredients constituting a finer level of matter. And so another possibility, should strings fail to be the final theory, is that they are one more layer in the cosmic onion, a layer that becomes visible at the Planck length, although not the final layer. In this case, strings could be made up of yet-smaller structures. String theorists have raised and continue to pursue this possibility. To date there are intriguing hints in theoretical studies that strings may have further substructure, but there is as yet no definitive evidence. Only time and intense research will supply the final word on this question.

Aside from a few speculations in Chapters 12 and 15, for our discussion here we approach strings in the manner proposed in the first answer—that is, we will take strings to be nature's most fundamental ingredient.

Unification through String Theory

Besides its inability to incorporate the gravitational force, the standard model has another shortcoming: There is no explanation for the details of its construction. Why did nature select the particular list of particles and forces outlined in previous chapters and recorded in Tables 1.1 and 1.2? Why do the 19 parameters that describe these ingredients quantitatively have the values that they do? You can't help feeling that their number and detailed properties seem so arbitrary. Is there a deeper understanding lurking behind these seemingly random ingredients, or were the detailed physical properties of the universe "chosen" by happenstance?

The standard model itself cannot possibly offer an explanation since it takes the list of particles and their properties as experimentally measured *input*. Just as the performance of the stock market cannot be used to determine the value of your portfolio without the input data of your initial investments, the standard model cannot be used to make any predictions without the input data of the fundamental particle properties.[6] After experimental particle physicists fastidiously measure these data, theorists can then use the standard model to make testable predictions, such as what should happen when particular particles are slammed together in an accelerator. But the standard model can no more explain the fundamental particle properties of Tables 1.1 and 1.2 than the Dow Jones average today can explain your initial investment in stocks ten years ago.

In fact, had experiments revealed a somewhat different particle content in the microscopic world, possibly interacting with somewhat different forces, these changes could have been fairly easily incorporated in the standard model by providing the theory with different input parameters. The structure of the standard model, in this sense, is too flexible to be able to explain the properties of the elementary particles, as it could have accommodated a range of possibilities.

String theory is dramatically different. It is a unique and inflexible theoretical edifice. It requires no input beyond a single number, described below, that sets the benchmark scale for measurements. All properties of the microworld are within the realm of its explanatory power. To understand this, let's first think about more familiar strings, such as those on a violin. Each such string can undergo a huge variety (in fact, infinite in number) of different vibrational patterns known as *resonances,* such as those shown in Figure 6.1. These are the wave patterns whose peaks and troughs are evenly spaced and fit perfectly between the string's two fixed endpoints. Our ears sense these different resonant vibrational patterns as different musical notes. The strings in string theory have similar properties. There are resonant vibrational patterns that the string can support by virtue of their evenly spaced peaks and troughs exactly fitting along its spatial extent. Some examples are given in Figure 6.2. Here's the central fact: Just as the different vibrational patterns of a violin string give rise to different musical notes, *the different vibrational patterns of a fundamental string give rise to different masses and force charges.* As this is a crucial point, let's say it again. According to string theory, the properties of an el-

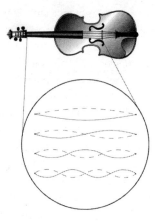

Figure 6.1 Strings on a violin can vibrate in resonant patterns in which a whole number of peaks and troughs exactly fit between the two ends.

ementary "particle"—its mass and its various force charges—are determined by the precise resonant pattern of vibration that its internal string executes.

It's easiest to understand this association for a particle's mass. The energy of a particular vibrational string pattern depends on its amplitude—the maximum displacement between peaks and troughs—and its wavelength—the separation between one peak and the next. The greater the amplitude and the shorter the wavelength, the greater the energy. This reflects what you would expect intuitively—more frantic vibrational pat-

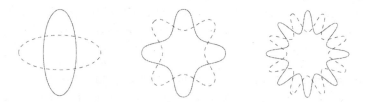

Figure 6.2 The loops in string theory can vibrate in resonance patterns—similar to those of violin strings—in which a whole number of peaks and troughs fit along their spatial extent.

terns have more energy, while less frantic ones have less energy. We give a couple of examples in Figure 6.3. This is again familiar, as violin strings that are plucked more vigorously will vibrate more wildly, while those plucked more gingerly will vibrate more gently. Now, from special relativity we know that energy and mass are two sides of the same coin: Greater energy means greater mass, and vice versa. Thus, according to string theory, the *mass* of an elementary particle is determined by the *energy* of the vibrational pattern of its internal string. Heavier particles have internal strings that vibrate more energetically, while lighter particles have internal strings that vibrate less energetically.

Since the mass of a particle determines its gravitational properties, we see that there is a direct association between the pattern of string vibration and a particle's response to the gravitational force. Although the reasoning involved is somewhat more abstract, physicists have found that a similar alignment exists between other detailed aspects of a string's pattern of vibration and its properties vis à vis other forces. The electric charge, the weak charge, and the strong charge carried by a particular string, for instance, are determined by the precise way it vibrates. Moreover, exactly the same idea holds for the messenger particles themselves. Particles like photons, weak gauge bosons, and gluons are yet other resonant patterns of string vibration. And of particular importance, among the vibrational string patterns, one matches perfectly the properties of the graviton, ensuring that gravity is an integral part of string theory.[7]

So we see that, according to string theory, the observed properties of each elementary particle arise because its internal string undergoes a particular resonant vibrational pattern. This perspective differs sharply from that espoused by physicists before the discovery of string theory; in the ear-

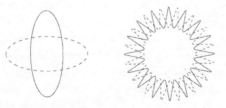

Figure 6.3 More frantic vibrational patterns have more energy than less frantic ones.

lier perspective the differences among the fundamental particles were explained by saying that, in effect, each particle species was "cut from a different fabric." Although each particle was viewed as elementary, the kind of "stuff" each embodied was thought to be different. Electron "stuff," for example, had negative electric charge, while neutrino "stuff" had no electric charge. String theory alters this picture radically by declaring that the "stuff" of all matter and all forces is the *same*. Each elementary particle is composed of a single string—that is, each particle *is* a single string—and all strings are absolutely identical. Differences between the particles arise because their respective strings undergo different resonant vibrational patterns. What appear to be different elementary particles are actually different "notes" on a fundamental string. The universe—being composed of an enormous number of these vibrating strings—is akin to a cosmic symphony.

This overview shows how string theory offers a truly wonderful unifying framework. Every particle of matter and every transmitter of force consists of a string whose pattern of vibration is its "fingerprint." Because every physical event, process, or occurrence in the universe is, at its most elementary level, describable in terms of forces acting between these elementary material constituents, string theory provides the promise of a single, all-inclusive, unified description of the physical universe: a theory of everything (T.O.E.).

The Music of String Theory

Even though string theory does away with the previous concept of structureless elementary particles, old language dies hard, especially when it provides an accurate description of reality down to the most minute of distance scales. Following the common practice of the field we shall therefore continue to refer to "elementary particles," yet we will always mean "what appear to be elementary particles but are actually tiny pieces of vibrating string." In the preceding section we proposed that the masses and the force charges of such elementary particles are the result of the way in which their respective strings are vibrating. This leads us to the following realization: If we can work out precisely the allowed resonant vibrational patterns of fundamental strings—the "notes," so to speak, that they can play—we should be able to explain the observed properties of the elementary particles. For the first time, therefore, string theory sets up a framework for *explaining* the properties of the particles observed in nature.

At this stage, then, we should "grab hold" of a string and "pluck" it in all sorts of ways to determine the possible resonant patterns of vibration. If string theory is right, we should find that the possible patterns yield exactly the observed properties of the matter and force particles in Tables 1.1 and 1.2. Of course, a string is too small to carry out this experiment literally as described. Rather, by using mathematical descriptions we can *theoretically* pluck a string. In the mid-1980s, many string adherents believed that the mathematical analysis required for doing this was on the verge of being able to explain every detailed property of the universe on its most microscopic level. Some enthusiastic physicists declared that the T.O.E. had finally been discovered. More than a decade of hindsight has shown that the euphoria generated by this belief was premature. String theory has the makings of a T.O.E., but a number of hurdles remain, preventing us from deducing the spectrum of string vibrations with the precision necessary to compare with experimental results. At the present time, therefore, we do not know if the fundamental characteristics of our universe, summarized in Tables 1.1 and 1.2, can be explained by string theory. As we will discuss in Chapter 9, under certain assumptions that we will clearly state, string theory can give rise to a universe with properties that are in qualitative agreement with the known particle and force data, but extracting detailed numerical predictions from the theory is currently beyond our abilities. And so, although the framework of string theory, unlike that of the point-particle standard model, is *capable* of giving an explanation for why the particles and forces have the properties they do, we have not, as yet, been able to extract it. But remarkably, string theory is so rich and far-reaching that, even though we cannot yet determine its most detailed properties, we *are* able to gain insight into a wealth of the new physical phenomena that follow from the theory, as we will see in subsequent chapters.

In the following chapters we shall also discuss the status of the hurdles in some detail, but it is instructive first to understand them at a general level. Strings in the world around us come with a variety of tensions. The string laced through a pair of shoes, for example, is usually quite slack compared to the string stretched from one end of a violin to another. Both of these, in turn, are under far less tension than the steel strings of a piano. The one number that string theory requires in order to set its over-all scale is the corresponding tension on its loops. How is this tension determined? Well, if we could pluck a fundamental string we would learn about its stiffness, and in this way we could measure its tension much as is done to measure the tension of more familiar everyday strings. But since fundamental strings are so tiny, this approach cannot be carried out and a more indirect method is called for. In 1974, when Scherk and Schwarz proposed that one particular pattern of string vibration was the graviton particle, they were able to exploit such an indirect approach and thereby predict the tension on the strings of string theory. Their calculations revealed that the strength of the force transmitted by the proposed graviton pattern of string vibration is inversely proportional to the string's tension. And since the graviton is supposed to transmit the gravitational force—a force that is intrinsically quite feeble—they found that this implies a colossal tension of a thousand billion billion billion billion (10^{39}) tons, the so-called *Planck tension*. Fundamental strings are therefore extremely stiff compared with more familiar examples. This has three important consequences.

Three Consequences of Stiff Strings

First, whereas the ends of a violin or a piano string are pinned down, ensuring that they have a fixed length, no analogous constraining frame pins down the size of a fundamental string. Instead, the huge string tension causes the loops of string theory to contract to a minuscule size. Detailed calculation reveals that being under Planck tension translates into a typical string having Planck length—10^{-33} centimeters—as previously mentioned.[8]

Second, because of the enormous tension, the typical energy of a vibrating loop in string theory is extremely high. To understand this, we note that the greater the tension a string is under, the harder it is to get it to vibrate. For instance, it's far easier to pluck a violin string and set it vibrating than it is to pluck a piano string. Two strings, therefore, that are under different tension and are vibrating in precisely the same way will not have the same energy. The string with higher tension will have more energy than the string with lower tension, since more energy must be exerted to set it in motion.

This alerts us to the fact that the energy of a vibrating string is determined by two things: the precise manner in which it vibrates (more frantic patterns corresponding to higher energies) and the tension of the string (higher tension corresponding to higher energy). At first, this description might lead you to think that by taking on ever gentler vibrational patterns—patterns with ever smaller amplitudes and fewer peaks and troughs—a string can embody less and less energy. But as we found in Chapter 4 in a different context, quantum mechanics tells us that this reasoning is not right. Like all vibrations or wavelike disturbances, quantum mechanics implies that they can exist only in discrete units. Roughly speaking, just as the money carried by a comrade in the warehouse is a *whole* number multiple of the monetary denomination with which he or she is entrusted, the energy embodied in a string vibrational pattern is a whole number multiple of a minimal energy denomination. In particular, this minimal energy denomination is proportional to the tension of the string (and it is also proportional to the number of peaks and troughs in the particular vibrational pattern), while the whole number multiple is determined by the amplitude of the vibrational pattern.

The key point for the present discussion is this: Since the minimal energy denominations are proportional to the string's tension, and since this tension is enormous, the fundamental minimal energies are, on the usual scales of elementary particle physics, similarly huge. They are multiples of what is known as the *Planck energy*. To get a sense of scale, if we translate the Planck energy into a mass using Einstein's famous conversion formula $E = mc^2$, they correspond to masses that are on the order of ten billion billion (10^{19}) times that of a proton. This gargantuan mass—by elementary particle standards—is known as the *Planck mass*; it's about equal to the mass of a grain of dust or a collection of a million average bacteria. And so, the typical mass-equivalent of a vibrating loop in string theory is generally some whole number (1, 2, 3, . . .) times the Planck mass. Physicists often express this by saying that the "natural" or "typical" energy scale (and hence mass scale) of string theory is the Planck scale.

This raises a crucial question directly related to the goal of reproducing the particle properties in Tables 1.1 and 1.2: If the "natural" energy scale of string theory is some ten billion billion times that of a proton, how can it possibly account for the far-lighter particles—electrons, quarks, photons, and so on—making up the world around us?

The answer, once again, comes from quantum mechanics. The uncertainty principle ensures that nothing is ever perfectly at rest. All objects undergo quantum jitter, for if they didn't we would know where they were and how fast they were moving with complete precision, in violation of Heisenberg's dictum. This holds true for the loops in string theory as well; no matter how placid a string appears it will always experience some amount of quantum vibration. The remarkable thing, as originally worked out in the 1970s, is that there can be energy *cancellations* between these quantum jitters and the more intuitive kind of string vibrations discussed above and illustrated in Figures 6.2 and 6.3. In effect, through the weirdness of quantum mechanics, the energy associated with the quantum jitters of a string is *negative,* and this *reduces* the overall energy content of a vibrating string by an amount that is roughly equal to Planck energy. This means that the lowest-energy vibrational string patterns, whose energies we would naively expect to be about equal to the Planck energy (i.e., 1 times the Planck energy), are largely canceled, thereby yielding relatively low net-energy vibrations—energies whose corresponding mass-equivalents are in the neighborhood of the matter and force particle masses shown in Tables 1.1 and 1.2. It is these *lowest* energy vibrational patterns, therefore, that should provide contact between the theoretical description of strings and the experimentally accessible world of particle physics. As an important example, Scherk and Schwarz found that for the vibrational pattern whose properties make it a candidate for the graviton messenger particle, the energy cancellations are *perfect,* resulting in a zero-mass gravitational-force particle. This is precisely what is expected for the graviton; the gravitational force is transmitted at light speed and only massless particles travel at this maximal velocity. But low-energy vibrational combinations are very much the exception rather than the rule. The more typical vibrating fundamental string corresponds to a particle whose mass is billions upon billions times greater than that of the proton.

This tells us that the comparatively light fundamental particles of Tables 1.1 and 1.2 should arise, in a sense, from the fine mist above the roaring ocean of energetic strings. Even a particle as heavy as the top quark, with a mass about 189 times that of the proton, can arise from a vibrating string only if the string's enormous characteristic Planck-scale energy is canceled by the jitters of quantum uncertainty to better than one part in a hundred million billion. It's as if you were playing *The Price Is Right* and Bob Barker gives you ten billion billion dollars and challenges you to purchase products that will cost—cancel, so to speak—all but 189 of the dollars, not a dollar more or less. Coming up with such an enormous yet precise expenditure, without being privy to the exact prices of the individual items, would severely tax the acumen of even the world's most expert shoppers. In string theory, where the currency is energy as opposed to money, approximate calculations have conclusively shown that analogous energy cancellations certainly *can* occur, but for reasons that will become increasingly clear in subsequent chapters, verifying the cancellations to such a high level of precision is generally beyond our theoretical ken at present. Even so, as indicated before, we shall see that many other properties of string theory that are less sensitive to these finest of details can be extracted and understood with confidence.

This takes us to the third consequence of the enormous value of the string tension. Strings can execute an infinite number of different vibrational patterns. For instance, in Figure 6.2 we showed the beginnings of a never-ending sequence of possibilities characterized by an ever greater number of peaks and troughs. Doesn't this mean that there would have to be a corresponding never-ending sequence of elementary particles, seemingly in conflict with the experimental situation summarized in Tables 1.1 and 1.2?

The answer is yes: If string theory is right, each of the infinitely many resonant patterns of string vibration should correspond to an elementary particle. An essential point, however, is that the high string tension ensures that all but a few of these vibrational patterns will correspond to extremely heavy particles (the few being the lowest-energy vibrations that have near-perfect cancellations with quantum string jitters). And again, the term "heavy" here means many times heavier than the Planck mass. As our most powerful particle accelerators can reach energies only on the order of a thousand times the proton mass, less than a millionth of a billionth of the Planck energy, we are very far from being able to search in the laboratory for any of these new particles predicted by string theory.

There are more indirect approaches by which we could search for them, though. For instance, the energies involved at the birth of the universe would have been high enough to produce these particles copiously. In general one would not expect them to survive to the present day, as such super-heavy particles are usually unstable, relinquishing their enormous mass by decaying into a cascade of ever lighter particles, ending with the familiar, relatively light particles in the world around us. However, it is possible that such a super-heavy vibrational string state—a relic from the big bang—did survive to the present. Finding such particles, as we discuss more fully in Chapter 9, would be a monumental discovery, to say the least.

Gravity and Quantum Mechanics in String Theory

The unified framework that string theory presents is compelling. But its real attraction is the ability to ameliorate the hostilities between the gravitational force and quantum mechanics. Recall that the problem in merging general relativity and quantum mechanics turns up when the central tenet of the former—that space and time constitute a smoothly curving geometrical structure—confronts the essential feature of the latter—that everything in the universe, including the fabric of space and time, undergoes quantum fluctuations that become increasingly turbulent when probed on smaller and smaller distance scales. On sub-Planck-scale distances, the quantum undulations are so violent that they destroy the notion of a smoothly curving geometrical space; this means that general relativity breaks down.

String theory softens the violent quantum undulations by "smearing" out the short-distance properties of space. There is a rough and a more precise answer to the question of what this really means and how it resolves the conflict. We discuss each in turn.

The Rough Answer

Although it sounds unsophisticated, one way that we learn about the structure of an object is by hurling other things at it and observing the precise way in which they are deflected. We are able to *see* things, for example, because our eyes collect and our brains decode information carried by photons as they bounce off of objects being viewed. Particle accelerators are based on the same principle: They hurl bits of matter such as electrons and protons at each other as well as at other targets, and elaborate detectors analyze the resulting spray of debris to determine the architecture of the objects involved.

As a general rule, the *size of the probe particle* that we use sets a lower limit to the length scale to which we are sensitive. To get a feel for what this important statement means, imagine that Slim and Jim decide to get some culture by enrolling in a drawing class. As the semester progresses, Jim becomes increasingly irritated by Slim's growing proficiency as an artist and challenges him to an unusual contest. He proposes that they each take a peach pit, secure it in a vise, and draw their most accurate "still life" renditions. The unusual feature of Jim's challenge is that neither he nor Slim will be allowed to look at the peach pits. Instead, each is allowed to learn about the size, shape, and features of his peach pit only by shooting things (other than photons!) at the pit and observing how they are deflected, as illustrated in Figure 6.4. Unbeknownst to Slim, Jim fills Slim's "shooter" with marbles (as in Figure 6.4(a)) but fills his own shooter with far smaller five-millimeter plastic pellets (as in Figure 6.4(b)). They both turn on their shooters, and the competition begins.

After a while, the best drawing Slim can come up with is that in Figure 6.4(a). By observing the trajectories of the deflected marbles he was able to learn that the pit is a small, hard-surfaced mass. But that's all he could learn. Marbles are just too large to be sensitive to the finer corrugated structure of the peach pit. When Slim takes a look at Jim's drawing (Figure 6.4(b)), he is surprised to see that he has been outdone. A momentary glance at Jim's shooter, though, reveals the trick: The smaller probe particles used by Jim are fine enough to have their angle of deflection affected by some of the largest features adorning the pit's surface. And so, by shooting many five-millimeter pellets at the pit and observing their deflected trajectories, Jim was able to draw a more detailed image. Slim, not to be outdone, goes back to his shooter, fills it with even smaller probe particles—half-millimeter pellets—that are tiny enough to enter and hence be deflected by the finest corrugations on the pit's surface. By observing how these impinging probe particles are deflected, he is able to draw the winning rendition shown in Figure 6.4(c).

The lesson taught by this little competition is clear: Useful probe par-

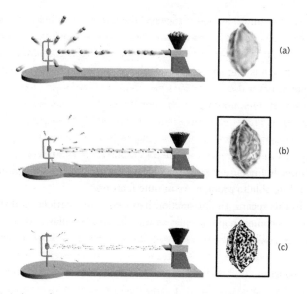

Figure 6.4 A peach pit is secured in a vise and it is drawn solely by observing how things—"probes"—thrown at it are deflected. By using ever smaller probes—(a) marbles, (b) five-millimeter pellets, (c) half-millimeter pellets—ever more detailed renditions can be drawn.

ticles cannot be substantially larger than the physical features being examined; otherwise, they will be insensitive to the structures of interest.

The same reasoning holds, of course, if one wants to probe the pit even more deeply to determine its atomic and subatomic structure. Half-millimeter pellets will not provide any useful information; they are clearly too big to have any sensitivity to structure on atomic scales. This is why particle accelerators use protons or electrons as probes, since their small size makes them much better suited to the task. On subatomic scales, where quantum concepts replace classical reasoning, the most appropriate measure of a particle's probing sensitivity is its quantum wavelength, which indicates the window of uncertainty in its position. This fact reflects our discussion of Heisenberg's uncertainty principle in Chapter 4, in which we found that the margin of error incurred when using a point particle as a probe (we focused on photon probes but the discussion applies to all other particles) is about equal to the probe particle's quantum wavelength. In somewhat looser language, the probing sensitivity of a point particle is smeared out by the jitteriness of quantum mechanics, in much the same way that the precision of a surgeon's scalpel is compromised if he or she has hands that shake. But recall that in Chapter 4 we also noted the important fact that a particle's quantum wavelength is inversely proportional to its momentum, which, roughly speaking, is its energy. And so, by increasing a point particle's energy, its quantum wavelength can be made shorter and shorter—quantum smearing can be decreased further and further—and hence we can use it to probe ever finer physical structures. Intuitively, higher-energy particles have greater penetrating power and are therefore able to probe more minute features.

In this regard, the distinction between point particles and strands of string becomes manifest. Just as was the case for plastic pellets probing the surface features of a peach pit, the string's inherent spatial extent prevents it from probing the structure of anything substantially smaller than its own size—in this case structures arising on length scales shorter than the Planck length. Somewhat more precisely, in 1988 David Gross, then of Princeton University, and his student Paul Mende showed that when quantum mechanics is taken into account, continually increasing the energy of a string does *not* continually increase its ability to probe finer structures, in direct contrast with what happens for a point particle. They found that when the energy of a string is increased, it is at first able to probe shorter-scale structures, just like an energetic point particle. But when its energy is increased beyond the value required for probing structures on the scale of the Planck length, the additional energy does not sharpen the string probe. Rather, the energy causes the string to *grow* in size, thereby *diminishing* its short-distance sensitivity. In fact, although the size of a typical string is the Planck length, if we pumped enough energy into a string—an amount of energy beyond our wildest imaginings but one that would likely have been attained by the big bang—we could cause it to grow to a *macroscopic* size, a clumsy probe of the microcosmos indeed! It's as if a string, unlike a point particle, has *two* sources of smearing: quantum jitters, as for a point particle, and also its own inherent spatial extent. Increasing a string's energy decreases the smearing from the first source but ultimately increases the smearing from the second. The upshot is that no matter how hard you try, the extended nature of a string prevents you from using it to probe phenomena on sub-Planck-length distances.

But the whole conflict between general relativity and quantum mechanics arises from the sub-Planck-length properties of the spatial fabric. *If the elementary constituent of the universe cannot probe sub-Planck-scale distances, then neither it nor anything made from it can be affected by the supposedly disastrous short-distance quantum undulations.* This is similar to what happens as we draw our hand across a highly polished granite surface. Although at a microscopic level the granite is discrete, grainy, and bumpy, our fingers are unable to detect these short-scale variations and the surface feels perfectly smooth. Our stumpy, extended fingers "smear" out the microscopic discreteness. Similarly, since the string has spatial extent, it also has limits on its short-distance sensitivity. It cannot detect variations on sub-Planck-distance scales. Like our fingers on granite, the string smears out the jittery ultramicroscopic fluctuations of the gravitational field. Although the resulting fluctuations are still substantial, this smearing smooths them out just enough to cure the incompatibility between general relativity and quantum mechanics. And, in particular, the pernicious infinities (discussed in the preceding chapter) that arise in the point-particle approach to forming a quantum theory of gravity are done away with by string theory.

An essential difference between the granite analogy and our real concern with the spatial fabric is that there *are* ways in which the microscopic discreteness of the granite's surface can be exposed: Finer, more precise probes than our fingers can be used. An electron microscope has the ability to resolve surface features to less than a millionth of a centimeter; this is sufficiently small to reveal the numerous surface imperfections. By contrast, in string theory there is no way to expose the sub-Planck-scale "imperfections" in the fabric of space. In a universe governed by the laws of string theory, the conventional notion that we can always dissect nature on ever smaller distances, without limit, is not true. There *is* a limit, and it comes into play before we encounter the devastating quantum foam of Figure 5.1. Therefore, in a sense that will be made more precise in later chapters, one can even say that the supposed tempestuous sub-Planckian quantum undulations *do not exist.* A positivist would say that something exists only if it can—at least in principle—be probed and measured. Since the string is supposed to be the most elementary object in the universe and since it is too large to be affected by the violent sub-Planck-length undulations of the spatial fabric, these fluctuations cannot be measured and hence, according to string theory, do not actually arise.

A Sleight of Hand?

This discussion may leave you feeling dissatisfied. Instead of showing that string theory tames the sub-Planck-length quantum undulations of space, we seem to have used the string's nonzero size to skirt the whole issue completely. Have we actually solved anything? We have. The following two points will serve to emphasize this.

First, what the preceding argument implies is that the supposedly problematic sub-Planck-length spatial fluctuations are an artifact of formulating general relativity and quantum mechanics in a point-particle framework. In a sense, therefore, the central conflict of contemporary theoretical physics has been a problem of our own making. Because we previously envisioned all matter particles and all force particles to be point-like objects with literally no spatial extent, we were obligated to consider properties of the universe on arbitrarily short distance scales. And on the tiniest of distances we ran into seemingly insurmountable problems. String theory tells us that we encountered these problems only because we did not understand the true rules of the game; the new rules tell us that there is a limit to how finely we can probe the universe—and, in a real sense, a limit to how finely our conventional notion of distance can even be applied to the ultramicroscopic structure of the cosmos. The supposed pernicious spatial fluctuations are now seen to have arisen in our theories because we were unaware of these limits and were thus led by a point-particle approach to grossly overstep the bounds of physical reality.

Given the apparent simplicity of this solution for overcoming the problem between general relativity and quantum mechanics, you might wonder why it took so long for someone to suggest that the point-particle description is merely an idealization and that in the real world elementary particles do have some spatial extent. This takes us to our second point. Long ago, some of the greatest minds in theoretical physics, such as Pauli, Heisenberg, Dirac, and Feynman, *did* suggest that nature's constituents might not actually be points but rather small undulating "blobs" or "nuggets." They and others found, however, that it is very hard to construct a theory, whose fundamental constituent is not a point particle, that is nonetheless consistent with the most basic of physical principles such as conservation of quantum-mechanical probability (so that physical objects do not suddenly vanish from the universe, without a trace) and the impossibility of faster-than-light-speed transmission of information. From a variety of perspectives, their research showed time and again that one or both of these principles were violated when the point-particle paradigm was discarded. For a long time, therefore, it seemed impossible to find a sensible quantum theory based on anything but point particles. The truly impressive feature of string theory is that more than twenty years of exacting research has shown that although certain features are unfamiliar, string theory *does* respect all of the requisite properties inherent in any sensible physical theory. And furthermore, through its graviton pattern of vibration, string theory is a quantum theory containing gravity.

The More Precise Answer

The rough answer captures the essence of why string theory prevails where previous point-particle theories failed. And so, if you like, you can go on to the next section without losing the logical flow of our discussion. But having developed the essential ideas of special relativity in Chapter 2, we already have the necessary tools for describing more accurately how string theory calms the violent quantum jitters.

In the more precise answer, we rely upon the same core idea as in the rough answer, but we express it directly at the level of strings. We do this by comparing, in some detail, point-particle and string probes. We will see how the extended nature of the string smears out the information that would be obtainable by point-particle probes, and therefore, again, how it happily does away with the ultra-short-distance behavior responsible for the central dilemma of contemporary physics.

We first consider the way in which point particles would interact, if they were actually to exist, and hence how they could be used as physical probes. The most basic interaction is between two point particles moving on a collision course so that their trajectories will intersect, as in Figure

Figure 6.5 Two particles interact—they "slam together"—and cause the path of each to be deflected.

6.5. If these particles were billiard balls they would collide, and each would be deflected onto a new trajectory. Point-particle quantum field theory shows that essentially the same thing happens when elementary particles collide—they scatter off one another and continue on deflected trajectories—but the details are a little different.

For concreteness and simplicity, imagine that one of the two particles is an electron and the other is its antiparticle, the positron. When matter and antimatter collide, they can annihilate in a flash of pure energy, producing, for example, a photon.[9] To distinguish the ensuing trajectory of the photon from the previous trajectories of the electron and positron, we follow a traditional physics convention and draw it with a wiggly line. The photon will typically travel for a bit and then release the energy derived from the initial electron-positron pair by producing another electron-positron pair with trajectories as indicated on the far right of Figure 6.6. In the end, two particles are fired at each other, they interact through the electromagnetic force, and finally they emerge on deflected trajectories, a sequence of events that bears some similarity to our description of colliding billiard balls.

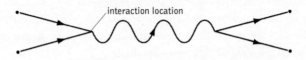

Figure 6.6 In quantum field theory, a particle and its antiparticle can momentarily annihilate one another, producing a photon. Subsequently, this photon can give rise to another particle and antiparticle traveling along different trajectories.

We are concerned with the details of the interaction—specifically, the point where the initial electron and positron annihilate and produce the photon. The central fact, as will become apparent, is that there is an unambiguous, completely identifiable time and place where this happens: It is marked in Figure 6.6.

How does this description change if, when we closely examine the objects we thought were zero-dimensional points, they turn out to be one-dimensional strings? The basic process of interaction is the same, but now the objects on a collision course are oscillating loops, as shown in Figure 6.7. If these loops are vibrating in just the right resonance patterns, they will correspond to an electron and a positron on collision course, just as in Figure 6.6. Only when examined at the most minute distance scales, far smaller than anything our present technology can access, is their true stringlike character apparent. As in the point-particle case, the two strings

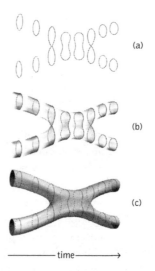

(a)

(b)

(c)

————— time —————→

Figure 6.7 (a) Two strings on a collision course can merge into a third string, which subsequently can split apart into two strings travelling along deflected trajectories. (b) The same process as shown in (a), emphasizing string motion. (c) A "time-lapse photograph" of two interacting strings sweeping out a "world-sheet."

collide and again annihilate each other in a flash. The flash, a photon, is itself a string in a particular vibrational pattern. Thus, the two incoming strings interact by merging together and producing a third string, as shown in Figure 6.7. Just as in our point-particle description, this string travels a bit, and then releases the energy derived from the two initial strings by dissociating into two strings that travel onward. Again, from any but the most microscopic perspective, this will look just like the point-particle interaction of Figure 6.6.

There is, however, a crucial difference between the two descriptions. We emphasized that the point-particle interaction occurs at an identifiable point in space and time, a location that all observers can agree on. As we shall now see, this is *not* true for interactions between strings. We will show this by comparing how George and Gracie, two observers in relative motion as in Chapter 2, would describe the interaction. We will see that they do not agree on where and when the two strings touch for the first time.

To do so, imagine that we view the interaction between two strings with a camera whose shutter is kept open so that the whole history of the process is captured on one piece of film.[10] We show the result—known as a *string world-sheet*—in Figure 6.7(c). By "slicing" the world-sheet into parallel pieces—much as one slices a loaf of bread—the moment-by-moment history of the string interaction can be recovered. We show an example of this slicing in Figure 6.8. Specifically, in Figure 6.8(a) we show George, intently focused on the two incoming strings, together with an attached plane that slices through *all events in space that occur at the same time,* according to his perspective. As we have done often in previous chapters, we have suppressed one spatial dimension in this diagram for visual clarity. In reality, of course, there is a three-dimensional array of events that occur at the same time according to any observer. Figures 6.8(b) and 6.8(c) give a couple of snapshots at subsequent times—subsequent "slices" of the world-sheet—showing how George sees the two strings approach each other. Of central importance, in Figure 6.8(c) we show the instant in time, according to George, when the two strings first touch and merge together, producing the third string.

Let's now do the same for Gracie. As discussed in Chapter 2, the relative motion of George and Gracie implies that they do not agree on what

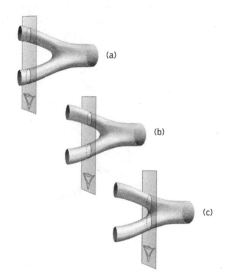

Figure 6.8 The two incoming strings from George's perspective at three consecutive moments in time. In (a) and (b) the strings are getting closer together; at (c) they touch for the first time, from his viewpoint.

events occur at the same time. From Gracie's perspective the events in space that occur simultaneously lie on a different plane, as shown in Figure 6.9. That is, from Gracie's perspective, the world-sheet of Figure 6.7(c) must be "sliced" into pieces at a different angle in order to reveal the moment-by-moment progression of the interaction.

In Figures 6.9(b) and 6.9(c) we show subsequent moments in time, now according to Gracie, including the moment when she sees the two incoming strings touch and produce the third string.

By comparing Figures 6.8(c) and 6.9(c), as we do in Figure 6.10, we see that George and Gracie do not agree on when and where the two initial strings first touch—where they interact. The string, being an extended object, ensures that *there is no unambiguous location in space or moment in time when the strings first interact*—rather, it depends upon the state of motion of the observer.

If we apply exactly the same reasoning to the interaction of point par-

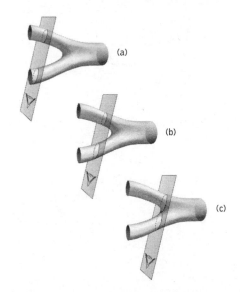

Figure 6.9 The two incoming strings from Gracie's perspective at three consecutive moments in time. In (a) and (b) the strings are getting closer together; at (c) they touch for the first time, from her viewpoint.

ticles, as summarized in Figure 6.11, we recover the conclusion stated earlier—there *is* a definite point in space and moment in time when the point particles interact. Point particles cram all of their interaction into a definite point. When the force involved in an interaction is the gravitational force—that is, when the messenger particle involved in the interaction is the graviton instead of the photon—this complete packing of the force's punch into a single point leads to disastrous results, such as the infinite answers we alluded to earlier. Strings, by contrast, "smear" out the place where interactions occur. Because different observers perceive that the interaction takes place at various locations along the left part of the surface of Figure 6.10, in a real sense this means that the interaction location is smeared out among all of them. This spreads out the force's punch and, in the case of the gravitational force, this smearing significantly dilutes its ultramicroscopic properties—so much so that calculations yield well-behaved finite answers in place of the previous infinities. This is a

Figure 6.10 George and Gracie do not agree on the location of the interaction.

more precise version of the smearing encountered in the rough answer of the last section. And once again, this smearing results in a smoothing of the ultramicroscopic jitteriness of space as sub-Planck-length distances are blurred together.

Like viewing the world through glasses that are too weak or too strong, fine sub-Planckian details that would be accessible to a point-particle probe are smeared together by string theory and rendered harmless. And unlike the case with poor eyesight, if string theory is the ultimate description of the universe, there is no corrective lens to bring the supposed sub-Planck-scale fluctuations into sharp focus. The incompatibility of general relativity and quantum mechanics—which would become apparent only on sub-Planck-scale distances—is avoided in a universe that has a lower limit on the distances that can be accessed, or even said to exist,

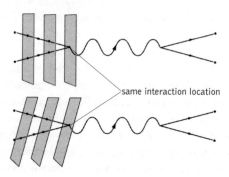

same interaction location

Figure 6.11 Observers in relative motion agree on when and where two point particles interact with another.

in the conventional sense. Such is the universe described by string theory, in which we see that the laws of the large and the small can be harmoniously merged together as the supposed catastrophe arising on ultramicroscopic distances is summarily done away with.

Beyond Strings?

Strings are special for two reasons. First, even though they are spatially extended they can be described consistently in the framework of quantum mechanics. Second, among the resonant vibrational patterns there is one that has the exact properties of the graviton, thus ensuring that the gravitational force is an intrinsic part of its structure. But just as string theory shows that the conventional notion of zero-dimensional point particles appears to be a mathematical idealization that is not realized in the real world, might it also be the case that an infinitely thin one-dimensional strand is similarly a mathematical idealization? Might it actually be the case that strings have some thickness—like the surface of a two-dimensional bicycle-tire inner tube or, even more realistically, like a thin three-dimensional doughnut? The seemingly insurmountable difficulties found by Heisenberg, Dirac, and others in their attempts to construct a quantum theory of three-dimensional nuggets have repeatedly stymied researchers following this natural chain of reasoning.

Quite unexpectedly, though, during the mid-1990s, string theorists realized, through indirect and rather shrewd reasoning, that such higher-dimensional fundamental objects actually do play an important and subtle role in string theory itself. Researchers have gradually realized that string theory is *not* a theory that contains only strings. A crucial observation, central to the second superstring revolution initiated by Witten and others in 1995, is that string theory actually includes ingredients with a variety of different dimensions: two-dimensional Frisbee-like constituents, three-dimensional blob-like constituents, and even more exotic possibilities to boot. These most recent realizations will be taken up in Chapters 12 and 13. For now we continue to follow the path of history and further explore the striking new properties of a universe built out of one-dimensional strings instead of zero-dimensional point-particles.

Chapter 7

The "Super" in Superstrings

When the success of Eddington's 1919 expedition to measure Einstein's prediction of the bending of starlight by the sun had been established, the Dutch physicist Hendrik Lorentz sent Einstein a telegram informing him of the good news. As word of the telegram's confirmation of general relativity spread, a student asked Einstein about what he would have thought if Eddington's experiment had not found the predicted bending of starlight. Einstein replied, "Then I would have been sorry for the dear Lord, for the theory *is* correct."[1] Of course, had experiments truly failed to confirm Einstein's predictions, the theory would not be correct and general relativity would not have become a pillar of modern physics. But what Einstein meant is that general relativity describes gravity with such a deep inner elegance, with such simple yet powerful ideas, that he found it hard to imagine that nature could pass it by. General relativity, in Einstein's view, was almost too beautiful to be wrong.

Aesthetic judgments do not arbitrate scientific discourse, however. Ultimately, theories are judged by how they fare when faced with cold, hard, experimental facts. But this last remark is subject to an immensely important qualification. While a theory is being constructed, its incomplete state of development often prevents its detailed experimental consequences from being assessed. Nevertheless, physicists must make choices and exercise judgments about the research direction in which to take their partially completed theory. Some of these decisions are dictated by internal logical consistency; we certainly require that any sensible theory avoid logical absurdities. Other decisions are guided by a sense of the qualitative experimental implications of one theoretical construct relative to another; we are generally not interested in a theory if it has no capacity to resemble anything we encounter in the world around us. But it is certainly the case that some decisions made by theoretical physicists are founded upon an aesthetic sense—a sense of which theories have an elegance and beauty of structure on par with the world we experience. Of course, nothing ensures that this strategy leads to truth. Maybe, deep down, the universe has a less elegant structure than our experiences have led us to believe, or maybe we will find that our current aesthetic criteria need significant refining when applied in ever less familiar contexts. Nevertheless, especially as we enter an era in which our theories describe realms of the universe that are increasingly difficult to probe experimentally, physicists do rely on such an aesthetic to help them steer clear of blind alleys and dead-end roads that they might otherwise pursue. So far, this approach has provided a powerful and insightful guide.

In physics, as in art, symmetry is a key part of aesthetics. But unlike the case in art, symmetry in physics has a very concrete and precise meaning. In fact, by diligently following this precise notion of symmetry to its mathematical conclusion, physicists during the last few decades have found theories in which matter particles and messenger particles are far more closely intertwined than anyone previously thought possible. Such theories, which unite not only the forces of nature but also the material constituents, have the greatest possible symmetry and for this reason have been called *supersymmetric*. Superstring theory, as we shall see, is both the progenitor and the pinnacle example of a supersymmetric framework.

The Nature of Physical Law

Imagine a universe in which the laws of physics are as ephemeral as the tastes of fashion—changing from year to year, from week to week, or even from moment to moment. In such a world, assuming that the changes do not disrupt basic life processes, you would never experience a dull moment, to say the least. The simplest acts would be an adventure, since random variations would prevent you or anyone else from using past experience to predict anything about future outcomes.

Such a universe is a physicist's nightmare. Physicists—and most everyone else as well—rely crucially upon the stability of the universe: The laws that are true today were true yesterday and will still be true tomorrow (even if we have not been clever enough to have figured them all out). After all, what meaning can we give to the term "law" if it can abruptly change? This does not mean that the universe is static; the universe certainly changes in innumerable ways from each moment to the next. Rather, it means that the laws governing such evolution are fixed and unchanging. You might ask whether we really know this to be true. In fact, we don't. But our success in describing numerous features of the universe, from a brief moment after the big bang right through to the present, assures us that if the laws are changing they must be doing so very slowly. The simplest assumption that is consistent with all that we know is that the laws are fixed.

Now imagine a universe in which the laws of physics are as parochial as local culture—changing unpredictably from place to place and defiantly resisting any outside influence to conform. Like the adventures of Gulliver, travels in such a world would expose you to an enormously rich array of unpredictable experiences. But from a physicist's perspective, this is yet another nightmare. It's hard enough, for instance, to live with the fact that laws that are valid in one country—or even one state—may not be valid in another. But imagine what things would be like if the laws of *nature* were as varied. In such a world experiments carried out in one locale would have no bearing on the physical laws relevant somewhere else. Instead, physicists would have to redo experiments over and over again in different locations to probe the local laws of nature that hold in each. Thankfully, everything we know points toward the laws of physics being the same everywhere. All experiments the world over converge on the same set of underlying physical explanations. Moreover, our ability to explain a vast number of astrophysical observations of far-flung regions of the cosmos using one, fixed set of physical principles leads us to believe that the same laws *do* hold true everywhere. Having never traveled to the opposite end of the universe, we can't definitively rule out the possibility that a whole new kind of physics prevails elsewhere, but everything points to the contrary.

Again, this does not mean that the universe looks the same—or has the same detailed properties—in different locations. An astronaut jumping on a pogo stick on the moon can do all sorts of things that are impossible to do on earth. But we recognize that the difference arises because the moon is far less massive than the earth; it does not mean that the law of gravity is somehow changing from place to place. Newton's, or more precisely, Einstein's, law of gravity is the same on earth as it is on the moon. The difference in the astronaut's experience is one of change in environmental detail, not variation of physical law.

Physicists describe these two properties of physical laws—that they do not depend on when or where you use them—as *symmetries* of nature. By this usage physicists mean that nature treats every moment in time and every location in space identically—symmetrically—by ensuring that the same fundamental laws are in operation. Much in the same manner that they affect art and music, such symmetries are deeply satisfying; they highlight an order and a coherence in the workings of nature. The elegance of rich, complex, and diverse phenomena emerging from a simple set of universal laws is at least part of what physicists mean when they invoke the term "beautiful."

In our discussions of the special and general theories of relativity, we came upon yet other symmetries of nature. Recall that the principle of relativity, which lies at the heart of special relativity, tells us that all physical laws must be the same regardless of the constant-velocity relative motion that individual observers might experience. This is a symmetry because it means that nature treats all such observers identically—symmetrically. Each such observer is justified in considering himself or herself to be at rest. Again, it's not that observers in relative motion will make identical observations; as we have seen earlier, there are all sorts of stunning *differences* in their observations. Instead, like the disparate experiences of the pogo-stick enthusiast on the earth and on the moon, the differences in observations reflect environmental details—the observers are in relative motion—even though their observations are governed by identical *laws*.

Through the equivalence principle of general relativity, Einstein significantly extended this symmetry by showing that the laws of physics are actually identical for all observers, even if they are undergoing complicated accelerated motion. Recall that Einstein accomplished this by realizing that an accelerated observer is also perfectly justified in declaring himself or herself to be at rest, and in claiming that the force he or she feels is due to a gravitational field. Once gravity is included in the framework, all possible observational vantage points are on a completely equal footing. Beyond the intrinsic aesthetic appeal of this egalitarian treatment of all motion, we have seen that these symmetry principles played a pivotal role in the stunning conclusions regarding gravity that Einstein found.

Are there any other symmetry principles having to do with space, time, and motion that the laws of nature should respect? If you think about this you might come up with one more possibility. The laws of physics should not care about the *angle* from which you make your observations. For instance, if you perform some experiment and then decide to rotate all of your equipment and do the experiment again, the same laws should apply. This is known as rotational symmetry, and it means that the laws of physics treat all possible *orientations* on equal footing. It is a symmetry principle that is on par with the previous ones discussed.

Are there others? Have we overlooked any symmetries? You might suggest the gauge symmetries associated with the nongravitational forces, as discussed in Chapter 5. These are certainly symmetries of nature, but they are of a more abstract sort; our focus here is on symmetries that have a direct link to space, time, or motion. With this stipulation, it's now likely that you can't think of any other possibilities. In fact, in 1967 physicists Sidney Coleman and Jeffrey Mandula were able to prove that no other symmetries associated with space, time, or motion could be combined with those just discussed and result in a theory bearing any resemblance to our world.

Subsequently, though, close examination of this theorem, based on insights of a number of physicists revealed precisely one subtle loophole: The Coleman-Mandula result did not exploit fully symmetries sensitive to something known as *spin*.

Spin

An elementary particle such as an electron can orbit an atomic nucleus in somewhat the same way that the earth orbits the sun. But, in the traditional point-particle description of an electron, it would appear that there is no analog of the earth's spinning around on its axis. When any object spins, points on the axis of rotation itself—like the *central point* of a spinning Frisbee—do not move. If something is truly pointlike, though, it has no "other points" that lie off of any purported spin axis. And so it would appear that there simply is no notion of a point object spinning. Many years ago, such reasoning fell prey to yet another quantum-mechanical surprise.

In 1925, the Dutch physicists George Uhlenbeck and Samuel Goudsmit realized that a wealth of puzzling data having to do with properties of light emitted and absorbed by atoms could be explained if electrons were assumed to have very particular *magnetic* properties. Some hundred years earlier, the Frenchman André-Marie Ampère had shown that magnetism arises from the motion of electric charge. Uhlenbeck and Goudsmit followed this lead and found that only one specific sort of electron motion could give rise to the magnetic properties suggested by the data: *rotational* motion—that is, *spin*. Contrary to classical expectations, Uhlenbeck and Goudsmit proclaimed that, somewhat like the earth, electrons both revolve *and* rotate.

Did Uhlenbeck and Goudsmit literally mean that the electron is spinning? Yes and no. What their work really showed is that there is a quantum-mechanical notion of spin that is somewhat akin to the usual image but inherently quantum mechanical in nature. It's one of those properties of the microscopic world that brushes up against classical ideas but injects an experimentally verified quantum twist. For instance, picture a spinning skater. As she pulls her arms in she spins more quickly; as she stretches out her arms she spins more slowly. And sooner or later, depending on how vigorously she threw herself into the spin, she will slow down and stop. Not so for the kind of spin revealed by Uhlenbeck and Goudsmit. According to their work and subsequent studies, every electron in the universe, always and forever, *spins at one fixed and never changing rate.* The spin of an electron is not a transitory state of motion as for more familiar objects that, for some reason or other, happen to be spinning. Instead, the spin of an electron is an *intrinsic* property, much like its mass or its electric charge. If an electron were not spinning, it would not be an electron.

Although early work focused on the electron, physicists have subsequently shown that these ideas about spin apply equally well to all of the matter particles that fill out the three families of Table 1.1. This is true down to the last detail: *All* of the matter particles (and their antimatter partners as well) have spin equal to that of the electron. In the language of the trade, physicists say that matter particles all have "spin-½," where the value ½ is, roughly speaking, a quantum-mechanical measure of how quickly the particles rotate.[2] Moreover, physicists have shown that the nongravitational force carriers—photons, weak gauge bosons, and gluons—also possess an intrinsic spinning characteristic that turns out to be *twice* that of the matter particles. They all have "spin-1."

What about gravity? Well, even before string theory, physicists were able to determine what spin the hypothesized graviton must have to be the transmitter of the gravitational force. The answer: twice the spin of photons, weak gauge bosons, and gluons—i.e., "spin-2."

In the context of string theory, spin—just like mass and force charges—is associated with the pattern of vibration that a string executes. As with point particles, it's a bit misleading to think of the spin carried by a string as arising from its spinning literally around in space, but this image does give a loose picture to have in mind. By the way, we can now clarify an important issue we encountered earlier. In 1974, when Scherk and Schwarz proclaimed that string theory should be thought of as a quantum theory incorporating the gravitational force, they did so because they had found that strings *necessarily* have a vibrational pattern in their repertoire that is *massless and has spin-2*—the hallmark features of the graviton. Where there is a graviton there is also gravity.

With this background on the concept of spin, let's now turn to the role it plays in revealing the loophole in the Coleman-Mandula result concerning the possible symmetries of nature, mentioned in the preceding section.

Supersymmetry and Superpartners

As we have emphasized, the concept of spin, although superficially akin to the image of a spinning top, differs in substantial ways that are rooted in quantum mechanics. Its discovery in 1925 revealed that there is another kind of rotational motion that simply would not exist in a purely classical universe.

This suggests the following question: Just as ordinary rotational motion allows for the symmetry principle of rotational invariance ("physics treats all spatial orientations on an equal footing"), could it be that the more subtle rotational motion associated with spin leads to another possible symmetry of the laws of nature? By 1971 or so, physicists showed that the answer to this question was yes. Although the full story is quite involved, the basic idea is that when spin is considered, there is precisely *one more symmetry of the laws of nature* that is mathematically possible. It is known as *supersymmetry*.[3]

Supersymmetry cannot be associated with a simple and intuitive change in observational vantage point; shifts in time, in spatial location, in angular orientation, and in velocity of motion exhaust these possibilities. But just as spin is "like rotational motion, with a quantum-mechanical twist," supersymmetry can be associated with a change in observational vantage point in a "quantum-mechanical extension of space and time." These quotes are especially important, as the last sentence is only meant to give a rough sense of where supersymmetry fits into the larger framework of symmetry principles.[4] Nevertheless, although understanding the origin of supersymmetry is rather subtle, we will focus on one of its primary *implications*—should the laws of nature incorporate its principles—and this is far easier to grasp.

In the early 1970s, physicists realized that if the universe is supersymmetric, the particles of nature must come in *pairs* whose respective spins differ by half a unit. Such pairs of particles—regardless of whether they are thought of as pointlike (as in the standard model) or as tiny vibrating loops—are called *superpartners*. Since matter particles have spin-½ while some of the messenger particles have spin-1, supersymmetry appears to result in a pairing—a partnering—of matter and force particles. As such, it seems like a wonderful unifying concept. The problem comes in the details.

By the mid-1970s, when physicists sought to incorporate supersymmetry into the standard model, they found that *none* of the known particles—those of Tables 1.1 and 1.2—could be superpartners of one another. Instead, detailed theoretical analysis showed that if the universe incorporates supersymmetry, then every known particle must have an as-yet-undiscovered superpartner particle, whose spin is half a unit less than its known counterpart. For instance, there should be a spin-0 partner of the electron; this hypothetical particle has been named the *selectron* (a contraction of supersymmetric-electron). The same should also be true for the other matter particles, with, for example, the hypothetical spin-0 superpartners of neutrinos and quarks being called *sneutrinos* and *squarks*. Similarly, the force particles should have spin-½ superpartners: For photons there should be *photinos*, for the gluons there should be *gluinos*, for the W and Z bosons there should be *winos* and *zinos*.

On closer inspection, then, supersymmetry seems to be a terribly uneconomical feature; it requires a whole slew of additional particles that wind up doubling the list of fundamental ingredients. Since none of the superpartner particles has ever been detected, you would be justified to take Rabi's remark from Chapter 1 regarding the discovery of the muon one step further, declare that "nobody ordered supersymmetry," and summarily reject this symmetry principle. For three reasons, however, many physicists believe strongly that such an out-of-hand dismissal of supersymmetry would be quite premature. Let's discuss these reasons.

The Case for Supersymmetry: Prior to String Theory

First, from an aesthetic standpoint, physicists find it hard to believe that nature would respect almost, but not quite all of the symmetries that are mathematically possible. Of course, it is possible that an incomplete utilization of symmetry is what actually occurs, but it would be such a shame. It would be as if Bach, after developing numerous intertwining voices to fill out an ingenious pattern of musical symmetry, left out the final, resolving measure.

Second, even within the standard model, a theory that ignores gravity, thorny technical issues that are associated with quantum processes are swiftly solved if the theory is supersymmetric. The basic problem is that every distinct particle species makes its own contribution to the microscopic quantum-mechanical frenzy. Physicists have found that in the bath of this frenzy, certain processes involving particle interactions remain consistent *only* if numerical parameters in the standard model are fine-tuned—to better than one part in a million billion—to cancel out the most pernicious quantum effects. Such precision is on par with adjusting the launch angle of a bullet fired from an enormously powerful rifle, so that it hits a specified target on the moon with a margin of error no greater than the thickness of an amoeba. Although numerical adjustments of an analogous precision can be made within the standard model, many physicists are quite suspect of a theory that is so delicately constructed that it falls apart if a number on which it depends is changed in the fifteenth digit after the decimal point.[5]

Supersymmetry changes this drastically because *bosons*—particles whose spin is a whole number (named after the Indian physicist Satyendra Bose)—and *fermions*—particles whose spin is half of a whole (odd) number (named after the Italian physicist Enrico Fermi)—tend to give cancelling quantum-mechanical contributions. Like opposite ends of a seesaw, when the quantum jitters of a boson are positive, those of a fermion tend to be negative, and vice versa. Since supersymmetry ensures that bosons and fermions occur in pairs, substantial cancellations occur from the outset—cancellations that significantly calm some of the frenzied quantum effects. It turns out that the consistency of the *supersymmetric standard model*—the standard model augmented by all of the superpartner particles—no longer relies upon the uncomfortably delicate numerical adjustments of the ordinary standard model. Although this is a highly technical issue, many particle physicists find that this realization makes supersymmetry very attractive.

The third piece of circumstantial evidence for supersymmetry comes from the notion of *grand unification*. One of the puzzling features of nature's four forces is the huge range in their intrinsic strengths. The electromagnetic force has less than 1 percent of the strength of the strong force, the weak force is some thousand times feebler than that, and the gravitational force is some hundred million billion billion billion (10^{-35}) times weaker still. Following the pathbreaking and ultimately Nobel Prize–winning work of Glashow, Salam, and Weinberg that established a deep connection between the electromagnetic and weak forces (discussed in Chapter 5), in 1974 Glashow, together with his Harvard colleague Howard Georgi, suggested that an analogous connection might be forged with the strong force. Their work, which proposed a "grand unification" of three of the four forces, differed in one essential way from that of the electroweak theory: Whereas the electromagnetic and weak forces crystallized out of a more symmetric union when the temperature of the universe dropped to about a million billion degrees above absolute zero (10^{15}

Kelvin), Georgi and Glashow showed that the union with the strong force would have been apparent only at a temperature some ten trillion times higher—around ten billion billion billion degrees above absolute zero (10^{28} Kelvin). From the point of view of energy, this is about a million billion times the mass of the proton, or about four orders of magnitude less than the Planck mass. Georgi and Glashow boldly took theoretical physics into an energy realm many orders of magnitude beyond that which anyone had previously dared explore.

Subsequent work at Harvard by Georgi, Helen Quinn, and Weinberg in 1974 made the potential unity of the nongravitational forces within the grand unified framework even more manifest. As their contribution continues to play an important role in unifying the forces and in assessing the relevance of supersymmetry to the natural world, let's spend a moment explaining it.

We are all aware that the electrical attraction between two oppositely charged particles or the gravitational attraction between two massive bodies gets stronger as the distance between the objects decreases. These are simple and well-known features of classical physics. There is a surprise, though, when we study the effect that quantum physics has on force strengths. Why should quantum mechanics have any effect at all? The answer, once again, lies in quantum fluctuations. When we examine the electric force field of an electron, for example, we are actually examining it through the "mist" of momentary particle-antiparticle eruptions and annihilations that are occurring all through the region of space surrounding it. Physicists some time ago realized that this seething mist of microscopic fluctuations obscures the full strength of the electron's force field, somewhat as a thin fog partially obscures the beacon of a lighthouse. But notice that as we get closer to the electron, we will have penetrated more of the cloaking particle-antiparticle mist and hence will be less subject to its diminishing influence. This implies that the strength of an electron's electric field will *increase* as we get closer to it.

Physicists distinguish this quantum-mechanical increase in strength as we get closer to the electron from that known in classical physics by saying that the *intrinsic* strength of the electromagnetic force increases on shorter distance scales. This reflects that the strength increases not merely because we are closer to the electron but also because more of the electron's intrinsic electric field becomes visible. In fact, although we have focused on the electron, this discussion applies equally well to all electrically charged particles and is summarized by saying that quantum effects drive the strength of the electromagnetic force to get larger when examined on shorter distance scales.

What about the other forces of the standard model? How do their intrinsic strengths vary with distance? In 1973, Gross and Frank Wilczek at Princeton, and, independently, David Politzer at Harvard, studied this question and found a surprising answer: The quantum cloud of particle eruptions and annihilations *amplifies* the strengths of the strong and weak forces. This implies that as we examine them on shorter distances, we penetrate more of this seething cloud and hence are subject to less of its amplification. And so, the strengths of these forces get *weaker* when they are probed on shorter distances.

Georgi, Quinn, and Weinberg took this realization and ran with it to a remarkable end. They showed that when these effects of the quantum frenzy are carefully accounted for, the net result is that the strengths of all three nongravitational forces are driven *together*. Whereas the strengths of these forces are very different on scales accessible to current technology, Georgi, Quinn, and Weinberg argued that this difference is actually due to the different effect that the haze of microscopic quantum activity has on each force. Their calculations showed that if this haze is penetrated by examining the forces not on everyday scales but as they act on distances of about a hundredth of a billionth of a billionth of a billionth (10^{-29}) of a centimeter (a mere factor of ten thousand larger than the Planck length), the three nongravitational force strengths appear to become equal.

Although far removed from the realm of common experience, the high energy necessary to be sensitive to such small distances was characteristic of the roiling, hot early universe when it was about a thousandth of a trillionth of a trillionth of a trillionth (10^{-39}) of a second old—when its temperature was on the order of 10^{28} Kelvin mentioned earlier. In somewhat the same way that a collection of disparate ingredients—pieces of metal, wood, rocks, minerals, and so on—all melt together and become a uniform, homogeneous plasma when heated to sufficiently high temperature, these theoretical works suggested that the strong, weak, and electromagnetic forces all merge into one grand force at such immense temperatures. This is shown schematically in Figure 7.1.[6]

Although we do not have the technology to probe such minute distance

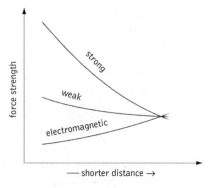

Figure 7.1 The strengths of the three nongravitational forces as they operate on ever shorter distance scales—equivalently, as they act on ever higher energy processes.

scales or to produce such scorching temperatures, since 1974 experimentalists have significantly refined the measured strengths of the three nongravitational forces under everyday conditions. These data—the starting points for the three force-strength curves in Figure 7.1—are the input data for the quantum-mechanical extrapolations of Georgi, Quinn, and Weinberg. In 1991, Ugo Amaldi of CERN, Wim de Boer and Hermann Fürstenau of the University of Karlsruhe, Germany, recalculated the Georgi, Quinn, and Weinberg extrapolations making use of these experimental refinements and showed two significant things. First, the strengths of the three nongravitational forces *almost agree, but not quite* at tiny distance scales (equivalently, high energy/high temperature) as shown in Figure 7.2. Second, this tiny but undeniable discrepancy in their strengths *vanishes* if supersymmetry is incorporated. The reason is that the new superpartner particles required by supersymmetry contribute additional quantum fluctuations, and these fluctuations are just right to nudge the strengths of the forces to converge with one another.

To many physicists, it is extremely difficult to believe that nature would choose the forces so that they *almost,* but not quite, have strengths that microscopically unify—microscopically become equal. It's like putting together a jigsaw puzzle in which the final piece is slightly misshapen and won't cleanly fit into its appointed position. Supersymmetry deftly refines its shape so that all pieces firmly lock into place.

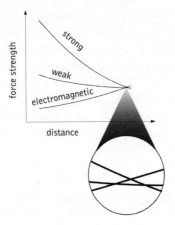

Figure 7.2 A refinement of the calculation of force strengths reveals that without supersymmetry they almost, but not quite, meet.

Another aspect of this latter realization is that it provides a possible answer to the question, Why haven't we discovered any of the superpartner particles? The calculations that lead to the convergence of the force strengths, as well as other considerations studied by a number of physicists, indicate that the superpartner particles must be a good deal heavier than the known particles. Although no definitive predictions can be made, studies show that the superpartner particles might be a thousand times as massive as a proton, if not heavier. As even our state-of-the-art accelerators cannot quite reach such energies, this provides an explanation for why these particles have not, as yet, been discovered. In Chapter 9, we will return to a discussion of the experimental prospects for determining in the near future whether supersymmetry truly is a property of our world.

Of course, the reasons we have given for believing in—or at least not yet rejecting—supersymmetry are far from airtight. We have described how supersymmetry elevates our theories to their most symmetric form— but you might suggest that the universe does not care about attaining the most symmetric form that is mathematically possible. We have noted the important technical point that supersymmetry relieves us from the delicate task of tuning numerical parameters in the standard model to avoid subtle quantum problems—but you might argue that the true theory describing nature may very well walk the fine edge between self-consistency and self-destruction. We have discussed how supersymmetry modifies the intrinsic strengths of the three nongravitational forces at tiny distances in just the right way for them to merge together into a grand unified force—but you might argue, again, that nothing in the design of nature dictates that these force strengths must exactly match on microscopic scales. And finally, you might suggest that a simpler explanation for why the super-partner particles have never been found is that our universe is not super-symmetric and, therefore, the superpartners do not exist.

No one can refute any of these responses. But the case for super-symmetry is strengthened immensely when we consider its role in string theory.

Supersymmetry in String Theory

The original string theory that emerged from Veneziano's work in the late 1960s incorporated all of the symmetries discussed at the beginning of this chapter, but it did not incorporate supersymmetry (which had not yet been discovered). This first theory based on the string concept was, more precisely, called the *bosonic string theory*. The name *bosonic* indicates that all of the vibrational patterns of the bosonic string have spins that are a whole number—there are no fermionic patterns, that is, no patterns with spins differing from a whole number by a half unit. This led to two problems.

First, if string theory was to describe all forces and all matter, it would somehow have to incorporate fermionic vibrational patterns, since the known matter particles all have spin-½. Second, and far more troubling, was the realization that there was one pattern of vibration in bosonic string theory whose mass (more precisely, whose mass squared) was *negative*— a so-called *tachyon*. Even before string theory, physicists had studied the possibility that our world might have tachyon particles, in addition to the familiar particles that all have positive masses, but their efforts showed that it is difficult if not impossible for such a theory to be logically sensi-ble. Similarly, in the context of bosonic string theory, physicists tried all sorts of fancy footwork to make sense of the bizarre prediction of a tachyon vibrational pattern, but to no avail. These features made it increasingly clear that although it was an interesting theory, the bosonic string was missing something essential.

In 1971, Pierre Ramond of the University of Florida took up the challenge of modifying the bosonic string theory to include fermionic patterns of vibration. Through his work and subsequent results of Schwarz and André Neveu, a new version of string theory began to emerge. And much to everyone's surprise, the bosonic and the fermionic patterns of vibration of this new theory appeared to come in pairs. For each bosonic pattern there was a fermionic pattern, and vice versa. By 1977, insights of Ferdinando Gliozzi of the University of Turin, Scherk, and David Olive of Imperial College put this pairing into the proper light. The new string theory incorporated supersymmetry, and the observed pairing of bosonic and fermionic vibrational patterns reflected this highly symmetric character. Supersymmetric string theory—superstring theory, that is—had been born. Moreover, the work of Gliozzi, Scherk, and Olive had one other crucial result: They showed that the troublesome tachyon vibration of the bosonic string does not afflict the superstring. Slowly, the pieces of the string puzzle were falling into place.

Nevertheless, the major initial impact of the work of Ramond, and also of Neveu and Schwarz, was not actually in string theory. By 1973, the physicists Julius Wess and Bruno Zumino realized that supersymmetry—the new symmetry emerging from the reformulation of string theory—was applicable even to theories based on point particles. They rapidly made important strides toward incorporating supersymmetry into the framework of point-particle quantum field theory. And since, at the time, quantum field theory was the central rage of the mainstream particle-physics community—with string theory increasingly becoming a subject on the fringe—the insights of Wess and Zumino launched a tremendous amount of subsequent research on what has come to be called *supersymmetric quantum field theory*. The supersymmetric standard model, discussed in the preceding section, is one of the crowning theoretical achievements of these pursuits; we now see that, through historical twists and turns, even this point-particle theory owes a great debt to string theory.

With the resurgence of superstring theory in the mid-1980s, supersymmetry has re-emerged in the context of its original discovery. And in this framework, the case for supersymmetry goes well beyond that presented in the preceding section. String theory is the only way we know of to merge general relativity and quantum mechanics. But it's only the supersymmetric version of string theory that avoids the pernicious tachyon problem and that has fermionic vibrational patterns that can account for the matter particles constituting the world around us. Supersymmetry therefore comes hand-in-hand with string theory's proposal for a quantum theory of gravity, as well as with its grand claim of uniting all forces and all of matter. If string theory is right, physicists expect that so is supersymmetry.

Until the mid-1990s, however, one particularly troublesome aspect plagued supersymmetric string theory.

A Super-Embarrassment of Riches

If someone tells you that they have solved the mystery of Amelia Earhart's fate, you might be skeptical at first, but if they have a well-documented, thoroughly pondered explanation, you would probably hear them out and, who knows, you might even be convinced. But what if, in the next breath, they tell you that they actually have a second explanation as well. You listen patiently and are surprised to find the alternate explanation to be as well documented and thought through as the first. And after finishing the second explanation, you are presented with a third, a fourth, and even a fifth explanation—each different from the others and yet equally convincing. No doubt, by the end of the experience you would feel no closer to Amelia Earhart's true fate than you did at the outset. In the arena of fundamental explanations, more is definitely less.

By 1985, string theory—notwithstanding the justified excitement it was engendering—was starting to sound like our overzealous Earhart expert. The reason is that by 1985 physicists realized that supersymmetry, by then a central element in the structure of string theory, could actually be incorporated into string theory in not one but *five* different ways. Each method results in a pairing of bosonic and fermionic vibrational patterns, but the details of this pairing as well as numerous other properties of the resulting theories differ substantially. Although their names are not all that important, it's worth recording that these five supersymmetric string theories are called the *Type I theory,* the *Type IIA theory,* the *Type IIB theory,* the *Heterotic type O(32) theory* (pronounced "oh-thirty-two"), and the

Heterotic type $E_8 \times E_8$ theory (pronounced "e-eight times e-eight"). All the features of string theory that we have discussed to this point are valid for each of these theories—they differ only in the finer details.

Having five different versions of what is supposedly the T.O.E.—possibly the ultimate unified theory—was quite an embarrassment for string theorists. Just as there is only one true explanation for whatever happened to Amelia Earhart (regardless of whether we will ever find it), we expect the same to be true regarding the deepest, most fundamental understanding of how the world works. We live in one universe; we expect one explanation.

One suggestion for resolving this problem might be that although there are five different superstring theories, four might be ruled out simply by experiment, leaving one true and relevant explanatory framework. But even if this were the case, we would still be left with the nagging question of why the other theories exist in the first place. In the wry words of Witten, "If one of the five theories describes our universe, who lives in the other four worlds?"[7] A physicist's dream is that the search for the ultimate answers will lead to a single, unique, absolutely inevitable conclusion. Ideally, the final theory—whether string theory or something else—should be the way it is because there simply is no other possibility. If we were to discover that there is only one logically sound theory incorporating the basic ingredients of relativity and quantum mechanics, many feel that we would have reached the deepest understanding of why the universe has the properties it does. In short, this would be unified-theory paradise.[8]

As we will see in Chapter 12, recent research has taken superstring theory one giant step closer to this unified utopia by showing that the five different theories are, remarkably, actually five different ways of describing *one and the same overarching theory.* Superstring theory *has* the uniqueness pedigree.

Things seem to be falling into place, but, as we will discuss in the next chapter, unification through string theory does require one more significant departure from conventional wisdom.

Chapter 8

More Dimensions Than
Meet the Eye

E instein resolved two of the major scientific conflicts of the past hundred years through special and then general relativity. Although the
initial problems that motivated his work did not portend the outcome,
each of these resolutions completely transformed our understanding of
space and time. String theory resolves the third major scientific conflict of
the past century and, in a manner that even Einstein would likely have
found remarkable, it requires that we subject our conceptions of space and
time to yet another radical revision. String theory so thoroughly shakes the
foundations of modern physics that even the generally accepted number
of dimensions in our universe—something so basic that you might think
it beyond questioning—is dramatically and convincingly overthrown.

The Illusion of the Familiar

Experience informs intuition. But it does more than that: Experience sets
the frame within which we analyze and interpret what we perceive. You
would no doubt expect, for instance, that the "wild child" raised by a pack
of wolves would interpret the world from a perspective that differs substantially from your own. Even less extreme comparisons, such as those

between people raised in very different cultural traditions, serve to underscore the degree to which our experiences determine our interpretive mindset.

Yet there are certain things that we *all* experience. And it is often the beliefs and expectations that follow from these universal experiences that can be the hardest to identify and the most difficult to challenge. A simple but profound example is the following. If you were to get up from reading this book, you could move in three independent directions—that is, through three independent spatial dimensions. Absolutely any path you follow—regardless of how complicated—results from some combination of motion through what we might call the "left-right dimension," the "back-forth dimension," and the "up-down dimension." Every time you take a step you implicitly make three separate choices that determine how you move through these three dimensions.

An equivalent statement, as encountered in our discussion of special relativity, is that any location in the universe can be fully specified by giving three pieces of data: where it is relative to these three spatial dimensions. In familiar language, you can specify a city address, say, by giving a street (location in the "left-right dimension"), a cross street or an avenue (location in the "back-forth dimension"), and a floor number (location in the "up-down dimension"). And from a more modern perspective, we have seen that Einstein's work encourages us to think about time as another dimension (the "future-past dimension"), giving us a total of four dimensions (three space dimensions and one time dimension). You specify events in the universe by telling where and when they occur.

This feature of the universe is so basic, so consistent, and so thoroughly pervasive that it really does seem beyond questioning. In 1919, however, a little-known Polish mathematician named Theodor Kaluza from the University of Königsberg had the temerity to challenge the obvious— he suggested that the universe might *not* actually have three spatial dimensions; it might have *more*. Sometimes silly-sounding suggestions are plain silly. Sometimes they rock the foundations of physics. Although it took quite some time to percolate, Kaluza's suggestion has revolutionized our formulation of physical law. We are still feeling the aftershocks of his astonishingly prescient insight.

Kaluza's Idea and Klein's Refinement

The suggestion that our universe might have more than three spatial dimensions may well sound fatuous, bizarre, or mystical. In reality, though, it is concrete and thoroughly plausible. To see this, it's easiest to shift our sights temporarily from the whole universe and think about a more familiar object, such as a long, thin garden hose.

Imagine that a few hundred feet of garden hose is stretched across a canyon, and you view it from, say, a quarter of a mile away, as in Figure 8.1(a). From this distance, you will easily perceive the long, unfurled, horizontal extent of the hose, but unless you have uncanny eyesight, the *thickness* of the hose will be difficult to discern. From your distant vantage point, you would think that if an ant were constrained to live on the hose, it would have only *one* dimension in which to walk: the left-right dimension along the hose's length. If someone asked you to specify where the ant was at a given moment, you would need to give only *one* piece of data: the distance of the ant from the left (or the right) end of the hose. The upshot

(a)

(b)

Figure 8.1 (a) A garden hose viewed from a substantial distance looks like a one-dimensional object. (b) When magnified, a second dimension—one that is in the shape of a circle and is curled around the hose—becomes visible.

is that from a quarter of a mile away, a long piece of garden hose appears to be a one-dimensional object.

In reality, we know that the hose *does* have thickness. You might have trouble resolving this from a quarter mile, but by using a pair of binoculars you can zoom in on the hose and observe its girth directly, as shown in Figure 8.1(b). From this magnified perspective, you see that a little ant living on the hose actually has *two* independent directions in which it can walk: along the left-right dimension spanning the length of the hose as already identified, *and* along the "clockwise-counterclockwise dimension" around the circular part of the hose. You now realize that to specify where the tiny ant is at any given instant, you must actually give *two* pieces of data: where the ant is along the length of the hose, and where the ant is along its circular girth. This reflects the fact the surface of the garden hose is two-dimensional.[1]

Nonetheless, there is a clear difference between these two dimensions. The direction along the length of the hose is long, extended, and easily visible. The direction circling around the thickness of the hose is short, "curled up," and harder to see. To become aware of the circular dimension, you have to examine the hose with significantly greater precision.

This example underscores a subtle and important feature of spatial dimensions: they come in two varieties. They can be large, extended, and therefore directly manifest, or they can be small, curled up, and much more difficult to detect. Of course, in this example you did not have to exert a great deal of effort to reveal the "curled-up" dimension encircling the thickness of the hose. You merely had to use a pair of binoculars. However, if you had a very thin garden hose—as thin as a hair or a capillary—detecting its curled-up dimension would be more difficult.

In a paper he sent to Einstein in 1919, Kaluza made an astounding suggestion. He proposed that the spatial fabric of the universe might possess more than the three dimensions of common experience. The motivation for this radical thesis, as we will discuss shortly, was Kaluza's realization that it provided an elegant and compelling framework for weaving together Einstein's general relativity and Maxwell's electromagnetic theory into a single, unified conceptual framework. But, more immediately, how can this proposal be squared with the apparent fact that we *see* precisely three spatial dimensions?

The answer, implicit in Kaluza's work and subsequently made explicit and refined by the Swedish mathematician Oskar Klein in 1926, is that *the spatial fabric of our universe may have both extended and curled-up dimensions.* That is, just like the horizontal extent of the garden hose, our universe has dimensions that are large, extended, and easily visible—the three spatial dimensions of common experience. But like the circular girth of a garden hose, the universe may also have additional spatial dimensions that are tightly curled up into a tiny space—a space so tiny that it has so far eluded detection by even our most refined experimental equipment.

To gain a clearer image of this remarkable proposal, let's reconsider the garden hose for a moment. Imagine that the hose is painted with closely spaced black circles along its girth. From far away, as before, the garden hose looks like a thin, one-dimensional line. But if you zoom in with binoculars, you can detect the curled-up dimension, even more easily after our paint job, and you see the image illustrated in Figure 8.2. This figure emphasizes that the surface of the garden hose is two-dimensional, with one large, extended dimension and one small, circular dimension. Kaluza and Klein proposed that our spatial universe is similar, but that it has three large, extended spatial dimensions and one small, circular dimension—for a total of four spatial dimensions. It is difficult to draw something with that many dimensions, so for visualization purposes we must settle for an illustration incorporating two large dimensions and one small, circular dimension. We illustrate this in Figure 8.3, in which we magnify the fabric

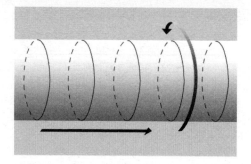

Figure 8.2 The surface of the garden hose is two-dimensional: one dimension (its horizontal extent), emphasized by the straight arrow, is long and extended; the other dimension (its circular girth), emphasized by the circular arrow, is short and curled up.

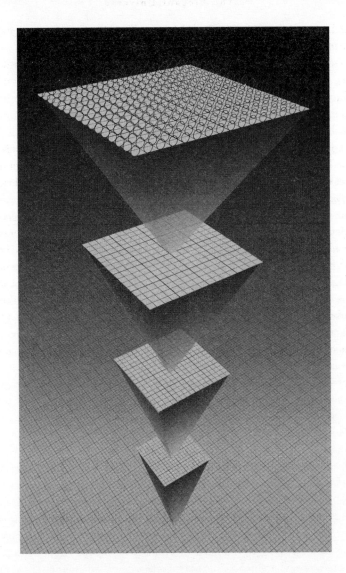

Figure 8.3 As in Figure 5.1, each subsequent level represents a huge magnification of the spatial fabric displayed in the previous level. Our universe may have extra dimensions—as we see by the fourth level of magnification—so long as they are curled up into a space small enough to have as yet evaded direct detection.

of space in much the same way that we zoomed in on the surface of the garden hose.

The lowest image in the figure shows the apparent structure of space—the ordinary world around us—on familiar distance scales such as meters. These distances are represented by the largest set of grid lines. In the subsequent images, we zoom in on the fabric of space by focusing our attention on ever smaller regions, which we sequentially magnify in order to make them easily visible. At first as we examine the fabric of space on shorter distance scales, not much happens; it appears to retain the same basic form as it has on larger scales, as we see in the first three levels of magnification. However, as we continue on our journey toward the most microscopic examination of space—the fourth level of magnification in Figure 8.3—a new, curled-up, circular dimension becomes apparent, much like the circular loops of thread making up the pile of a tightly woven piece of carpet. Kaluza and Klein suggested that the extra circular dimension exists at *every* point in the extended dimensions, just as the circular girth of the garden hose exists at every point along its unfurled, horizontal extent. (For visual clarity, we have drawn only an illustrative sample of the circular dimension at regularly spaced points in the extended dimensions.) We show a close-up of the Kaluza-Klein vision of the microscopic structure of the spatial fabric in Figure 8.4.

The similarity with the garden hose is manifest, although there are some important differences. The universe has three large, extended space dimensions (only two of which we have actually drawn), compared with the garden hose's one, and, more important, we are now describing the spatial fabric of the *universe* itself, not just an object, like the garden hose, that exists *within* the universe. But the basic idea is the same: Like the circular girth of the garden hose, if the additional curled-up, circular dimension of the universe is extremely small, it is much harder to detect than the manifest, large, extended dimensions. In fact, if its size is small enough, it will be beyond detection by even our most powerful magnifying instruments. And, of utmost importance, the circular dimension is *not* merely a circular bump within the familiar extended dimensions as the illustration might lead you to believe. Rather, the circular dimension is a *new* dimension, one that exists at every point in the familiar extended dimensions just as each of the up-down, left-right, and back-forth dimen-

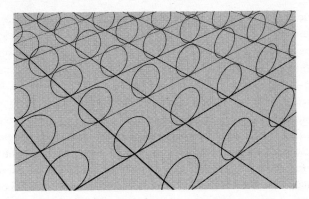

Figure 8.4 The grid lines represent the extended dimensions of common experience, whereas the circles are a new, tiny, curled-up dimension. Like the circular loops of thread making up the pile of a carpet, the circles exist at every point in the familiar extended dimensions—but for visual clarity we draw them as spread out on intersecting grid lines.

sions exists at every point as well. It is a new and independent direction in which an ant, if it were small enough, could move. To specify the spatial location of such a microscopic ant, we would need to say where it is in the three familiar extended dimensions (represented by the grid) and *also* where it is in the circular dimension. We would need *four* pieces of spatial information; if we add in time, we get a total of five pieces of spacetime information—one more than we normally would expect.

And so, rather surprisingly, we see that although we are aware of only three extended spatial dimensions, Kaluza's and Klein's reasoning shows that this does not preclude the existence of additional curled-up dimensions, at least if they are very small. The universe may very well have more dimensions than meet the eye.

How small is "small?" Cutting-edge equipment can detect structures as small as a billionth of a billionth of a meter. So long as an extra dimension is curled up to a size less than this tiny distance, it is too small for us to detect. In 1926 Klein combined Kaluza's initial suggestion with some ideas from the emerging field of quantum mechanics. His calculations indicated that the additional circular dimension might be as small as the

Planck length, far shorter than experimental accessibility. Since then, physicists have called the possibility of extra tiny space dimensions *Kaluza-Klein theory*.[2]

Comings and Goings on a Garden Hose

The tangible example of the garden hose and the illustration in Figure 8.3 are meant to give you some sense of how it is possible that our universe has extra spatial dimensions. But even for researchers in the field, it is quite difficult to visualize a universe with more than three spatial dimensions. For this reason, physicists often hone their intuition about these extra dimensions by contemplating what life would be like if we lived in an imaginary *lower*-dimensional universe—following the lead of Edwin Abbott's enchanting 1884 classic popularization *Flatland*[3]—in which we slowly realize that the universe has more dimensions than those of which we are directly aware. Let's try this by imagining a two-dimensional universe shaped like our garden hose. Doing so requires that you relinquish an "outsider's" perspective that views the garden hose as an object in our universe. Rather, you must leave the world as we know it and enter a new Garden-hose universe in which the surface of a very long garden hose (you can think of it as being infinitely long) is *all there is* as far as spatial extent. Imagine that you are a tiny ant living your life on its surface.

Let's start by making things even a little more extreme. Imagine that the length of the circular dimension of the Garden-hose universe is very short—so short that neither you nor any of your fellow Hose-dwellers are even aware of its existence. Instead, you and everyone else living in the Hose universe take one basic fact of life to be so evident as to be beyond questioning: the universe has *one* spatial dimension. (If the Garden-hose universe had produced its own ant-Einstein, Hose-dwellers would say that the universe has one spatial and one time dimension.) In fact, this feature is so self-evident that Hose-dwellers have named their home *Lineland,* directly emphasizing its having one spatial dimension.

Life in Lineland is very different from life as we know it. For example, the body with which you are familiar *cannot fit* in Lineland. No matter how much effort you may put into body reshaping, one thing you can't get around is that you definitely have length, width, and breadth—spatial extent in three dimensions. In Lineland there is no room for such an extravagant design. Remember, although your mental image of Lineland may still be tied to a long, threadlike object existing in our space, you really need to think of Lineland as a *universe*—all there is. As an inhabitant of Lineland you must fit within its spatial extent. Try to imagine it. Even if you take on an ant's body, you still will not fit. You must squeeze your ant body to look more like a worm, and then further squeeze it until you have no thickness at all. To fit in Lineland you must be a being that has *only* length.

Imagine further that you have an eye on each end of your body. Unlike your human eyes, which can swivel around to look in all three dimensions, your eyes as a Linebeing are forever locked into position, each staring off into the one-dimensional distance. This is *not* an anatomical limitation of your new body. Instead, you and all other Linebeings recognize that since Lineland has but one dimension, there simply isn't another direction in which your eyes can look. Forward and backward exhaust the extent of Lineland.

We can try to go further in imagining life in Lineland, but we quickly realize that there's not much more to it. For instance, if another Linebeing is on one or the other side of you, picture how it will appear: you will see one of her eyes—the one facing you—but unlike human eyes, hers will be a single dot. Eyes in Lineland have no features and display no emotion—there is just no room for these familiar characteristics. Moreover, you will be forever stuck with this dotlike image of your neighbor's eye. If you wanted to pass her and explore the realm of Lineland on the other side of her body, you would be in for a great disappointment. *You can't pass by her.* She is fully "blocking the road," and there is no space in Lineland to go around her. The order of Linebeings as they are sprinkled along the extent of Lineland is fixed and unchanging. What drudgery.

A few thousand years after a religous epiphany in Lineland, a Linebeing named Kaluza K. Line offers some hope for the downtrodden Linedwellers. Either from divine inspiration or from the sheer exasperation of years of staring at his neighbor's dot-eye, he suggests that Lineland may not be one-dimensional after all. What if, he theorizes, Lineland is actually two-dimensional, with the second space dimension being a very small circular direction that has, as yet, evaded direct detection because of its tiny spatial extent. He goes on to paint a picture of a vastly new life, if only this curled-up space direction would expand in size—something that is at least possible according to the recent work of his colleague, Linestein. Kaluza K. Line describes a universe that amazes you and your comrades and fills everyone with hope—a universe in which Linebeings can move freely past one another by making use of the second dimension: the end of spatial enslavement. We realize that Kaluza K. Line is describing life in a "thickened" Garden-hose universe.

In fact, if the circular dimension were to grow, "inflating" Lineland into the Garden-hose universe, your life would change in profound ways. Take your body, for example. As a Linebeing, anything between your two eyes constitutes the interior of your body. Your eyes, therefore, play the same role for your linebody as skin plays for an ordinary human body: They constitute the barrier between the inside of your body and the outside world. A doctor in Lineland can access the interior of your linebody only by puncturing its surface—in other words, "surgery" in Lineland takes place through the eyes.

But now imagine what happens if Lineland does, à la Kaluza K. Line, have a secret, curled-up dimension, and if this dimension expands to an observably large size. Now one Linebeing can view your body at an angle and thereby directly see into its interior, as we illustrate in Figure 8.5. Using this second dimension, a doctor can operate on your body by reaching directly inside your exposed interior. Weird! In time, Linebeings, no doubt, would develop a skinlike cover to shield the newly exposed interior

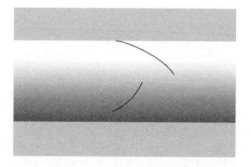

Figure 8.5 One Linebeing can see directly into the interior of another's body when Lineland expands into the Garden-hose universe.

of their bodies from contact with the outside world. And moreover, they would undoubtedly evolve into beings with length as well as breadth: Flat-beings sliding along the two-dimensional Garden-hose universe as illustrated in Figure 8.6. If the circular dimension were to grow very large, this two-dimensional universe would be closely akin to Abbott's Flatland—an imaginary two-dimensional world Abbott suffused with a rich cultural heritage and even a satirical caste system based upon one's geometrical shape. Whereas it's hard to imagine *anything* interesting happening in Lineland—there is just not enough room—life on a Garden-hose becomes replete with possibilities. The evolution from one to two observably large space dimensions is dramatic.

And now the refrain: Why stop there? The two-dimensional universe might itself have a curled-up dimension and therefore secretly be three-dimensional. We can illustrate this with Figure 8.4, so long as we recognize that we are now imagining that there *are* only two extended space dimensions (whereas when we first introduced this figure we were imagining the flat grid to represent three extended dimensions). If the circular dimension should expand, a two-dimensional being would find itself in a vastly new world in which movement is not limited just to left-right and back-forth along the extended dimensions. Now, a being can also move in a third dimension—the "up-down" direction along the circle. In fact, if the circular dimension were to grow to a large enough size, this could be *our* three-dimensional universe. We do not know at present whether any of our three spatial dimensions extends outward forever, or in fact curls back on

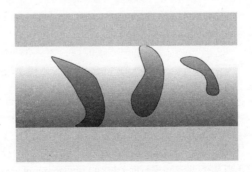

Figure 8.6 Flat, two-dimensional beings living in the Garden-hose universe.

itself in the shape of a giant circle, beyond the range of our most powerful telescopes. If the circular dimension in Figure 8.4 got big enough—billions of light-years in extent—the figure could very well be a drawing of our world.

But the refrain replays: Why stop there? This takes us to Kaluza's and Klein's vision: that our three-dimensional universe might have a previously unanticipated curled-up fourth spatial dimension. If this striking possibility, or its generalization to numerous curled-up dimensions (to be discussed shortly) is true, and if these curled-up dimensions were themselves to expand to a macroscopic size, the lower-dimensional examples discussed make it clear that life as we know it would change immensely.

Surprisingly, though, even if they should always stay curled up and small, the existence of extra curled-up dimensions has profound implications.

Unification in Higher Dimensions

Although Kaluza's 1919 suggestion that our universe might have more spatial dimensions than those of which we are directly aware was a remarkable possibility in its own right, something else really made it compelling. Einstein had formulated general relativity in the familiar setting of a universe with three spatial dimensions and one time dimension. The mathematical formalism of his theory, however, could be extended fairly directly to write down analogous equations for a universe with additional space dimensions. Under the "modest" assumption of one extra space dimension, Kaluza carried out the mathematical analysis and explicitly derived the new equations.

He found that in the revised formulation the equations pertaining to the three ordinary dimensions were essentially identical to Einstein's. But because he included an extra space dimension, not surprisingly Kaluza found extra equations beyond those Einstein originally derived. After studying the extra equations associated with the new dimension, Kaluza realized that something amazing was going on. The extra equations were none other than those Maxwell had written down in the 1880s for describing the electromagnetic force! By adding another space dimension,

Kaluza had united Einstein's theory of gravity with Maxwell's theory of light.

Before Kaluza's suggestion, gravity and electromagnetism were thought of as two unrelated forces; nothing had even hinted that there might be a relation between them. By having the bold creativity to imagine that our universe has an additional space dimension, Kaluza suggested that there was a deep connection, indeed. His theory argued that both gravity and electromagnetism are associated with ripples in the fabric of space. Gravity is carried by ripples in the familiar three space dimensions, while electromagnetism is carried by ripples involving the new, curled-up dimension.

Kaluza sent his paper to Einstein, and at first Einstein was quite intrigued. On April 21, 1919, Einstein wrote back to Kaluza and told him that it had never occurred to him that unification might be achieved "through a five-dimensional [four space and one time] cylinder-world." He added, "At first glance, I like your idea enormously."[4] About a week later, though, Einstein wrote Kaluza again, this time with some skepticism: "I have read through your paper and find it really interesting. Nowhere, so far, can I see an impossibility. On the other hand, I have to admit that the arguments brought forward so far do not appear convincing enough."[5] But then, on October 14, 1921, more than two years later, Einstein wrote to Kaluza again, having had time to digest Kaluza's novel approach more fully: "I am having second thoughts about having restrained you from publishing your idea on a unification of gravitation and electricity two years ago. . . . If you wish, I shall present your paper to the academy after all."[6] Belatedly, Kaluza had received the master's stamp of approval.

Although it was a beautiful idea, subsequent detailed study of Kaluza's proposal, augmented by Klein's contributions, showed that it was in serious conflict with experimental data. The simplest attempts to incorporate the electron into the theory predicted relations between its mass and its charge that were vastly different from their measured values. Because there did not seem to be any obvious way of getting around this problem, many of the physicists who had taken notice of Kaluza's idea lost interest. Einstein and others continued, now and then, to dabble with the possibility of extra curled-up dimensions, but it quickly came to be an enterprise on the outskirts of theoretical physics.

In a real sense, Kaluza's idea was way ahead of its time. The 1920s marked the start of a bull market for theoretical and experimental physics concerned with understanding the basic laws of the microworld. Theorists had their hands full as they sought to develop the structure of quantum mechanics and quantum field theory. Experimentalists had the detailed properties of the atom as well as numerous other elementary material constituents to discover. Theory guided experiment and experiment refined theory as physicists pushed forward for half a century, ultimately to reveal the standard model. It is no wonder that speculations on extra dimensions took a distant backseat during these productive and heady times. With physicists exploring powerful quantum methods, the implications of which gave rise to experimentally testable predictions, there was little interest in the mere possibility that the universe might be a vastly different place on length scales far too small to be probed by even the most powerful of instruments.

But sooner or later, bull markets lose steam. By the late 1960s and early 1970s the theoretical structure of the standard model was in place. By the late 1970s and early 1980s many of its predictions had been verified experimentally, and most particle physicists concluded that it was just a matter of time before the rest were confirmed as well. Although a few important details remained unresolved, many felt that the major questions concerning the strong, weak, and electromagnetic forces had been answered.

The time was finally ripe to return to the grandest question of all: the enigmatic conflict between general relativity and quantum mechanics. The success in formulating a quantum theory of three of nature's forces emboldened physicists to try to bring the fourth, gravity, into the fold. Having pursued numerous ideas that all ultimately failed, the mind-set of the community became more open to comparatively radical approaches. After being left for dead in the late 1920s, Kaluza-Klein theory was resuscitated.

Modern Kaluza-Klein Theory

The understanding of physics had significantly changed and substantially deepened in the six decades since Kaluza's original proposal. Quantum mechanics had been fully formulated and experimentally verified. The strong and the weak forces, unknown in the 1920s, had been discovered and were largely understood. Some physicists suggested that Kaluza's original proposal had failed because he was unaware of these other forces and had therefore been too *conservative* in his revamping of space. More forces meant the need for even more dimensions. It was argued that a single new, circular dimension, although able to show hints of a connection between general relativity and electromagnetism, was just not enough.

By the mid-1970s, an intense research effort was underway, focusing on higher-dimensional theories with numerous curled-up spatial directions. Figure 8.7 illustrates an example with two extra dimensions that are curled up into the surface of a ball—that is, a sphere. As in the case of the single circular dimension, these extra dimensions are tacked on to *every point* of the familiar extended dimensions. (For visual clarity we again have drawn only an illustrative sample of the spherical dimensions at regularly spaced grid points in the extended dimensions.) Beyond proposing a different number of extra dimensions, one can also imagine other shapes for the extra dimensions. For instance, in Figure 8.8 we illustrate a possibility in which there are again two extra dimensions, now in the shape of a hollow doughnut—that is, a torus. Although they are beyond our ability

Figure 8.7 Two extra dimensions curled up into the shape of a sphere.

Figure 8.8 Two extra dimensions curled up in the shape of a hollow doughnut, or torus.

to draw, more complicated possibilities can be imagined in which there are three, four, five, essentially any number of extra spatial dimensions, curled up into a wide spectrum of exotic shapes. The essential requirement, again, is that all of these dimensions have a spatial extent smaller than the smallest length scales we can probe, since no experiment has yet revealed their existence.

The most promising of the higher-dimensional proposals were those that also incorporated supersymmetry. Physicists hoped that the partial cancelling of the most severe quantum fluctuations, arising from the pairing of superpartner particles, would help to soften the hostilities between gravity and quantum mechanics. They coined the name *higher-dimensional supergravity* to describe those theories encompassing gravity, extra dimensions, and supersymmetry.

As had been the case with Kaluza's original attempt, various versions of higher-dimensional supergravity looked quite promising at first. The new equations resulting from the extra dimensions were strikingly reminiscent of those used in the description of electromagnetism, and the strong and the weak forces. But detailed scrutiny showed that the old conundrums persisted. Most importantly, the pernicious short-distance quantum undulations of space were lessened by supersymmetry, but not sufficiently to yield a sensible theory. Physicists also found it difficult to find a single, sensible, higher-dimensional theory incorporating all features of forces and matter.[7]

It gradually became clear that bits and pieces of a unified theory were surfacing, but that a crucial element capable of tying them all together in a quantum-mechanically consistent manner was missing. In 1984 this missing piece—string theory—dramatically entered the story and took center stage.

More Dimensions and String Theory

By now you should be convinced that our universe *may* have additional curled-up spatial dimensions; certainly, so long as they are small enough, nothing rules them out. But extra dimensions may strike you as an artifice. Our inability to probe distances smaller than a billionth of a billionth of a meter permits not only extra tiny dimensions but all manner of whimsical possibilities as well—even a microscopic civilization populated by even tinier green people. While the former certainly seems more rationally motivated than the latter, the act of postulating either of these experimentally untested—and, at present, untestable—possibilities might seem equally arbitrary.

Such was the case until string theory. Here is a theory that resolves the central dilemma confronting contemporary physics—the incompatibility between quantum mechanics and general relativity—and that unifies our understanding of all of nature's fundamental material constituents and forces. But to accomplish these feats, it turns out that string theory *requires* that the universe have extra space dimensions.

Here's why. One of the main insights of quantum mechanics is that our predictive power is fundamentally limited to asserting that such-and-such outcome will occur with such-and-such probability. Although Einstein felt that this was a distasteful feature of our modern understanding, and you may agree, it certainly appears to be fact. Let's accept it. Now, we all know that probabilities are always numbers between 0 and 1—equivalently, when expressed as percentages, probabilities are numbers between 0 and 100. Physicists have found that a key signal that a quantum-mechanical theory has gone haywire is that particular calculations yield "probabilities" that are *not* within this acceptable range. For instance, we mentioned earlier that a sign of the grinding incompatibility between general relativity and quantum mechanics in a point-particle framework is that calculations result in infinite probabilities. As we have discussed, string theory cures these infinities. But what we have not as yet mentioned is that a residual, somewhat more subtle problem still remains. In the early days of string theory physicists found that certain calculations yielded *negative* probabilities, which are also outside of the acceptable range. So, at first sight, string theory appeared to be awash in its own quantum-mechanical hot water.

With stubborn determination, physicists sought and found the cause of this unacceptable feature. The explanation begins with a simple observation. If a string is constrained to lie on a two-dimensional surface—such as the surface of a table or a garden hose—the number of independent directions in which it can vibrate is reduced to *two*: the left-right and back-forth dimensions along the surface. Any vibrational pattern that remains on the surface involves some combination of vibrations in these two directions. Correspondingly, we see that this also means that a string in Flatland, the Garden-hose universe, or in any other two-dimensional universe, is also constrained to vibrate in a total of two independent spatial directions. If, however, the string is allowed to leave the surface, the number of independent vibrational directions increases to three, since the string then can also oscillate in the up-down direction. Equivalently, in a universe with three spatial dimensions, a string can vibrate in three independent directions. Although it gets harder to envision, the pattern continues: In a universe with ever more spatial dimensions, there are ever more independent directions in which it can vibrate.

We emphasize this fact of string vibrations because physicists found that the troublesome calculations were highly sensitive to the number of independent directions in which a string can vibrate. The negative probabilities arose from a *mismatch* between what the theory required and what reality seemed to impose: The calculations showed that if strings could vibrate in *nine* independent spatial directions, all of the negative probabilities would cancel out. Well, that's great in theory, but so what? If string theory is meant to describe our world with three spatial dimensions, we still seem to be in trouble.

But are we? Taking a more than half-century-old lead, we see that

Kaluza and Klein provide a loophole. Since strings are so small, not only can they vibrate in large, extended dimensions, they can also vibrate in ones that are tiny and curled up. And so we *can* meet the nine-space-dimension requirement of string theory in *our* universe, by assuming—à la Kaluza and Klein—that in addition to our familiar three extended spatial dimensions there are six other curled-up spatial dimensions. In this manner, string theory, which appeared to be on the brink of elimination from the realm of physical relevance, is saved. Moreover, rather than just postulating the existence of extra dimensions, as had been done by Kaluza, Klein, and their followers, string theory *requires* them. For string theory to make sense, the universe should have nine space dimensions and one time dimension, for a total of ten dimensions. In this way, Kaluza's 1919 proposal finds its most convincing and powerful forum.

Some Questions

This raises a number of questions. First, why does string theory require the particular number of nine space dimensions to avoid nonsensical probability values? This is probably the hardest question in string theory to answer without appealing to mathematical formalism. A straightforward string theory calculation reveals this answer, but no one has an intuitive, nontechnical explanation for the particular number that emerges. The physicist Ernest Rutherford once said, in essence, that if you can't explain a result in simple, nontechnical terms, then you don't really understand it. He wasn't saying that this means your result is wrong; rather, he was saying that it means you do not fully understand its origin, meaning, or implications. Perhaps this is true regarding the extradimensional character of string theory. (In fact, let's take this opportunity to brace—parenthetically—for a central aspect of the second superstring revolution that we will discuss in Chapter 12. The calculation underlying the conclusion that there are ten spacetime dimensions—nine space and one time—turns out to be *approximate*. In the mid-1990s, Witten, based on his own insights and previous work by Michael Duff from Texas A&M University and Chris Hull and Paul Townsend from Cambridge University, gave convincing evidence that the approximate calculation actually *misses* one space dimension: String theory, he argued to most string theorists' amazement, actually requires *ten* space dimensions and one time dimension, for a total of *eleven* dimensions. We will ignore this important result until Chapter 12, as it will have little direct bearing on the material we develop before then.)

Second, if the equations of string theory (or, more precisely, the approximate equations guiding our pre–Chapter 12 discussion) show that the universe has nine space dimensions and one time dimension, why is it that three space (and one time) dimensions are large and extended while all of the others are tiny and curled up? Why aren't they *all* extended, or all curled up, or some other possibility in between? At present no one knows the answer to this question. If string theory is right, we should eventually be able to extract the answer, but as yet our understanding of the theory is not refined enough to reach this goal. That's not to say that there haven't been valiant attempts to explain it. For instance, from a cosmological perspective, we can imagine that all of the dimensions start out being tightly curled up and then, in a big bang–like explosion, three spatial dimensions and one time dimension unfurl and expand to their present large extent while the other spatial dimensions remain small. Rough arguments have been put forward as to why only three space dimensions grow large, as we will discuss in Chapter 14, but it's fair to say that these explanations are only in the formative stages. In what follows, we will assume that all but three space dimensions are curled up, in accordance with what we see around us. A primary goal of modern research is to establish that this assumption emerges from the theory itself.

Third, given the requirement of numerous extra dimensions, is it possible that some are additional *time* dimensions, as opposed to additional space dimensions? If you think about this for a moment, you will see that it's a truly bizarre possibility. We all have a visceral understanding of what it means for the universe to have multiple space dimensions, since we live in a world in which we constantly deal with a plurality—three. But what would it mean to have multiple times? Would one align with time as we presently experience it psychologically while the other would somehow be "different?"

It gets even stranger when you think about a curled-up time dimension. For instance, if a tiny ant walks around an extra space dimension that is curled up like a circle, it will find itself returning to the same position over and over again as it traverses complete circuits. This holds little mystery as we are familiar with the ability to return, should we so choose, to the same location in space as often as we like. But, if a curled-up dimension is a time dimension, traversing it means returning, after a temporal lapse, to *a prior instant in time*. This, of course, is well beyond the realm of our experience. Time, as we know it, is a dimension we can traverse in only one direction with absolute inevitability, never being able to return to an instant after it has passed. Of course, it might be that curled-up time dimensions have very different properties from the familiar, vast time dimension that we imagine reaching back to the creation of the universe and forward to the present moment. But, in contrast to extra spatial dimensions, new and previously unknown time dimensions would clearly require an even more monumental restructuring of our intuition. Some theorists have been exploring the possibility of incorporating extra time dimensions into string theory, but as yet the situation is inconclusive. In our discussion of string theory, we will stick to the more "conventional" approach in which all of the curled-up dimensions are space dimensions, but the intriguing possibility of new time dimensions could well play a role in future developments.

The Physical Implications of Extra Dimensions

Years of research, dating back to Kaluza's original paper, have shown that even though any extra dimensions that physicists propose must be smaller than we or our equipment can directly "see" (since we haven't seen them), they do have important *indirect* effects on the physics that we observe. In string theory, this connection between the microscopic properties of space and the physics we observe is particularly transparent.

To understand this, you need to recall that masses and charges of particles in string theory are determined by the possible resonant vibrational string patterns. Picture a tiny string as it moves and oscillates, and you will realize that the resonant patterns are influenced by its spatial surroundings. Think, for example, of ocean waves. Out in the grand expanse of the open ocean, isolated wave patterns are relatively free to form and travel this way or that. This is much like the vibrational patterns of a string as it moves through large, extended spatial dimensions. As in Chapter 6, such a string is equally free to oscillate in any of the extended directions at any moment. But if an ocean wave passes through a more cramped spatial environment, the detailed form of its wave motion will surely be affected by, for example, the depth of the water, the placement and shape of the rocks encountered, the canals through which the water is channeled, and so on. Or, think of an organ pipe or a French horn. The sounds that each of these instruments can produce are a direct consequence of the resonant patterns of vibrating air streams in their interior; these are determined by the precise size and shape of the spatial surroundings within the instrument through which the air streams are channeled. Curled-up spatial dimensions have a similar impact on the possible vibrational patterns of a string. Since tiny strings vibrate through all of the spatial dimensions, the precise way in which the extra dimensions are twisted up and curled back on each other strongly influences and tightly constrains the possible resonant vibrational patterns. These patterns, largely determined by the extradimensional geometry, constitute the array of possible particle properties observed in the familiar extended dimensions. This means that *extradimensional geometry determines fundamental physical attributes like particle masses and charges that we observe in the usual three large space dimensions of common experience.*

This is such a deep and important point that we say it once again, with feeling. According to string theory, the universe is made up of tiny strings whose resonant patterns of vibration are the microscopic origin of particle masses and force charges. String theory also requires extra space dimensions that must be curled up to a very small size to be consistent with our never having seen them. But a tiny string can probe a tiny space. As a string moves about, oscillating as it travels, the geometrical form of the extra dimensions plays a critical role in determining resonant patterns of vibration. Because the patterns of string vibrations appear to us as the masses and charges of the elementary particles, we conclude that these fundamental properties of the universe are determined, in large measure, by the geometrical size and shape of the extra dimensions. That's one of the most far-reaching insights of string theory.

Since the extra dimensions so profoundly influence basic physical properties of the universe, we should now seek—with unbridled vigor— an understanding of what these curled-up dimensions look like.

What Do the Curled-Up Dimensions Look Like?

The extra spatial dimensions of string theory cannot be "crumpled" up any which way; the equations that emerge from the theory severely restrict the geometrical form that they can take. In 1984, Philip Candelas of the University of Texas at Austin, Gary Horowitz and Andrew Strominger of the University of California at Santa Barbara, and Edward Witten showed that a particular class of six-dimensional geometrical shapes can meet these conditions. They are known as *Calabi-Yau spaces* (or *Calabi-Yau shapes*) in honor of two mathematicians, Eugenio Calabi from the University of Pennsylvania and Shing-Tung Yau from Harvard University, whose research in a related context, but prior to string theory, plays a central role in understanding these spaces. Although the mathematics describing Calabi-Yau spaces is intricate and subtle, we can get an idea of what they look like with a picture.[8]

In Figure 8.9 we show an example of a Calabi-Yau space.[9] As you view this figure, you must bear in mind that the image has built-in limitations. We are trying to represent a six-dimensional shape on a two-dimensional piece of paper, and this introduces significant distortions. Nevertheless, the image does convey the rough idea of what a Calabi-Yau space looks

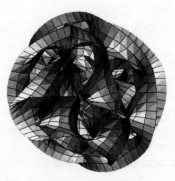

Figure 8.9 One example of a Calabi-Yau space.

like.[10] The shape in Figure 8.9 is but one of many tens of thousands of examples of Calabi-Yau shapes that meet the stringent requirements for the extra dimensions that emerge from string theory. Although belonging to a club with tens of thousands of members might not sound very exclusive, you must compare it with the infinite number of shapes that are mathematically possible; by this measure Calabi-Yau spaces are rare indeed.

To put it all together, you should now imagine replacing each of the spheres in Figure 8.7—which represented two curled-up dimensions— with a Calabi-Yau space. That is, at every point in the three familiar extended dimensions, string theory claims that there are six hitherto unanticipated dimensions, tightly curled up into one of these rather complicated-looking shapes, as illustrated in Figure 8.10. These dimensions are an integral and ubiquitous part of the spatial fabric; they exist everywhere. For instance, if you sweep your hand in a large arc, you are moving not only through the three extended dimensions, but also through these curled-up dimensions. Of course, because the curled-up dimensions are so small, as you move your hand you circumnavigate them an enormous number of times, repeatedly returning to your starting point. Their tiny extent means that there is not much room for a large object like your hand to move—it all averages out so that after sweeping your arm, you are completely unaware of the journey you took through the curled-up Calabi-Yau dimensions.

Figure 8.10 According to string theory, the universe has extra dimensions curled up into a Calabi-Yau shape.

This is a stunning feature of string theory. But if you are practically minded, you are bound to bring the discussion back to an essential and concrete issue. Now that we have a better sense of what the extra dimensions look like, what are the physical properties that emerge from strings that vibrate through them, and how do these properties compare with experimental observations? This is string theory's $64,000 question.

Chapter 9

The Smoking Gun:
Experimental Signatures

Nothing would please string theorists more than to proudly present the world with a list of detailed, experimentally testable predictions. Certainly, there is no way to establish that any theory describes our world without subjecting its predictions to experimental verification. And no matter how compelling a picture string theory paints, if it does not accurately describe our universe, it will be no more relevant than an elaborate game of Dungeons and Dragons.

Edward Witten is fond of declaring that string theory has already made a dramatic and experimentally confirmed prediction: "String theory has the remarkable property of *predicting gravity.*"[1] What Witten means by this is that both Newton and Einstein developed theories of gravity because their observations of the world clearly showed them that gravity exists, and that, therefore, it required an accurate and consistent explanation. On the contrary, a physicist studying string theory—even if he or she was completely unaware of general relativity—would be inexorably led to it by the string framework. Through its massless spin-2 graviton pattern of vibration, string theory has gravity thoroughly sewn into its theoretical fabric. As Witten has said, "the fact that gravity is a consequence of string theory is one of the greatest theoretical insights ever."[2] In acknowledging that this "prediction" is more precisely labeled a "postdiction" because physicists had discovered theoretical descriptions of gravity before they knew of string theory, Witten points out that this is a mere accident of history on earth. In other advanced civilizations in the universe, Witten fancifully argues, it is quite possible that string theory was discovered first, and a theory of gravity found as a stunning consequence.

Since we are bound to the history of science on our planet, there are many who find this postdiction of gravity unconvincing experimental confirmation of string theory. Most physicists would be far happier with one of two things: a bona fide prediction from string theory that experimentalists could confirm, or a postdiction of some property of the world (like the mass of the electron or the existence of three families of particles) for which there is currently no explanation. In this chapter we will discuss how far string theorists have gone toward reaching these goals.

Ironically, we will see that although string theory has the potential to be *the* most predictive theory that physicists have ever studied—a theory that has the capacity to explain the most fundamental of nature's properties—physicists have not as yet been able to make predictions with the precision necessary to confront experimental data. Like a child who receives his or her dream gift for Christmas but can't quite get it to work because a few pages of the instructions are missing, today's physicists are in possession of what may well be the Holy Grail of modern science, but they can't unleash its full predictive power until they succeed in *writing* the full instruction manual. Nevertheless, as we discuss in this chapter, with a bit of luck, one central feature of string theory could receive experimental verification within the next decade. And with a good deal more luck, indirect fingerprints of the theory could be confirmed at any moment.

Crossfire

Is string theory right? We don't know. If you share the belief that the laws of physics should not be fragmented into those that govern the large and those that govern the small, and if you also believe that we should not rest until we have a theory whose range of applicability is limitless, string theory is the only game in town. You might well argue, though, that this highlights only physicists' lack of imagination rather than some fundamental uniqueness of string theory. Perhaps. You might further argue that, like the man searching for his lost keys solely under a street light, physicists are huddled around string theory merely because the vagaries of scientific history have shed one random ray of insight in this direction. Maybe. And, if you're either relatively conservative or fond of playing devil's advocate, you might even say that physicists have no business wasting time on a theory that postulates a new feature of nature some hundred million billion times smaller than anything we can directly probe experimentally.

If you voiced these complaints in the 1980s when string theory first made its splash, you would have been joined by some of the most respected physicists of our age. For instance, in the mid-1980s Nobel Prize–winning Harvard physicist Sheldon Glashow, together with physicist Paul Ginsparg, then also at Harvard, publicly disparaged string theory's lack of experimental accessibility:

> In lieu of the traditional confrontation between theory and experiment, superstring theorists pursue an inner harmony, where elegance, uniqueness and beauty define truth. The theory depends for its existence upon magical coincidences, miraculous cancellations and relations among seemingly unrelated (and possibly undiscovered) fields of mathematics. Are these properties reasons to accept the reality of superstrings? Do mathematics and aesthetics supplant and transcend mere experiment?[3]

Elsewhere, Glashow went on to say,

> Superstring theory is so ambitious that it can only be totally right, or totally wrong. The only problem is that the mathematics is so new and difficult that we won't know which for decades to come.[4]

And he even questioned whether string theorists should "be paid by physics departments and allowed to pervert impressionable students," warning that string theory was was undermining science, much as medieval theology did during the Middle Ages.[5]

Richard Feynman, shortly before he died, made it clear that he did not believe that string theory was the *unique* cure for the problems—the pernicious infinities, in particular—besetting a harmonious merger of gravity and quantum mechanics:

My feeling has been—and I could be wrong—that there is more than one way to skin a cat. I don't think that there's only one way to get rid of the infinities. The fact that a theory gets rid of infinities is to me not a sufficient reason to believe its uniqueness.[6]

And Howard Georgi, Glashow's eminent Harvard colleague and collaborator, was also a vociferous string critic in the late 1980s:

If we allow ourselves to be beguiled by the siren call of the "ultimate" unification at distances so small that our experimental friends cannot help us, then we are in trouble, because we will lose that crucial process of pruning of irrelevant ideas which distinguishes physics from so many other less interesting human activities.[7]

As with many issues of great importance, for each of these naysayers, there is an enthusiastic supporter. Witten has said that when he learned how string theory incorporates gravity and quantum mechanics, it was "the greatest intellectual thrill" of his life.[8] Cumrun Vafa, a leading string theorist from Harvard University, has said that "string theory is definitely revealing the deepest understanding of the universe which we have ever had."[9] And Nobel Prize–winner Murray Gell-Mann has said that string theory is "a fantastic thing" and that he expects that some version of string theory will someday be the theory of the whole world.[10]

As you can see, the debate is fueled in part by physics and in part by distinct philosophies about how physics should be done. The "traditionalists" want theoretical work to be closely tied to experimental observation, largely in the successful research mold of the last few centuries. But others think that we are ready to tackle questions that are beyond our present technological ability to test directly.

Different philosophies notwithstanding, during the past decade much of the criticism of string theory has subsided. Glashow attributes this to two things. First, he notes that in the mid-1980s,

String theorists were enthusiastically and exuberantly proclaiming that they would shortly answer all questions in physics. As they are now more prudent with their enthusiasm, much of my criticism in the 1980s is no longer that relevant.[11]

Second, he also points out,

> We non–string theorists have not made any progress whatsoever in the
> last decade. So the argument that string theory is the only game in
> town is a very strong and powerful one. There are questions that will
> not be answered in the framework of conventional quantum field the-
> ory. That much is clear. They may be answered by something else, and
> the only something else I know of is string theory.[12]

Georgi reflects back on the 1980s in much the same way:

> At various times in its early history, string theory has gotten oversold.
> In the intervening years I have found that some of the ideas of string
> theory have led to interesting ways of thinking about physics which
> have been useful to me in my own work. I am much happier now to see
> people spending their time on string theory since I can now see how
> something useful will come out of it.[13]

Theorist David Gross, a leader in both conventional and string physics, has
eloquently summed up the situation in the following way:

> It used to be that as we were climbing the mountain of nature the ex-
> perimentalists would lead the way. We lazy theorists would lag behind.
> Every once in a while they would kick down an experimental stone
> which would bounce off our heads. Eventually we would get the idea
> and we would follow the path that was broken by the experimentalists.
> Once we joined our friends we would explain to them what the view
> was and how they got there. That was the old and easy way (at least for
> theorists) to climb the mountain. We all long for the return of those
> days. But now we theorists might have to take the lead. This is a much
> more lonely enterprise.[14]

String theorists have no desire for a solo trek to the upper reaches of
Mount Nature; they would far prefer to share the burden and the excitement with experimental colleagues. It is merely a technological mismatch
in our current situation—a historical asynchrony—that the theoretical
ropes and crampons for the final push to the top have at least been par-

tially fashioned, while the experimental ones do not yet exist. But this does *not* mean that string theory is fundamentally divorced from experiment. Rather, string theorists have high hopes of "kicking down a *theoretical* stone" from the ultra-high-energy mountaintop to experimentalists working at a lower base camp. This is a prime goal of present-day research in string theory. No stones have as yet been dislodged from the summit to be sent hurtling down, but, as we now discuss, a few tantalizing and promising pebbles certainly have.

The Road to Experiment

Without monumental technological breakthroughs, we will never be able to focus on the tiny length scales necessary to see a string directly. Physicists can probe down to a billionth of a billionth of a meter with accelerators that are roughly a few miles in size. Probing smaller distances requires higher energies and this means larger machines capable of focusing that energy on a single particle. As the Planck length is some 17 orders of magnitude smaller than what we can currently access, using today's technology we would need an accelerator the size of the *galaxy* to see individual strings. In fact, Shmuel Nussinov of Tel Aviv University has shown that this rough estimate based on straightforward scaling is likely to be overly optimistic; his more careful study indicates that we would require an accelerator the size of the whole *universe*. (The energy required to probe matter at the Planck length is roughly equal to a thousand kilowatt-hours—the energy needed to run an average air conditioner for about one hundred hours—and so is not particularly outlandish. The seemingly insurmountable technological challenge is to focus all of this energy on a single particle, that is, on a single string.) As the U.S. Congress ultimately canceled funding for the Superconducting Supercollider—an accelerator a "mere" 54 miles in circumference—don't hold your breath while waiting for the money for a Planck-probing accelerator. If we are going to test string theory experimentally, it will have to be in an indirect manner. We will have to determine physical implications of string theory that can be observed on length scales that are far larger than the size of a string itself.[15]

In their groundbreaking paper, Candelas, Horowitz, Strominger, and Witten took the first steps toward this goal. They not only found that the extra dimensions in string theory must be curled up into a Calabi-Yau shape, but they also worked out some of the implications this has on the possible patterns of string vibrations. One central result they found highlights the amazingly unexpected solutions string theory offers to long-standing particle-physics problems.

Recall that the elementary particles that physicists have found fall into three families of identical organization, with the particles in each successive family being increasingly massive. The puzzling question for which there was no insight prior to string theory is, Why *families* and why *three*? Here is string theory's proposal. A typical Calabi-Yau shape contains *holes* that are analogous to those found at the center of a phonograph record, or a doughnut, or a "multidoughnut", as shown in Figure 9.1. In the higher-dimensional Calabi-Yau context, there are actually a variety of different types of holes that can arise—holes which themselves can have a variety of dimensions ("multidimensional holes")—but Figure 9.1 conveys the basic idea. Candelas, Horowitz, Strominger, and Witten closely examined the effect that these holes have on the possible patterns of string vibration, and here is what they found.

There is a *family* of lowest-energy string vibrations associated with each *hole* in the Calabi-Yau portion of space. Because the familiar elementary particles should correspond to the lowest-energy oscillatory patterns, the existence of multiple holes—somewhat like those in the multidoughnut—means that the patterns of string vibrations will fall into multiple families. If the curled-up Calabi-Yau has three holes, then we will find three families of elementary particles.[16] And so, string theory proclaims that the

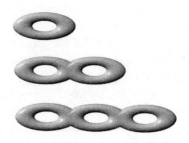

Figure 9.1 A doughnut, or torus, and its multihandled cousins.

family organization observed experimentally, rather than being some un-explainable feature of either random or divine origin, is a reflection of the number of holes in the geometrical shape comprising the extra dimensions! This is the kind of result that makes a physicist's heart skip a beat.

You might think that the number of holes in the curled-up Planck-sized dimensions—mountaintop physics *par excellence*—has now kicked an experimentally testable stone down to accessible energies. After all, experimentalists can establish—in fact, already have established—the number of particle families: 3. Unfortunately, the number of holes contained in each of the tens of thousands of known Calabi-Yau shapes spans a wide range. Some have 3. But others have 4, 5, 25, and so on—some even have as many as 480 holes. *The problem is that at present no one knows how to deduce from the equations of string theory which of the Calabi-Yau shapes constitutes the extra spatial dimensions.* If we could find the principle that allows the selection of one Calabi-Yau shape from the numerous possibilities, then indeed a stone from the mountaintop would go tumbling down into the experimentalists' camp. If the particular Calabi-Yau shape singled out by the equations of the theory were to have three holes, we would have found an impressive postdiction from string theory explaining a known feature of the world that is otherwise completely mysterious. But finding the principle for choosing among Calabi-Yau shapes is a problem that as yet remains unsolved. Nevertheless—and this is the important point—we see that string theory provides the potential for answering this basic puzzle of particle physics, and this in itself is substantial progress.

The number of families is but one experimental consequence of the geometrical form of the extra dimensions. Through their effect on possible patterns of string vibrations, other consequences of the extra dimensions include the detailed properties of the force and matter particles. As one primary example, subsequent work by Strominger and Witten showed that the masses of the particles in each family depend upon—hang on, this is a bit tricky—the way in which the boundaries of the various multidimensional holes in the Calabi-Yau shape intersect and overlap with one another. It's hard to visualize, but the idea is that as strings vibrate through the extra curled-up dimensions, the precise arrangement of the various holes and the way in which the Calabi-Yau shape folds around them has a direct impact on the possible resonant patterns of vibration. Although the details get difficult to follow and are really not all that essential, what is important is that, as with the number of families, string theory can provide us with a framework for answering questions—such as why the electron and other particles have the masses they do—on which previous theories are completely silent. Once again, though, carrying through with such calculations requires that we know which Calabi-Yau space to take for the extra dimensions.

The preceding discussion gives some idea of how string theory may one day explain the properties of the matter particles recorded in Table 1.1. String theorists believe that a similar story will one day also explain the properties of the messenger particles of the fundamental forces, listed in Table 1.2. That is, as strings twist and vibrate while meandering through the extended and curled-up dimensions, a small subset of their vast oscillatory repertoire consists of vibrations with spin equal to 1 or 2. These are the candidate force-carrying string-vibrational states. Regardless of the shape of the Calabi-Yau space, there is always one vibrational pattern that is massless and has spin-2; we identify this pattern as the graviton. The precise list of spin-1 messenger particles—their number, the strength of the force they convey, the gauge symmetries they respect—though, does depend crucially on the precise geometrical form of the curled-up dimensions. And so, once again, we come to realize that string theory provides a framework for explaining the observed messenger-particle content of our universe, that is, for explaining the properties of the fundamental forces, but that without knowing exactly which Calabi-Yau shape the extra dimensions are curled into, we cannot make any definitive predictions or postdictions (beyond Witten's remark regarding the postdiction of gravity).

Why can't we figure out which is the "right" Calabi-Yau shape? Most string theorists blame this on the inadequacy of the theoretical tools currently being used to analyze string theory. As we shall discuss in some detail in Chapter 12, the mathematical framework of string theory is so complicated that physicists have been able to perform only approximate calculations through a formalism known as *perturbation theory*. In this approximation scheme, each possible Calabi-Yau shape appears to be on equal footing with every other; none is fundamentally singled out by the equations. And since the physical consequences of string theory depend sensitively on the precise form of the curled-up dimensions, without the ability to select one Calabi-Yau space from the many, no definitive experimentally testable conclusions can be drawn. A driving force behind present-day research is to develop theoretical methods that transcend the approximate approach in the hope that, among other benefits, we will be led to a unique Calabi-Yau shape for the extra dimensions. We will discuss progress along these lines in Chapter 13.

Exhausting Possibilities

So you might ask: Even though we can't as yet figure out which Calabi-Yau shape string theory selects, does *any* choice yield physical properties that agree with what we observe? In other words, if we were to work out the corresponding physical properties associated with each and every Calabi-Yau shape and collect them in a giant catalog, would we find any that match reality? This is an important question, but, for two main reasons, it is also a hard one to answer fully.

A sensible start is to focus only on those Calabi-Yau shapes that yield three families. This cuts down the list of viable choices considerably, although many still remain. In fact, notice that we can deform a multihandled doughnut from one shape to a slew of others—an infinite variety, in fact—without changing the number of holes it contains. In Figure 9.2 we illustrate one such deformation of the bottom shape from Figure 9.1. In much the same way, we can start with a three-holed Calabi-Yau space and smoothly deform its shape without changing the number of holes, again through what amounts to an infinite sequence of shapes. (When we mentioned earlier that there were tens of thousands of Calabi-Yau shapes, we were already grouping together all those shapes that can be changed into one another by such smooth deformations, and we were counting

Figure 9.2 The shape of a multihandled doughnut can be deformed in many ways, one of which is illustrated here, without changing the number of holes it contains.

the whole group as one Calabi-Yau space.) The problem is that the detailed physical properties of string vibrations, their masses and their response to forces, *are* very much affected by such detailed changes in shape, but once again, we have no means of selecting one possibility over any other. And no matter how many graduate students physics professors might set to work, it's just not possible to figure out the physics corresponding to an infinite list of different shapes.

This realization has led string theorists to examine the physics resulting from a sample of possible Calabi-Yau shapes. Even here, however, life is not completely smooth sailing. The approximate equations that string theorists currently use are not powerful enough to work out the resulting physics fully for any given choice of Calabi-Yau shape. They can take us a long way toward understanding, in the sense of a ballpark estimate, the properties of the string vibrations that we hope will align with the particles we observe. But precise and definitive physical conclusions, such as the mass of the electron or the strength of the weak force, require equations that are far more exact than the present approximate framework. Recall from Chapter 6—and the *Price is Right* example—that the "natural" energy scale of string theory is the Planck energy, and it is only through extremely delicate cancellations that string theory yields vibrational patterns with masses in the vicinity of those of the known matter and force particles. Delicate cancellations require precise calculations because even small errors have a profound impact on accuracy. As we will discuss in Chapter 12, during the mid-1990s physicists have made significant progress toward transcending the present approximate equations, although there is still far to go.

So where do we stand? Well, even with the stumbling block of having no fundamental criteria by which to select one Calabi-Yau shape over any other, as well as not having all the theoretical tools necessary to extract the observable consequences of such a choice fully, we can still ask whether *any* of the choices in the Calabi-Yau catalog gives rise to a world that is in even rough agreement with observation. The answer to this question is quite encouraging. Although most of the entries in the Calabi-Yau catalog yield observable consequences significantly different from our world (different numbers of particle families, different number and types of fundamental forces, among other substantial deviations), a few entries in the catalog yield physics that *is* qualitatively close to what we actually observe. That is, there *are* examples of Calabi-Yau spaces that, when chosen for the curled-up dimensions required by string theory, give rise to string vibrations that are closely akin to the particles of the standard model. And, of prime importance, string theory successfully stitches the gravitational force into this quantum-mechanical framework.

With our present level of understanding, this situation is the best we could have hoped for. If many of the Calabi-Yau shapes were in rough agreement with experiment, the link between a specific choice and the physics we observe would be less compelling. Many choices could fit the bill and hence none would appear to be singled out, even from an experimental perspective. On the other hand, if none of the Calabi-Yau shapes came even remotely close to yielding observed physical properties, it would seem that string theory, although a beautiful theoretical framework, could have no relevance for our universe. Finding a small number of Calabi-Yau shapes that, with our present, fairly coarse ability to determine detailed physical implications, appear to be well within the ballpark of acceptability is an extremely encouraging outcome.

Explaining the elementary matter and force particle properties would be among the greatest—if not *the* greatest—of scientific achievements. Nevertheless, you might ask whether there are any string theoretic *pre*dictions—as opposed to *post*dictions—that experimental physicists could attempt to confirm, either now or in the foreseeable future. There are.

Superparticles

The theoretical hurdles currently preventing us from extracting detailed string predictions force us to search for *generic*, rather than specific, aspects of a universe consisting of strings. Generic in this context refers to characteristics that are so fundamental to string theory that they are fairly insensitive to, if not completely independent of, those detailed properties of the theory that are now beyond our theoretical purview. Such characteristics can be discussed with confidence, even with an incomplete understanding of the full theory. In subsequent chapters we shall return to other examples, but for now we focus on one: supersymmetry.

As we have discussed, a fundamental property of string theory is that it is highly symmetric, incorporating not only intuitive symmetry principles but respecting, as well, the maximal mathematical extension of these principles, supersymmetry. This means, as discussed in Chapter 7, that patterns of string vibrations come in pairs—superpartner pairs—differing from each other by a half unit of spin. If string theory is right, then some of the string vibrations will correspond to the known elementary particles. And due to the supersymmetric pairing, string theory makes the *prediction* that each such known particle will have a superpartner. We can determine the force charges that each of these superpartner particles should carry, but we do not currently have the ability to predict their masses. Even so, the prediction that superpartners *exist* is a generic feature of string theory; it is a property of string theory that is true, independent of those aspects of the theory we haven't yet figured out.

No superpartners of the known elementary particles have ever been observed. This might mean that they do not exist and that string theory is wrong. But many particle physicists feel that it means that the superpartners are very heavy and are thus beyond our current capacity to observe experimentally. Physicists are now constructing a mammoth accelerator in Geneva, Switzerland, called the Large Hadron Collider. Hopes run high that this machine will be powerful enough to find the superpartner particles. The accelerator should be ready for operation before 2010, and shortly thereafter supersymmetry may be confirmed experimentally. As Schwarz has said, "Supersymmetry ought to be discovered before too long. And when that happens, it's going to be dramatic."[17]

You should bear in mind two things, though. Even if superpartner particles are found, this fact alone will not establish that string theory is correct. As we have seen, although supersymmetry was discovered by studying string theory, it has also been successfully incorporated into point-particle theories and is therefore not unique to its stringy origins. Conversely, even if superpartner particles are not found by the Large Hadron Collider, this fact alone will not rule out string theory, since it might be that the superpartners are so heavy that they are beyond the reach of this machine as well.

Having said this, if in fact the superpartner particles are found, it will most definitely be strong and exciting circumstantial evidence for string theory.

Fractionally Charged Particles

Another experimental signature of string theory, having to do with electric charge, is somewhat less generic than superpartner particles but equally dramatic. The elementary particles of the standard model have a very limited assortment of electric charges: The quarks and antiquarks have electric charges of one-third or two-thirds, and their negatives, while the other particles have electric charges of zero, one, or negative one. Combinations of these particles account for all known matter in the universe. In string theory, however, it is possible for there to be resonant vibrational patterns corresponding to particles of significantly different electric charges. For instance, the electric charge of a particle can take on exotic fractional values such as $\frac{1}{5}$, $\frac{1}{11}$, $\frac{1}{13}$, or $\frac{1}{53}$, among a variety of other possibilities. These unusual charges can arise if the curled-up dimensions have a certain geometrical property: Holes with the peculiar property that strings encircling them can disentangle themselves only by wrapping around a specified number of times.[18] The details are not particularly important, but it turns out that the number of windings required to get disentangled manifests itself in the allowed patterns of vibration by determining the denominator of the fractional charges.

Some Calabi-Yau shapes have this geometrical property while others do not, and for this reason the possibility of unusual electric-charge fractions is not as generic as the existence of superpartner particles. On the other hand, whereas the prediction of superpartners is not a unique property of string theory, decades of experience have shown that there is no compelling reason for such exotic electric-charge fractions to exist in *any* point-particle theory. They can be forced into a point-particle theory, but doing so would be as natural as the proverbial bull in a china shop. Their possible emergence from simple geometrical properties that the extra dimensions can have makes these unusual electric charges a natural experimental signature for string theory.

As with the situation with superpartners, no such exotically charged particles have ever been observed, and our understanding of string theory does not allow for a definitive prediction of their masses should the extra dimensions have the correct properties to generate them. One explanation for not seeing them, again, is that if they do exist, their masses must be beyond our present technological means—in fact, it is likely that their masses would be on the order of the Planck mass. But should a future experiment come across such exotic electric charges, it would constitute very strong evidence for string theory.

Some Longer Shots

There are yet other ways in which evidence for string theory might be found. For example, Witten has pointed out the long-shot possibility that astronomers might one day see a direct signature of string theory in the data they collect from observing the heavens. As encountered in Chapter 6, the size of a string is typically the Planck length, but strings that are more energetic can grow substantially larger. The energy of the big bang, in fact, would have been high enough to produce a few macroscopically large strings that, through cosmic expansion, might have grown to astronomical scales. We can imagine that now or sometime in the future, a string of this sort might sweep across the night sky, leaving an unmistakable and measurable imprint on data collected by astronomers (such as a small shift in the cosmic microwave background temperature; see Chapter 14). As Witten says, "Although somewhat fanciful, this is my favorite scenario for confirming string theory as nothing would settle the issue quite as dramatically as seeing a string in a telescope."[19]

Closer to earth, there are other possible experimental signatures of string theory that have been put forward. Here are five examples. First, in Table 1.1 we noted that we do not know if neutrinos are just very light, or if in fact they are exactly massless. According to the standard model, they are massless, but not for any particularly deep reason. A challenge to string theory is to provide a compelling explanation of present and future neutrino data, especially if experiments ultimately show that neutrinos do have a tiny but nonzero mass. Second, there are certain hypothetical processes that are forbidden by the standard model, but that may be allowed by string theory. Among these are the possible disintegration of the proton (don't worry, such disintegration, if true, would happen very slowly) and the possible transmutations and decays of various combinations of quarks, in violation of certain long-established properties of point-particle quantum field theory.[20] These kinds of processes are especially interesting because their absence from conventional theory makes them sensitive signals of physics that cannot be accounted for without invoking new theoretical principles. If observed, any one of these processes would provide fertile ground for string theory to offer an explanation. Third, for certain choices of the Calabi-Yau shape there are particular patterns of string vibration that can effectively contribute new, tiny, long-range force fields. Should the effects of any such new forces be discovered, they might well reflect some of the new physics of string theory. Fourth, as we note in the next chapter, astronomers have collected evidence that our galaxy and possibly the whole of the universe is immersed in a bath of *dark matter*, the identity of which has yet to be determined. Through its many possible patterns of resonant vibration, string theory suggests a number of candidates for the dark matter; the verdict on these candidates must await future experimental results establishing the detailed properties of the dark matter.

And finally, a fifth possible means of connecting string theory to observations involves the cosmological constant—remember, as discussed in Chapter 3, this is the modification Einstein temporarily imposed on his original equations of general relativity to ensure a static universe. Although the subsequent discovery that the universe is expanding led Einstein to retract the modification, physicists have since realized that there is no explanation for *why* the cosmological constant should be zero. In fact, the cosmological constant can be interpreted as a kind of overall energy stored in the vacuum of space, and hence its value should be theoretically calculable and experimentally measurable. But, to date, such calculations and measurements lead to a colossal mismatch: Observations show that the cosmological constant is either zero (as Einstein ultimately suggested) or quite small; calculations indicate that quantum-mechanical fluctuations in the vacuum of empty space tend to *generate* a nonzero cosmological constant whose value is some 120 orders of magnitude (a 1 followed by 120 zeros) larger than experiment allows! This presents a wonderful challenge and opportunity for string theorists: Can calculations in string theory improve on this mismatch and explain why the cosmological constant is zero, or if experiments do ultimately establish that its value is small but nonzero, can string theory provide an explanation? Should string theorists be able to rise to this challenge—as yet they have not—it would provide a compelling piece of evidence in support of the theory.

An Appraisal

The history of physics is filled with ideas that when first presented seemed completely untestable but, through various unforeseen developments, were ultimately brought within the realm of experimental verifiability. The notion that matter is made of atoms, Pauli's hypothesis that there are ghostly neutrino particles, and the possibility that the heavens are dotted with neutron stars and black holes are three prominent ideas of precisely this sort—ideas that we now embrace fully but that, at their inception, seemed more like musings of science fiction than aspects of science fact.

The motivation for introducing string theory is at least as compelling as any of these three ideas—in fact, string theory has been hailed as the most important and exciting development in theoretical physics since the discovery of quantum mechanics. This comparison is particularly apt because the history of quantum mechanics teaches us that revolutions in physics can easily take many decades to reach maturity. And compared to today's string theorists, the physicists working out quantum mechanics had a great advantage: Quantum mechanics, even when only partially formulated, could make direct contact with experimental results. Even so, it took close to 30 years for the logical structure of quantum mechanics to be worked out, and about another 20 years to incorporate special relativity fully into the theory. We are now incorporating general relativity, a far more challenging task, and, moreover, one that makes contact with experiment much more difficult. Unlike those who worked out quantum theory, today's string theorists do not have the shining light of nature—through detailed experimental results—to guide them from one step to the next.

This means that it's conceivable that one or more generations of physicists will devote their lives to the investigation and development of string theory without getting a shred of experimental feedback. The substantial number of physicists the world over who are vigorously pursuing string theory know that they are taking a risk: that a lifetime of effort might result in an inconclusive outcome. Undoubtedly, significant theoretical progress will continue, but will it be sufficient to overcome present hurdles and yield definitive, experimentally testable predictions? Will the indirect tests we have discussed above result in a true smoking gun for string theory? These questions are of central concern to all string theorists, but they are also questions about which nothing can really be said. Only the passage of time will reveal the answers. The beautiful simplicity of string theory, the way in which it tames the conflict between gravity and quantum mechanics, its ability to unify all of nature's ingredients, and its potential of limitless predictive power all serve to provide rich inspiration that makes the risk worth taking.

These lofty considerations have been continually reinforced by the ability of string theory to uncover remarkably new physical characteristics of a string-based universe—characteristics that reveal a subtle and deep coherence in the workings of nature. In the language introduced above, many of these are generic features that, regardless of currently unknown details, will be basic properties of a universe built of strings. Of these, the most astonishing have had a profound effect on our ever evolving understanding of space and time.

Part IV

String Theory and the Fabric of Spacetime

Chapter 10

Quantum Geometry

In the course of about a decade, Einstein singlehandedly overthrew the centuries-old Newtonian framework and gave the world a radically new and demonstrably deeper understanding of gravity. It does not take much to get experts and nonexperts alike to gush over the sheer brilliance and monumental originality of Einstein's accomplishment in fashioning general relativity. Nevertheless, we should not lose sight of the favorable historical circumstances that strongly contributed to Einstein's success. Foremost among these are the nineteenth-century mathematical insights of Georg Bernhard Riemann that firmly established the geometrical apparatus for describing curved spaces of arbitrary dimension. In his famous 1854 inaugural lecture at the University of Göttingen, Riemann broke the chains of flat-space Euclidean thought and paved the way for a democratic mathematical treatment of geometry on all varieties of curved surfaces. It is Riemann's insights that provide the mathematics for quantitatively analyzing warped spaces such as those illustrated in Figures 3.4 and 3.6. Einstein's genius lay in recognizing that this body of mathematics was tailor-made for implementing his new view of the gravitational force. He boldly declared that the mathematics of Riemann's geometry aligns perfectly with the physics of gravity.

But now, almost a century after Einstein's tour-de-force, string theory gives us a quantum-mechanical description of gravity that, by necessity, modifies general relativity when the distances involved become as short as the Planck length. Since Riemannian geometry is the mathematical core of general relativity, this means that it too must be modified in order to reflect faithfully the new short-distance physics of string theory. Whereas general relativity asserts that the curved properties of the universe are described by Riemannian geometry, string theory asserts that this is true only if we examine the fabric of the universe on large enough scales. On scales as small as the Planck length a new kind of geometry must emerge, one that aligns with the new physics of string theory. This new geometrical framework is called *quantum geometry*.

Unlike the case of Riemannian geometry, there is no ready-made geometrical opus sitting on some mathematician's shelf that string theorists can adopt and put in the service of quantum geometry. Instead, physicists and mathematicians are now vigorously studying string theory and, little by little, piecing together a new branch of physics and mathematics. Although the full story has yet to be written, these investigations have already uncovered many new geometrical properties of spacetime entailed by string theory—properties that would almost certainly have thrilled even Einstein.

The Heart of Riemannian Geometry

If you jump on a trampoline, the weight of your body causes it to warp by stretching its elastic fibers. This stretching is most severe right under your body and becomes less noticeable toward the trampoline's edge. You can see this clearly if a familiar image such as the Mona Lisa is painted on the trampoline. When the trampoline is not supporting any weight, the Mona Lisa looks normal. But when you stand on the trampoline, the image of the Mona Lisa becomes distorted, especially the part directly under your body, as illustrated in Figure 10.1.

This example cuts to the heart of Riemann's mathematical framework for describing warped shapes. Riemann, building on earlier insights of the mathematicians Carl Friedrich Gauss, Nikolai Lobachevsky, Janos Bolyai, and others, showed that a careful analysis of the *distances* between all locations on or in an object provides a means of quantifying the extent of its curvature. Roughly speaking, the greater the (nonuniform) stretching—the greater the deviation from the distance relations on a flat shape—the greater the curvature of the object. For example, the trampoline is most

Figure 10.1 When standing on the Mona Lisa trampoline, the image becomes most distorted under your weight.

significantly stretched right under your body and therefore the distance relations between points in this area are most severely distorted. This region of the trampoline, therefore, has the largest amount of curvature, in line with what you expect, since this is where the Mona Lisa suffers the greatest distortion, yielding the hint of a grimace at the corner of her customary enigmatic smile.

Einstein adopted Riemann's mathematical discoveries by giving them a precise physical interpretation. He showed, as we discussed in Chapter 3, that the curvature of spacetime embodies the gravitational force. But let's now think about this interpretation a little more closely. Mathematically, the curvature of spacetime—like the curvature of the trampoline—reflects the distorted distance relations between its *points*. Physically, the gravitational force felt by an object is a direct reflection of this distortion. In fact, by making the object smaller and smaller, the physics and the mathematics align ever more precisely as we get closer and closer to physically realizing the abstract mathematical concept of a point. But string theory limits how precisely Riemann's geometrical formalism can be realized by the physics of gravity, because there is a limit to how small we can make any object. Once you get down to strings, you can't go any further. The traditional notion of a point particle does not exist in string theory—an essential element in its ability to give us a quantum theory of gravity. This concretely shows us that Riemann's geometrical framework, which relies fundamentally upon distances between points, is modified on ultra-microscopic scales by string theory.

This observation has a very small effect on ordinary macroscopic applications of general relativity. In cosmological studies, for example, physicists routinely model whole galaxies as if they are points, since their size, in relation to the whole of the universe, is extremely tiny. For this reason, implementing Riemann's geometrical framework in this crude manner proves to be a very accurate approximation, as evidenced by the success of general relativity in a cosmological context. But in the ultramicroscopic realm, the extended nature of the string ensures that Riemann's geometry simply will not be the right mathematical formalism. Instead, as we will now see, it must be replaced by the quantum geometry of string theory, leading to dramatically new and unexpected properties.

A Cosmological Playground

According to the big bang model of cosmology, the whole of the universe violently emerged from a singular cosmic explosion, some 15 or so billion years ago. Today, as originally discovered by Hubble, we can see that the "debris" from this explosion, in the form of many billions of galaxies, is still streaming outward. The universe is expanding. We do not know whether this cosmic growth will continue forever or if there will come a time when the expansion slows to a halt and then reverses itself, leading to a cosmic implosion. Astronomers and astrophysicists are trying to settle this question experimentally, since the answer turns on something that in principle can be measured: the average density of matter in the universe.

If the average matter density exceeds a so-called *critical density* of about a hundredth of a billionth of a billionth of a billionth (10^{-29}) of a gram per cubic centimeter—about five hydrogen atoms for every cubic meter of the universe—then a large enough gravitational force will permeate the cosmos to halt and reverse the expansion. If the average matter density is less than the critical value, the gravitational attraction will be too weak to stop the expansion, which will continue forever. (Based upon your own observations of the world, you might think that the average mass density of the universe greatly exceeds the critical value. But bear in mind that matter—like money—tends to clump. Using the average mass density of the earth, or the solar system, or even the Milky Way galaxy as an indicator for that of the whole universe would be like using Bill Gates's net worth as an indicator of the average earthling's finances. Just as there are many people whose net worth pales in comparison to that of Bill Gates, thereby diminishing the average enormously, there is a lot of nearly empty space between the galaxies that drastically lowers the overall average matter density.)

By carefully studying the distribution of galaxies throughout space, astronomers can get a pretty good handle on the average amount of visible matter in the universe. This turns out to be significantly less than the critical value. But there is strong evidence, of both theoretical and experimental origin, that the universe is permeated with dark matter. This is matter that does not participate in the processes of nuclear fusion that powers stars and hence does not give off light; it is therefore invisible to the astronomer's telescope. No one has figured out the identity of the dark matter, let alone the precise amount that exists. The fate of our presently expanding universe, therefore, is as yet unclear.

Just for argument's sake, let's assume that the mass density does exceed the critical value and that someday in the distant future the expansion will stop and the universe will begin to collapse upon itself. All galaxies will start to approach one another slowly, and then as time goes by, their speed of approach will increase until they rush together at blinding speed. You need to picture the whole of the universe squeezing together into an ever shrinking cosmic mass. As in Chapter 3, from a maximum size of many billions of light-years, the universe will shrink to millions of light-years, every moment gaining speed as *everything* is crushed together to the size of a single galaxy, and then to the size of a single star, a planet, and down to the size of an orange, a pea, a grain of sand, and further, according to general relativity, to the size of a molecule, an atom, and in a final inexorable cosmic crunch to *no size at all*. According to conventional theory, the universe began with a bang from an initial state of zero size, and if it has enough mass, it will end with a crunch to a similar state of ultimate cosmic compression.

But when the distance scales involved are around the Planck length or less, quantum mechanics invalidates the equations of general relativity, as we are now well aware. We must instead make use of string theory. And so, whereas Einstein's general relativity allows the geometrical form of the universe to get arbitrarily small—in exactly the same way that the mathematics of Riemannian geometry allows an abstract shape to take on as small a size as the intellect can imagine—we are led to ask how string theory modifies the picture. As we shall now see, there is evidence that string theory once again sets a lower limit to physically accessible distance scales and, in a remarkably novel way, proclaims that the universe cannot be squeezed to a size shorter than the Planck length in any of its spatial dimensions.

Based on the familiarity you now have with string theory, you might be tempted to hazard a guess as to how this comes about. After all, you might argue that no matter how many points you pile up on top of each other—point particles that is—their combined volume is still zero. By contrast, if these particles are really strings, collapsed together in completely random orientations, they will fill out a nonzero-sized blob, roughly like a Planck-sized ball of entangled rubber bands. If you made this argument, you would be on the right track, but you would be missing significant, subtle features that string theory elegantly employs to suggest a minimum size to the universe. These features serve to emphasize, in a concrete manner, the new stringy physics that comes into play and its resultant impact on the geometry of spacetime.

To explain these important aspects, let's first call upon an example that pares away extraneous details without sacrificing the new physics. Instead of considering all ten of the spacetime dimensions of string theory—or even the four extended spacetime dimensions we are familiar with—let's go back to the Garden-hose universe. We originally introduced this two-spatial-dimension universe in Chapter 8 in a prestring context to explain aspects of Kaluza's and Klein's insights in the 1920s. Let's now use it as a "cosmological playground" to explore the properties of string theory in a simple setting; we will shortly use the insights we gain to better understand all of the spatial dimensions string theory requires. Toward this end, we imagine that the circular dimension of the Garden-hose universe starts out nice and plump but then shrinks to shorter and shorter size, approaching the form of Lineland—a simplified, partial version of the big crunch.

The question we seek to answer is whether the geometrical and physical properties of this cosmic collapse have features that markedly differ between a universe based on strings and one based on point particles.

The Essential New Feature

We do not have to search far to find the essential new string physics. A point particle moving in this two-dimensional universe can execute the kinds of motion illustrated in Figure 10.2: It can move along the extended dimension of the Garden-hose, it can move along the curled-up part of the Garden-hose, or any combination of the two. A loop of string can undergo similar motion, with one difference being that it oscillates as it moves around on the surface, as shown in Figure 10.3(a). This is a distinction we have already discussed in some detail: The oscillations of the string imbue it with characteristics such as mass and force charges. Although a crucial aspect of string theory, this is not our present focus, since we already understand its physical implications.

Instead, our present interest is in another difference between point-particle and string motion, a difference directly dependent on the *shape* of the space through which the string is moving. Since the string is an extended object, there is another possible configuration beyond those already mentioned: It can *wrap around*—lasso, so to speak—the circular part of the Garden-hose universe, as shown in Figure 10.3(b).[1] The string will continue to slide around and oscillate, but it will do so in this extended configuration. In fact, the string can wrap around the circular part of the space any number of times, as also shown in Figure 10.3(b), and again will execute oscillatory motion as it slides around. When a string is in such a wrapped configuration, we say that it is in a *winding mode* of motion. Clearly, being in a winding mode is a possibility inherent to strings. There is no point-particle counterpart. We now seek to understand the implications of this qualitatively new kind of string motion on the string itself as well as on the geometrical properties of the dimension it wraps.

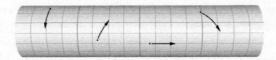

Figure 10.2 Point particles moving on a cylinder.

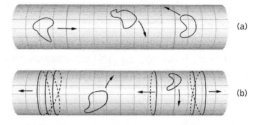

(a)

(b)

Figure 10.3 Strings can move on a cylinder in two different ways—in "unwrapped" or "wrapped" configurations.

The Physics of Wound Strings

Throughout our previous discussion of string motion, we have focused on unwound strings. Strings that wrap around a circular component of space share almost all of the same properties as the strings we have studied. Their oscillations, just as those of their unwound counterparts, contribute strongly to their observed properties. The essential difference is that a wrapped string has a *minimum* mass, determined by the *size* of the circular dimension and the number of times it wraps around. The string's oscillatory motion determines a contribution in excess of this minimum.

It is not difficult to understand the origin of this minimum mass. A wound string has a minimum length determined by the circumference of the circular dimension and the number of times the string encircles it. The minimum length of a string determines its minimum mass: The longer this length, the greater the mass, since there is more of it. Since the circumference of a circle is proportional to its radius, the minimum winding-mode masses are proportional to the radius of the circle being wrapped. By using Einstein's $E = mc^2$ relating mass to energy, we can also say that the energy bound in a wound string is proportional to the radius of the circular dimension. (Unwrapped strings also have a tiny minimum length since if they didn't, we would be back in the realm of point particles. The same reasoning might lead to the conclusion that even unwrapped strings have a minuscule yet nonzero minimum mass. In a sense this is true, but the quantum-mechanical effects encountered in Chapter 6—remember *The Price Is Right,* again—are able to exactly cancel this contribution to the mass. This is how, we recall, unwrapped strings can yield the zero-mass photon, graviton, and the other massless or near-massless particles, for example. Wrapped strings are different in this regard.)

How does the existence of wrapped string configurations affect the *geometrical* properties of the dimension around which the strings wind? The answer, first recognized in 1984 by the Japanese physicists Keiji Kikkawa and Masami Yamasaki, is bizarre and remarkable.

Let's consider the last cataclysmic stages of our variant on the big crunch in the Garden-hose universe. As the radius of the circular dimension shrinks to the Planck length and, in the mold of general relativity, continues to shrink to yet smaller lengths, string theory insists upon a radical reinterpretation of what actually happens. String theory claims that *all* physical processes in the Garden-hose universe in which the radius of the circular dimension is shorter than the Planck length and is decreasing are absolutely identical to physical processes in which the circular dimension is longer than the Planck length and increasing! This means that as the circular dimension tries to collapse through the Planck length and head toward ever smaller size, its attempts are made futile by string theory, which turns the tables on geometry. String theory shows that this evolution can be rephrased—exactly reinterpreted—as the circular dimension shrinking down to the Planck length and then proceeding to expand. String theory rewrites the laws of short-distance geometry so that what previously appeared to be complete cosmic collapse is now seen to be a cosmic *bounce*. The circular dimension can shrink to the Planck-length. But because of the winding modes, attempts to shrink further actually result in expansion. Let's see why.

The Spectrum of String States*

The new possibility of wound-string configurations implies that the energy of a string in the Garden-hose universe comes from *two* sources: vibrational motion and winding energy. From the legacy of Kaluza and Klein,

---

*Some of the ideas in this and the next few sections are rather subtle, so don't be put off if you have trouble following every link in the explanatory chain—especially in a single reading.

each depends on the geometry of the hose, that is, on the radius of its curled-up circular component, but with a distinctly stringy twist, since point particles cannot wrap around dimensions. Our first task, then, will be to determine precisely how the winding and vibrational contributions to the energy of a string depend on the size of the circular dimension. For this purpose, it proves convenient to separate the vibrational motion of strings into two categories: *uniform* and *ordinary* vibrations. Ordinary vibrations refer to the usual oscillations we have discussed repeatedly, such as those illustrated in Figure 6.2; uniform vibrations refer to even simpler motion: the overall motion of string as it slides from one position to another without changing its shape. All string motion is a combination of sliding and oscillating—of uniform and ordinary vibrations—but for the present discussion it is easier to separate them in this manner. In fact, the ordinary vibrations will not play a central part in our reasoning, and we will therefore include their effects only after we have finished giving the gist of the argument.

Here are the two essential observations. First, uniform vibrational excitations of a string have energies that are *inversely* proportional to the radius of the circular dimension. This is a direct consequence of the quantum-mechanical uncertainty principle: A smaller radius more strictly confines a string and therefore, through quantum-mechanical claustrophobia, increases the amount of energy in its motion. So, as the radius of the circular dimension decreases, the energy of motion of the string necessarily increases—the hallmark feature of an inverse proportionality. Second, as found in the preceding section, the winding mode energies are *directly*—not inversely—proportional to the radius. Remember, this is because the minimum length of wound strings, and hence their minimum energy, is proportional to the radius. These two observations establish that large values of the radius imply large winding energies and small vibration energies, whereas small values of the radius imply small winding energies and large vibration energies.

This leads us to the key fact: For any large circular radius of the Garden-hose universe, there is a corresponding small circular radius for which the winding energies of strings in the former universe equal the vibration energies of strings in the latter, and vibration energies of strings in the former equal winding energies of strings in the latter. As physical properties are sensitive to the *total* energy of a string configuration—and not to how the energy is divided between vibration and winding contributions—there is *no physical distinction* between these *geometrically distinct* forms for the Garden-hose universe. And so, strangely enough, string theory claims that there is no difference whatsoever between a "fat" Garden-hose universe and a "thin" one.

It's a cosmic hedging of bets, somewhat akin to what you, as a smart investor, should do if faced with the following puzzle. Imagine you learn that the fate of two stocks trading on Wall Street—say, a company making fitness machines and a company making heart-bypass valves—are inextricably connected. They each closed trading today valued at one dollar per share, and you are told by a reliable source that if one company's stock goes up the other's will go down, and vice versa. Moreover, your source—who is completely trustworthy (but whose guidance might be crossing over legal boundaries)—tells you that the next day's closing prices of these two companies are absolutely certain to be inversely related to one another. That is, if one stock closes at $2 per share, the other will close at $1/2 (50 cents) per share; if one stock closes at $10 per share, the other will close at $1/10 (10 cents) per share, and so on. But the one thing your source can't tell you is which stock will close high and which will close low. What do you do?

Well, you immediately invest all of your money in the stock market, equally divided between the shares of these two companies. As you can easily check by working out a few examples, no matter what happens on the next day, your investment cannot lose value. At worse it can remain the same (if both companies again close at $1), but any movement of share prices—consistent with your insider information—will increase your holdings. For instance, if the fitness company closes at $4 and the heart-valve company closes at $1/4 (25 cents), their combined value is $4.25 (for each pair of shares), compared with $2 the previous day. Furthermore, from the perspective of net worth, it does not matter one bit whether the fitness company closes high and the heart-valve company low, or vice versa. If you care only about the total amount of money, these two distinct circumstances are financially indistinguishable.

The situation in string theory is analogous in that the energy in string configurations comes from two sources—vibrations and windings—whose contributions to the total energy of a string are generally different. But, as we shall see in more detail below, certain pairs of distinct geometrical circumstances—leading to high-winding-energy/low-vibration-energy or low-winding-energy/high-vibration-energy—are *physically* indistinguishable. And, unlike the financial analogy for which considerations beyond total wealth can distinguish between the two types of stock holdings, there is absolutely no physical distinction between the two string scenarios.

Actually, we shall see that to make the analogy with string theory tighter, we should consider what would happen if you did not divide your money equally between the two companies in your initial investment, but bought, say, 1,000 shares of the fitness company and 3,000 shares of the heart-valve company. Now the total value of your holdings does depend on which company closes high and which closes low. For instance, if the stocks close at $10 (fitness) and 10 cents (heart-valve), your initial investment of $4,000 will now be worth $10,300. If the reverse happens—the stocks close at 10 cents (fitness) and $10 (heart-valve)—your holdings will be worth $30,100—significantly more.

Nevertheless, the inverse relationship between the closing stock prices does ensure the following. If a friend of yours invests exactly "opposite" to you—3,000 shares of the fitness company and 1,000 shares of the heart-valve company—then the value of her holdings will be $10,300 if stocks close valves-high/fitness-low (the same as your holdings in the fitness-high/valves-low closing) and $30,100 if they close with fitness-high/valves-low (again, the same as your holdings in the reciprocal situation). That is, from the point of view of total stock value, interchanging which stock closes high and which closes low is exactly compensated by interchanging the number of shares you own of each company.

Hold this last observation in mind as we now return to string theory and think about the possible string energies in a specific example. Imagine that the radius of the circular Garden-hose dimension is, say, ten times the Planck length. We will write this as $R = 10$. A string can wrap around this circular dimension one time, two times, three times, and so forth. The number of times a string wraps around the circular dimension is called its *winding number*. The energy from winding, being determined by the length of wound string, is proportional to the *product* of the radius and the winding number. Additionally, for any amount of winding, the string can undergo vibrational motion. As the uniform vibrations we are currently focusing on have energies that are inversely dependent on the radius, they are proportional to whole-number multiples of the *reciprocal* of the radius—1/R—which in this case is one-tenth of the Planck length. We call this whole number multiple the *vibration number*.[2]

As you can see, this situation is very similar to what we encountered on Wall Street, with the winding and vibration numbers being direct analogs of the shares held in the two companies, while R and 1/R are the analogs of the closing prices per share in each. Now, just as you can easily calculate the total value of your investment from the number of shares held in each company and the closing prices, we can calculate the total energy carried by a string in terms of its vibration number, its winding number, and the radius. In Table 10.1 we give a partial list of these total energies for various string configurations, which we specify by their winding and vibration numbers, in a Garden-hose universe with radius R = 10.

A complete table would be infinitely long, since the winding and vibration numbers can take on arbitrary whole-number values, but this representative piece of the table is adequate for our discussion. We see from the table and our remarks that we are in a high-winding-energy/low-vibration-energy situation: Winding energies come in multiples of 10, while vibrational energies come in multiples of the smaller number 1/10.

Now imagine that the radius of the circular dimension shrinks, say, from 10 to 9.2 to 7.1 and on down to 3.4, 2.2, 1.1, .7, all the way to .1 (1/10), where, for our present discussion, it stops. In this geometrically distinct form of the Garden-hose universe we can compile an analogous table of string energies: Winding energies are now multiples of 1/10 while vibration energies are multiples of its reciprocal, 10. The results are shown in Table 10.2.

At first glance, the two tables might appear to be different. But closer inspection reveals that although arranged in a different order, the "total energy" columns of both tables have *identical* entries. To find the corresponding entry in Table 10.2 for a chosen entry in Table 10.1, one must simply interchange the vibration and winding numbers. That is, vibration and winding contributions play complementary roles when the radius of the circular dimension changes from 10 to 1/10. And so, as far as total string energies go, there is *no distinction* between these different sizes for the circular dimension. Just as the interchange of fitness-high/valves-low

Vibration number	Winding number	Total energy
1	1	$1/10 + 10 = 10.1$
1	2	$1/10 + 20 = 20.1$
1	3	$1/10 + 30 = 30.1$
1	4	$1/10 + 40 = 40.1$
2	1	$2/10 + 10 = 10.2$
2	2	$2/10 + 20 = 20.2$
2	3	$2/10 + 30 = 30.2$
2	4	$2/10 + 40 = 40.2$
3	1	$3/10 + 10 = 10.3$
3	2	$3/10 + 20 = 20.3$
3	3	$3/10 + 30 = 30.3$
3	4	$3/10 + 40 = 40.3$
4	1	$4/10 + 10 = 10.4$
4	2	$4/10 + 20 = 20.4$
4	3	$4/10 + 30 = 30.4$
4	4	$4/10 + 40 = 40.4$

Table 10.1 Sample vibration and winding configurations of a string moving in a universe shown in Figure 10.3, with radius $R = 10$. The vibration energies contribute in multiples of 1/10 and the winding energies contribute in multiples of 10, yielding the total energies listed. The energy unit is the Planck energy, so for example, 10.1 in the last column means 10.1 times the Planck energy.

with valves-high/fitness-low is exactly compensated by an interchange of the number of shares held in each company, interchange of radius 10 and radius 1/10 is exactly compensated by the interchange of vibration and

Vibration number	Winding number	Total energy
1	1	$10 + 1/10 = 10.1$
1	2	$10 + 2/10 = 10.2$
1	3	$10 + 3/10 = 10.3$
1	4	$10 + 4/10 = 10.4$
2	1	$20 + 1/10 = 20.1$
2	2	$20 + 2/10 = 20.2$
2	3	$20 + 3/10 = 20.3$
2	4	$20 + 4/10 = 20.4$
3	1	$30 + 1/10 = 30.1$
3	2	$30 + 2/10 = 30.2$
3	3	$30 + 3/10 = 30.3$
3	4	$30 + 4/10 = 30.4$
4	1	$40 + 1/10 = 40.1$
4	2	$40 + 2/10 = 40.2$
4	3	$40 + 3/10 = 40.3$
4	4	$40 + 4/10 = 40.4$

Table 10.2 As in Table 10.1, except that the radius is now taken to be 1/10.

winding numbers. Moreover, while for simplicity we have focused on an initial radius of $R = 10$ and its reciprocal 1/10, the conclusions drawn are the same for any choice of the radius and its reciprocal.[3]

Tables 10.1 and 10.2 are incomplete for two reasons. First, as mentioned, we have listed only a few of the infinite possibilities for winding/vibration numbers that a string can assume. This, of course, poses no problem—we could make the tables as long as our patience allows and would find that the relation between them will continue to hold. Second, beyond winding energy, we have so far considered only energy contributions arising from the uniform-vibrational motion of a string. We should now include the ordinary vibrations as well, since these give additional contributions to the string's total energy and also determine the force charges it carries. The important point, however, is that investigations have revealed that these contributions do not depend on the size of the radius. Thus, even if we were to include these more detailed features of string attributes in Tables 10.1 and 10.2, the tables would still correspond exactly, since the ordinary vibrational contributions affect each table identically. We therefore conclude that the masses and the charges of particles in a Garden-hose universe with radius R are completely identical to those in a Garden-hose universe with radius $1/R$. And since these masses and force charges govern fundamental physics, there is no way to distinguish physically these two geometrically distinct universes. Any experiment done in one such universe has a corresponding experiment that can be done in the other, leading to exactly the same results.

A Debate

George and Gracie, after being flattened out into two-dimensional beings, take up residence as physics professors in the Garden-hose universe. After setting up their competing laboratories, each claims to have determined the size of the circular dimension. Surprisingly, although each has a reputation for carrying out research with great precision, their conclusions do not agree. George claims that the circular radius is $R = 10$ times the Planck length, while Gracie claims that the circular radius is $R = 1/10$ times the Planck length.

"Gracie," says George, "based on my string theory calculations, I know that if the circular dimension has radius 10, then I should expect to see strings whose energies are listed in Table 10.1. I have done extensive experiments using the new Planck energy accelerator and they have revealed that this prediction is precisely confirmed. Therefore, with confidence, I claim that the circular dimension has radius $R = 10$." Gracie, in defense of her claims, makes exactly the same remarks except for her conclusion that the list of energies in Table 10.2 is found, confirming that the radius is $R = 1/10$.

In a flash of insight, Gracie shows George that the two tables, although arranged differently, are actually identical. Now George, who, as is well known, reasons a bit more slowly than Gracie, replies, "How can this be? I know that different values for the radius give rise, through basic quantum mechanics and the properties of wound strings, to different possible values for string energies and string charges. If we agree on the latter, then we must agree on the radius."

Gracie, using her newfound insight into string physics replies, "What you say is almost, but not quite, correct. It is *usually* true that two different values for the radius give rise to different allowed energies. However, in the special circumstance when the two values for the radius are inversely related to one another—like 10 and 1/10—then the allowed energies and charges are actually identical. You see, what you would call a winding mode I would call a vibration mode, and what you would call a vibration mode I would call a winding mode. But nature does not care about the language we use. Instead, physics is governed by the properties of the *fundamental ingredients*—the particle masses (energies) and the force charges they carry. And whether the radius is R or $1/R$, the complete list of these properties for the fundamental ingredients in string theory is identical."

In a moment of bold comprehension, George responds, "I think I understand. Although the detailed description you and I might give for strings may differ—whether they are wound around the circular dimension, or the particulars of their vibrational behavior—the complete list of physical characteristics they can attain is the same. Therefore, since the physical properties of the universe depend upon these properties of the basic constituents, there is no distinction, no way to differentiate, between radii that are inversely related to one another." Exactly.

Three Questions

At this point you might say, "Look, if I was a little being in the Garden-hose universe I would simply measure the circumference of the hose with a tape measure and thereby unambiguously determine the radius—

no ifs, ands, or buts. So what is this nonsense about two indistinguishable possibilities with different radii? Furthermore, doesn't string theory do away with sub-Planck distances, so why are we even talking about circular dimensions with radii that are a fraction of the Planck length? And finally, while we are at it, who really cares about the two-dimensional Garden-hose universe—what does all this add up to when we include *all* dimensions?"

Let's begin with the last question, as the answer will force us to come face to face with the first two.

Although our discussion has taken place in the Garden-hose universe, we restricted ourselves to one extended and one curled-up spatial dimension merely for simplicity. If we have three extended spatial dimensions and six circular dimensions—the latter being the simplest of all Calabi-Yau spaces—the conclusion is exactly the same. Each of the circles has a radius that, if interchanged with its reciprocal, yields a physically identical universe.

We can even take this conclusion one giant step further. In our universe, we observe three spatial dimensions, each of which, according to astronomical observations, appears to extend for about 15 billion light-years (a light-year is about 6 trillion miles, so this distance is about 90 billion trillion miles). As mentioned in Chapter 8, nothing tells us what happens after that. We do not know whether they continue on indefinitely or perhaps curve back on themselves in the shape of an enormous circle, beyond the visual sensitivity of state-of-the-art telescopes. If the latter is the case, an astronaut travelling out into space, continuously going in a fixed direction, would ultimately circle around the universe—like Magellan travelling around the earth—and wind up back at the initial starting point.

The familiar extended dimensions, therefore, may very well also be in the shape of circles and hence subject to the R and $1/R$ physical identification of string theory. To put some rough numbers in, if the familiar dimensions are circular then their radii must be about as large as the 15 billion light-years mentioned above, which is about ten trillion trillion trillion trillion trillion ($R=10^{61}$) times the Planck length, and growing as the universe expands. If string theory is right, this is physically identical to the familiar dimensions being circular with incredibly tiny radii of about $1/R=1/10^{61} = 10^{-61}$ times the Planck length! *These are our well-known familiar dimensions in an alternate description provided by string theory.* In fact, in this reciprocal language, these tiny circles are getting ever smaller as time goes by, since as R grows, $1/R$ shrinks. Now we seem to have really gone off the deep end. How can this possibly be true? How can a six-foot tall human being "fit" inside such an unbelievably microscopic universe? How can such a speck of a universe be physically identical to the great expanse we view in the heavens? Furthermore, we are now led forcefully to the second of our initial three questions: String theory was supposed to eliminate the ability to probe sub-Planck distances. But if a circular dimension has radius R whose length is larger than the Planck length, its reciprocal $1/R$ is necessarily a fraction of the Planck length. So what is going on? The answer, which will also address the first of our three questions, highlights an important and subtle aspect of space and distance.

Two Interrelated Notions of Distance in String Theory

Distance is such a basic concept in our understanding of the world that it is easy to underestimate the depth of its subtlety. With the surprising effects that special and general relativity have had on our notions of space and time, and the new features arising from string theory, we are led to be a bit more careful even in our definition of distance. The most meaningful definitions in physics are those that are operational—that is, definitions that provide a means, at least in principle, for measuring whatever is being defined. After all, no matter how abstract a concept is, having an operational definition allows us to boil down its meaning to an experimental procedure for measuring its value.

How can we give an operational definition of the concept of distance? The answer to this question in the context of string theory is rather surprising. In 1988, the physicists Robert Brandenberger of Brown University and Cumrun Vafa of Harvard University pointed out that if the spatial shape of a dimension is circular, there are two different yet related operational definitions of distance in string theory. Each lays out a distinct experimental procedure for measuring distance and is based, roughly speaking, on the simple principle that if a probe travels at a fixed and known speed then we can measure a given distance by determining how long the probe takes to traverse it. The difference between the two procedures is the choice of probe used. The first definition uses strings that are *not* wound around a circular dimension, whereas the second definition uses strings that *are* wound. We see that the extended nature of the fundamental probe is responsible for there being two natural operational definitions of distance in string theory. In a point-particle theory, for which there is no notion of winding, there would be only one such definition.

How do the results of each procedure differ? The answer found by Brandenberger and Vafa is as surprising as it is subtle. The rough idea underlying the result can be understood by appealing to the uncertainty principle. Unwound strings can move around freely and probe the full circumference of the circle, a length proportional to R. By the uncertainty principle, their energies are proportional to $1/R$ (recall from Chapter 6 the inverse relation between the energy of a probe and the distances to which it is sensitive). On the other hand, we have seen that wound strings have minimum energy proportional to R; as probes of distances the uncertainty principle tells us that they are therefore sensitive to the reciprocal of this value, $1/R$. The mathematical embodiment of this idea shows that if each is used to measure the radius of a circular dimension of space, unwound string probes will measure R while wound strings will measure $1/R$, where, as before, we are measuring distances in multiples of the Planck length. The result of each experiment has an equal claim to being the radius of the circle—what we learn from string theory is that using different probes to measure distance can result in different answers. In fact, this property extends to all measurements of lengths and distances, not just to determining the size of a circular dimension. The results obtained by wound and unwound string probes will be inversely related to one another.[4]

If string theory describes our universe, why have we not encountered these two possible notions of distance in any of our day-to-day or scientific endeavors? Any time we talk about distance, we do so in a manner that conforms to our experience of there being one concept of distance without any hint of there being a second notion. Why have we missed the alternative possibility? The answer is that although there is a high degree of symmetry in our discussion, whenever R (and hence $1/R$ as well) differ significantly from the value 1 (meaning, again, 1 times the Planck length), then one of our operational definitions proves extremely difficult to carry out while the other proves extremely easy to carry out. In essence, we have always carried out the easy approach, completely unaware of there being another possibility.

The discrepancy in difficulty between the two approaches is due to the very different masses of the probes used—high-winding-energy/low-vibration-energy, and vice versa—if the radius R (and hence $1/R$ as well) differs significantly from the Planck length (that is, $R = 1$). "High" energy here, for radii that are vastly different from the Planck length, corresponds to incredibly massive probes—billions and billions of times heavier than the proton, for instance—while "low" energy corresponds to probe masses at most a speck above zero. In such circumstances, there is a monumental difference in difficulty between the two approaches, since even producing the heavy-string configurations is an undertaking that, at present, is beyond our technological prowess. In practice, then, only one of the two approaches is technologically feasible—the one involving the lighter of the two types of string configurations. This is the one used implicitly in all of our discussions involving distance encountered to this point. This is the one that informs and hence meshes with our intuition.

Putting issues of practicality aside, in a universe governed by string theory one is free to measure distances using either of the two approaches. When astronomers measure the "size of the universe" they do so by examining photons that have traveled across the cosmos and have happened to enter their telescopes. No pun intended, photons are the *light* string modes in this situation. The result obtained is the 10^{61} times the Planck length quoted earlier. If the three familiar spatial dimensions are in fact circular and string theory is right, astronomers using vastly different (and currently nonexistent) equipment, in principle, should be able to measure the extent of the heavens with heavy wound-string modes and find a result that is the reciprocal of this huge distance. It is in this sense that we can think of the universe as being either huge, as we normally do, or terribly minute. According to the light string modes, the universe is large and expanding; according to the heavy modes it is tiny and contracting. There is no contradiction here; instead, we have two distinct but equally sensible definitions of distance. We are far more familiar with the first definition due to technological limitations, but, nevertheless, each is an equally valid concept.

Now we can answer our earlier question about big humans in a little universe. When we measure the height of a human and find six feet, for instance, we necessarily use the light string modes. To compare their size to that of the universe, we must use the same measuring procedure and, as above, this yields 15 billion light-years for the size of the universe, a result that is much larger than six feet. Asking how such a person can fit into the "tiny" universe as measured by the heavy string modes is asking a meaningless question—it's comparing apples and oranges. Since we now have two concepts of distance—using light or heavy string probes—we must compare measurements made in the same manner.

A Minimum Size

It's been a bit of a trek, but we are now set for the key point. If one does stick to measuring distances "the easy way"—that is, using the lightest of the string modes instead of the heavy ones—the results obtained will *always* be larger than the Planck length. To see this, let's think through the hypothetical big crunch for the three extended dimensions, assuming them to be circular. For argument's sake, let's say that at the beginning of our thought experiment, unwound string modes are the light ones and by using them it is determined that the universe has an enormously large radius and that it is shrinking in time. As it shrinks, these unwound modes get heavier and the winding modes get lighter. When the radius shrinks all the way to the Planck length—that is, when R takes on the value 1—the winding and vibration modes have comparable mass. The two approaches to measuring distance become equally difficult to carry out and, moreover, each would yield the same result since 1 is its own reciprocal.

As the radius continues to shrink, the winding modes become lighter than the unwound modes and hence, since we are always opting for the "easier approach," *they* should now be used to measure distances. According to this method of measurement, which yields the *reciprocal* of that measured by the unwound modes, *the radius is larger than one times the Planck length and increasing*. This simply reflects that as R—the quantity measured by unwound strings—shrinks to 1 and continues to get smaller, $1/R$—the quantity measured by wound strings—grows to 1 and gets larger. Therefore, if one takes care to always use the light string modes—the "easy" approach to measuring distance—the minimal value encountered is the Planck length.

In particular, a big crunch to zero size is avoided, as the radius of the universe as measured using light string-mode probes is always larger than the Planck length. Rather than heading through the Planck length on to ever smaller size, the radius, as measured by the lightest string modes, decreases to the Planck length and then immediately starts to increase. The crunch is replaced by a bounce.

Using light string modes to measure distances aligns with our conventional notion of length—the one that was around long before the discovery of string theory. It is according to *this* notion of distance, as seen in Chapter 5, that we encountered insurmountable problems with violent quantum undulations if sub-Planck-scale distances play a physical role. We once again see, from this complementary perspective, that the ultra-short distances are avoided by string theory. In the physical framework of general relativity and in the corresponding mathematical framework of Riemannian geometry there is a single concept of distance, and it can acquire arbitrarily small values. In the physical framework of string theory, and, correspondingly, in the realm of the emerging discipline of quantum geometry, there are two notions of distance. By judiciously making use of both we find a concept of distance that meshes with both our intuition and with general relativity when distance scales are large, but that differs from them dramatically when distance scales get small. Specifically, sub-Planck-scale distances are inaccessible.

As this discussion is quite subtle, let's re-emphasize one central point. If we were to spurn the distinction between "easy" and "hard" approaches to measuring length and, say, continue to use the unwound modes as R shrinks through the Planck length, it might seem that we would indeed be able to encounter a sub-Planck-length distance. But the paragraphs above inform us that the word "distance" in the last sentence must be carefully interpreted, since it can have two different meanings, only one of which conforms to our traditional notion. And in this case, when R shrinks to sub-Planck length but we continue to use the unwound strings (even though they have now become heavier than the wound strings), we are employing the "hard" approach to measuring distance, and hence the meaning of "distance" does *not* conform to our standard usage. However, the discussion is far more than one of semantics or even of convenience or practicality of measurement. Even if we choose to use the nonstandard notion of distance and thereby describe the radius as being shorter than the Planck length, the *physics* we encounter—as discussed in previous sections—will be identical to that of a universe in which the radius, in the conventional sense of distance, is larger than the Planck length (as attested to, for example, by the exact correspondence between Tables 10.1 and 10.2). And it is physics, not language, that really matters.

Brandenberger, Vafa, and other physicists have made use of these ideas to suggest a rewriting of the laws of cosmology in which both the big bang and the possible big crunch do not involve a zero-size universe, but rather one that is Planck-length in all dimensions. This is certainly a very appealing proposal for avoiding the mathematical, physical, and logical conundrums of a universe that emanates from or collapses to an infinitely dense point. Although it is conceptually difficult to imagine the whole of the universe compressed together into a tiny Planck-sized nugget, it is truly beyond the pale to imagine it crushed to a point of no size at all. String cosmology, as we shall discuss in Chapter 14, is a field very much in its infancy but one that holds great promise, and may very well provide us with this easier-to-swallow alternative to the standard big bang model.

How General Is This Conclusion?

What if the spatial dimensions are not circular in shape? Do these remarkable conclusions about minimum spatial extent in string theory still hold? No one knows for sure. The essential aspect of circular dimensions is that they permit the possibility of wound strings. As long as the spatial dimensions—regardless of the details of their shape—allow strings to wind around them, most of the conclusions we have drawn should still apply. But what if, say, two of the dimensions are in the shape of a sphere? In this case, strings cannot get "trapped" in a wound configuration, because they can always "slip off" much as a stretched rubber band can pop off a basketball. Does string theory nevertheless limit the size to which these dimensions can shrink?

Numerous investigations seem to show that the answer depends on whether a full spatial dimension is being shrunk (as in the examples in this chapter) or (as we shall encounter and explain in Chapters 11 and 13) an isolated "chunk" of space is collapsing. The general belief among string theorists is that, regardless of shape, there *is* a minimum limiting size, much as in the case of circular dimensions, so long as we are shrinking a full spatial dimension. Establishing this expectation is an important goal for further research because it has a direct impact on a number of aspects of string theory, including its implications for cosmology.

Mirror Symmetry

Through general relativity, Einstein forged a link between the physics of gravity and the geometry of spacetime. At first blush, string theory strengthens and broadens the link between physics and geometry, since the properties of vibrating strings—their mass and the force charges they carry—are largely determined by the properties of the curled-up component of space. We have just seen, though, that quantum geometry—the geometry-physics association in string theory—has some surprising twists. In general relativity, and in "conventional" geometry, a circle of radius R is different from one whose radius is $1/R$, pure and simple; yet, in string theory they are physically indistinguishable. This leads us to be bold enough to go further and ask whether there might be geometrical forms of space that differ in more drastic ways—not just in overall size, but possibly also in shape—but that are nevertheless physically indistinguishable in string theory.

In 1988, Lance Dixon of the Stanford Linear Accelerator Center made a pivotal observation in this regard that was further amplified by Wolfgang Lerche of CERN, Vafa at Harvard, and Nicholas Warner, then of the Massachusetts Institute of Technology. Based upon aesthetic arguments rooted in considerations of symmetry, these physicists made the audacious suggestion that it might be possible for two different Calabi-Yau shapes, chosen for the extra curled-up dimensions in string theory, to give rise to identical physics.

To give you an idea of how this rather far-fetched possibility might actually occur, recall that the number of holes in the extra Calabi-Yau dimensions determines the number of families into which string excitations will arrange themselves. These holes are analogous to the holes one finds in a torus or its multihandled cousins, as illustrated in Figure 9.1. One deficiency of the two-dimensional figure that we must show on the printed page is that it cannot show that a six-dimensional Calabi-Yau space can have holes of a variety of dimensions. Although such holes are harder to picture, they can be described with well-understood mathematics. A key fact is that the number of families of particles arising from string vibrations is sensitive only to the total number of holes, not to the number of holes of each particular dimension (that's why, for instance, we did not worry about drawing distinctions between the different types of holes in our discussion in Chapter 9). Imagine, then, two Calabi-Yau spaces in which the number of holes in various dimensions differs, but in which the total number of holes is the same. Since the number of holes in each dimension is not the same, the two Calabi-Yaus have different shapes. But since they have the same total number of holes, each yields a universe with the *same number of families*. This, of course, is but one physical property. Agreement on *all* physical properties is a far more restrictive requirement, but this at least gives the flavor of how the Dixon-Lerche-Vafa-Warner conjecture could possibly be true.

In the fall of 1987, I joined the physics department at Harvard as a postdoctoral fellow and my office was just down the hall from Vafa's. As my thesis research had focused on the physical and mathematical properties of curled-up Calabi-Yau dimensions in string theory, Vafa kept me closely apprised of his work in this area. When he stopped by my office in the fall of 1988 and told me of the conjecture that he, Lerche, and Warner had come upon, I was intrigued but also skeptical. The intrigue arose from the realization that if their conjecture was true, it might open a new avenue of research on string theory; the skepticism arose from the realization that guesses are one thing, established properties of a theory are quite another.

During the following months, I thought frequently about their conjecture and, frankly, half convinced myself that it wasn't true. Surprisingly, though, a seemingly unrelated research project I had undertaken in collaboration with Ronen Plesser, then a graduate student at Harvard and now on the faculty of the Weizmann Institute and Duke University, was soon to change my mind completely. Plesser and I had become interested in developing methods for starting with an initial Calabi-Yau shape and mathematically manipulating it to produce hitherto unknown Calabi-Yau shapes. We were particularly drawn to a technique known as *orbifolding*, which was pioneered by Dixon, Jeffrey Harvey of the University of Chicago, Vafa, and Witten in the mid-1980s. Roughly speaking, this is a procedure in which different points on an initial Calabi-Yau shape are glued together according to mathematical rules that ensure that a new Calabi-Yau shape is produced. This is schematically illustrated in Figure 10.4. The mathematics underlying the manipulations illustrated in Figure 10.4 is formidable, and for this reason string theorists had thoroughly investigated this procedure only as applied to the simplest of shapes— higher-dimensional versions of the doughnut shapes shown in Figure 9.1. Plesser and I realized, though, that some beautiful new insights of Doron Gepner, then of Princeton University, might give a powerful theoretical framework for applying the orbifolding technique to full-fledged Calabi-Yau shapes, such as the one in Figure 8.9.

After a few months of intensive pursuit of this idea we came to a surprising realization. If we glued particular groups of points together in just the right way, the Calabi-Yau shape we produced differed from the one we started with in a startling manner: The number of *odd*-dimensional holes in the new Calabi-Yau shape equaled the number of *even*-dimensional holes in the original, and vice versa. In particular, this means that the

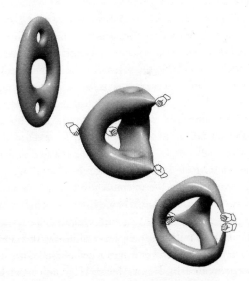

Figure 10.4 Orbifolding is a procedure in which a new Calabi-Yau shape is produced by gluing together various points on an initial Calabi-Yau shape.

total number of holes—and therefore the number of particle families—in each is the *same* even though the even-odd interchange means that their shapes and fundamental geometrical structures are quite different.[5]

Excited by the apparent contact we had made with the Dixon-Lerche-Vafa-Warner guess, Plesser and I pressed on to the linchpin question: Beyond the number of families of particles, do the two different Calabi-Yau spaces agree on the rest of their physical properties? After a couple more months of detailed and arduous mathematical analysis during which we received valuable inspiration and encouragement from Graham Ross, my thesis advisor at Oxford, and also from Vafa, Plesser and I were able to argue that the answer was, most definitely, yes. For mathematical reasons having to do with the even-odd interchange, Plesser and I coined the term *mirror manifolds* to describe the physically equivalent yet geometrically distinct Calabi-Yau spaces.[6] The individual spaces in a mirror pair of Calabi-Yau spaces are not literally mirror images of one another, in the sense of everyday usage. But even though they have different geometrical properties, they give rise to one and the same physical universe when used for the extra dimensions in string theory.

The weeks after finding this result were an extremely anxious time. Plesser and I knew that we were sitting on an important new piece of string physics. We had shown that the tight association between geometry and physics originally set down by Einstein was substantially modified by string theory: Drastically different geometrical shapes that would imply different physical properties in general relativity were giving rise to identical physics in string theory. But what if we had made a mistake? What if their physical implications did differ in some subtle way that we had missed? When we showed our results to Yau, for example, he politely but firmly claimed that we must have made an error; he asserted that from a mathematical standpoint our results were far too outlandish to be true. His assessment gave us substantial pause. It's one thing to make a mistake in a small or modest claim that attracts little attention. Our result, though, was suggesting an unexpected step in a new direction that would certainly engender a strong response. If it were wrong, everyone would know.

Finally, after much checking and rechecking, our confidence grew and we sent our paper off for publication. A few days later, I was sitting in my office at Harvard and the phone rang. It was Philip Candelas from the University of Texas, and he immediately asked me if I was seated. I was.

He then told me that he and two of his students, Monika Lynker and Rolf Schimmrigk, had found something that was going to knock me off of my chair. By carefully examining a huge sample set of Calabi-Yau spaces that they had generated by computer, they found that almost all came in pairs differing precisely by the interchange of the number of even and odd holes. I told him that I was still seated—that Plesser and I had found the same result. Candelas's and our work turned out to be complementary; we had gone one step further by showing that all of the resulting physics in a mirror pair was identical, whereas Candelas and his students had shown that a significantly larger sample of Calabi-Yau shapes fell into mirror pairs. Through the two papers, we had discovered the *mirror symmetry* of string theory.[7]

The Physics and the Mathematics of Mirror Symmetry

The loosening of Einstein's rigid and unique association between the geometry of space and observed physics is one of the striking paradigm shifts of string theory. But these developments entail far more than a change in philosophical stance. Mirror symmetry, in particular, provides a powerful tool for understanding both the physics of string theory and the mathematics of Calabi-Yau spaces.

Mathematicians working in a field called algebraic geometry had been studying Calabi-Yau spaces for purely mathematical reasons long before string theory was discovered. They had worked out many of the detailed properties of these geometrical spaces without an inkling of a future physical application. Certain aspects of Calabi-Yau spaces, however, had proven difficult—essentially impossible—for mathematicians to unravel fully. But the discovery of mirror symmetry in string theory changed this significantly. In essence, mirror symmetry proclaims that particular pairs of Calabi-Yau spaces, pairs that were previously thought to be completely unrelated, are now intimately connected by string theory. They are linked by the common physical universe each implies if either is the one selected for the extra curled-up dimensions. This previously unsuspected interconnection provides an incisive new physical and mathematical tool.

Imagine, for instance, that you are busily calculating the physical properties—particle masses and force charges—associated with one possible Calabi-Yau choice for the extra dimensions. You are not particularly concerned with matching your detailed results with experiment, since as we have seen a number of theoretical and technological obstacles make doing this quite difficult at present. Instead, you are working through a thought experiment concerned with what the world *would* look like if a particular Calabi-Yau space *were* selected. For a while, everything is going along fine, but then, in the midst of your work, you come upon a mathematical calculation of insurmountable difficulty. No one, not even the world's most expert mathematicians, can figure out how to proceed. You are stuck. But then you realize that this Calabi-Yau has a mirror partner. Since the resulting string physics associated with each member of a mirror pair is identical, you recognize that you are free to do your calculations making use of either. And so, you rephrase the difficult calculation on the original Calabi-Yau space in terms of a calculation on its mirror, assured that the result of the calculation—the physics—will be the same. At first sight you might think that the rephrased version of the calculation will be as difficult as the original. But here you come upon a pleasant and powerful surprise: You discover that although the result will be the same, the detailed form of the calculation is very different, and in some cases the horribly difficult calculation you started with turns into an extremely easy calculation on the mirror Calabi-Yau space. There is no simple explanation for why this happens, but—at least for certain calculations—it most definitely does, and the decrease in level of difficulty can be dramatic. The implication, of course, is clear: You are no longer stuck.

It's somewhat as if someone requires you to count exactly the number of oranges that are haphazardly jumbled together in an enormous bin, some 50 feet on each side and 10 feet deep. You start to count them one by one, but soon realize that the task is just too laborious. Luckily, though, a friend comes along who was present when the oranges were delivered. He tells you that they arrived neatly packed in smaller boxes (one of which he just happens to be holding) that when stacked were 20 boxes long, by 20 boxes deep, by 20 boxes high. You quickly calculate that they arrived in 8,000 boxes, and that all you need to do is figure out how many oranges are packed in each. This you easily do by borrowing your friend's box and filling it with oranges, allowing you to finish your huge counting task with almost no effort. In essence, by cleverly reorganizing the calculation, you were able to make it substantially easier to accomplish.

The situation with numerous calculations in string theory is similar. From the perspective of one Calabi-Yau space, a calculation might involve an enormous number of difficult mathematical steps. By translating the calculation to its mirror, though, the calculation is reorganized in a far more efficient manner, allowing it to be completed with relative ease. This point was made by Plesser and me, and was impressively put into practice in subsequent work by Candelas with his collaborators Xenia de la Ossa and Linda Parkes, from the University of Texas, and Paul Green, from the University of Maryland. They showed that calculations of almost unimaginable difficulty could be accomplished by using the mirror perspective, with a few pages of algebra and a desktop computer.

This was an especially exciting development for mathematicians, because some of these calculations were precisely the ones they had been stuck on for many years. String theory—or so the physicists claimed—had beaten them to the solution.

Now you should bear in mind that there is a good deal of healthy and generally good-natured competition between mathematicians and physicists. And as it turns out, two Norwegian mathematicians—Geir Ellingsrud and Stein Arild Strømme—happened to be working on one of numerous calculations that Candelas and his collaborators had successfully conquered with mirror symmetry. Roughly speaking, it amounted to calculating the number of spheres that could be "packed" inside a particular Calabi-Yau space, somewhat like our analogy of counting oranges in a large bin. At a meeting of physicists and mathematicians in Berkeley in 1991, Candelas announced the result reached by his group using string theory and mirror symmetry: 317,206,375. Ellingsrud and Strømme announced the result of their very difficult mathematical calculation: 2,682,549,425. For days, mathematicians and physicists debated: Who was right? The question turned into a real litmus test of the quantitative reliability of string theory. A number of people even commented—somewhat in jest—that this test was the next best thing to being able to compare string theory with experiment. Moreover, Candelas's results went far beyond the single numerical result that Ellingsrud and Strømme claimed to have calculated. He and his collaborators claimed to have also answered many other questions that were tremendously more difficult—so difficult in fact, that no mathematician had ever even attempted to address them. But could the string theory results be trusted? The meeting ended with a great deal of fruitful exchange between mathematicians and physicists, but no resolution of the discrepancy.

About a month later, an e-mail message was widely circulated among participants in the Berkeley meeting with the subject heading *Physics Wins!* Ellingsrud and Strømme had found an error in their computer code that, when corrected, confirmed Candelas's result. Since then, there have been many mathematical checks on the quantitative reliability of the mirror symmetry of string theory: It has passed all with flying colors. Even more recently, almost a decade after physicists discovered mirror symmetry, mathematicians have made great progress in revealing its inherent mathematical foundations. By utilizing substantial contributions of the mathematicians Maxim Kontsevich, Yuri Manin, Gang Tian, Jun Li, and Alexander Givental, Yau and his collaborators Bong Lian and Kefeng Liu have finally found a rigorous mathematical proof of the formulas used to count spheres inside Calabi-Yau spaces, thereby solving problems that have puzzled mathematicians for hundreds of years.

Beyond the particulars of this success, what these developments really highlight is the role that physics has begun to play in modern mathematics. For quite some time, physicists have "mined" mathematical archives in search of tools for constructing and analyzing models of the physical world. Now, through the discovery of string theory, physics is beginning to repay the debt and to provide mathematicians with powerful new approaches to their unsolved problems. String theory not only provides a unifying framework for physics, but it may well forge an equally deep union with mathematics as well.

Chapter 11

Tearing the Fabric of Space

If you relentlessly stretch a rubber membrane, sooner or later it will tear. This simple fact has inspired numerous physicists over the years to ask whether the same might be true of the spatial fabric making up the universe. That is, can the fabric of space rip apart, or is this merely a misguided notion that arises from taking the rubber membrane analogy too seriously?

Einstein's general relativity says no, the fabric of space cannot tear.[1] The equations of general relativity are firmly rooted in Riemannian geometry and, as we noted in the preceding chapter, this is a framework that analyzes distortions in the distance relations between nearby locations in space. In order to speak meaningfully about these distance relations, the underlying mathematical formalism requires that the substrate of space is *smooth*—a term with a technical mathematical meaning, but whose everyday usage captures its essence: no creases, no punctures, no separate pieces "stuck" together, and no tears. Were the fabric of space to develop such irregularities, the equations of general relativity would break down, signaling some or other variety of cosmic catastrophe—a disastrous outcome that our apparently well-behaved universe avoids.

This has not kept imaginative theorists over the years from pondering the possibility that a new formulation of physics that goes beyond Einstein's classical theory and incorporates quantum physics might show that rips, tears, and mergers of the spatial fabric can occur. In fact, the realization that quantum physics leads to violent short-distance undulations led some to speculate that rips and tears might be a commonplace microscopic feature of the spatial fabric. The concept of *wormholes* (a notion with which any fan of *Star Trek: Deep Space Nine* is familiar) makes use of such musings. The idea is simple: Imagine you're the CEO of a major corporation with headquarters on the ninetieth floor of one of New York City's World Trade Center towers. Through the vagaries of corporate history, an arm of your company with which you need to have ever increasing contact is ensconced on the ninetieth floor of the other tower. As it is impractical to move either office, you come up with a natural suggestion: Build a bridge from one office to the other, connecting the two towers. This allows employees to move freely between the offices without having to go down and then up ninety floors.

A wormhole plays a similar role: It is a bridge or tunnel that provides a shortcut from one region of the universe to another. Using a two-dimensional model, imagine that a universe is shaped as in Figure 11.1.

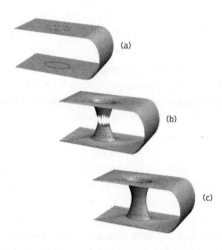

Figure 11.1 (a) In a "U-shaped" universe, the only way to get from one end to the other is by traversing the whole cosmos. (b) The fabric of space tears, and two ends of a wormhole start to grow. (c) The two wormhole ends merge together, forming a new bridge—a shortcut—from one end of the universe to the other.

If your corporate headquarters are located near the lower circle in 11.1(a), you can get to your field office, located near the upper circle, only by traversing the entire U-shaped path, taking you from one end of the universe to another. But if the fabric of space can tear, developing punctures as in 11.1(b), and if these punctures can "grow" tentacles that merge together as in 11.1(c), a spatial bridge would connect the previously remote regions. This is a wormhole. You should note that the wormhole has some similarity to the World Trade Center bridge, but there is one essential difference: The World Trade Center bridge would traverse a region of *existing* space—the space between the two towers. On the contrary, the wormhole creates a *new* region of space, since the curved two-dimensional space in Figure 11.1(a) is *all* there is (in the setting of our two-dimensional analogy). Regions lying off of the membrane merely reflect the inadequacy of the illustration, which depicts the U-shaped universe as if it were an object within our higher-dimensional universe. The wormhole creates new space and therefore blazes new spatial territory.

Do wormholes exist in the universe? No one knows. And if they do, it is far from clear whether they would take on only a microscopic form or if they could span vast regions of the universe (as in *Deep Space Nine*). But one essential element in assessing whether they are fact or fiction is determining whether or not the fabric of space can tear.

Black holes provide another compelling example in which the fabric of space is stretched to its limits. In Figure 3.7, we saw that the enormous gravitational field of a black hole results in such extreme curvature that the fabric of space *appears* to be pinched or punctured at the black hole's center. Unlike in the case of wormholes, there is strong experimental evidence supporting the existence of black holes, so the question of what really happens at their central point is one of science, not speculation. Once again, the equations of general relativity break down under such extreme conditions. Some physicists have suggested that there really is a puncture, but that we are protected from this cosmic "singularity" by the event horizon of the black hole, which prevents anything from escaping its gravitational grip. This reasoning led Roger Penrose of Oxford University to speculate on a "cosmic censorship hypothesis" that allows these kinds of spatial irregularities to occur only if they are deeply hidden from our view behind the shroud of an event horizon. On the other hand, prior to the discovery of string theory, some physicists surmised that a proper merger of quantum mechanics and general relativity would show that the apparent puncture of space is actually smoothed out—"sewn up," so to speak—by quantum considerations.

With the discovery of string theory and the harmonious merger of quantum mechanics and gravity, we are finally poised to study these issues. As yet, string theorists have not been able to answer them fully, but during the last few years closely related issues *have* been solved. In this chapter we discuss how string theory, for the first time, definitively shows that there are physical circumstances—differing from wormholes and black holes in certain ways—in which the fabric of space *can* tear.

A Tantalizing Possibility

In 1987, Shing-Tung Yau and his student Gang Tian, now at the Massachusetts Institute of Technology, made an interesting mathematical observation. They found, using a well-known mathematical procedure, that certain Calabi-Yau shapes could be transformed into others by puncturing their surface and then sewing up the resulting hole according to a precise mathematical pattern.[2] Roughly speaking, they identified a particular kind of two-dimensional sphere—like the surface of a beach ball—sitting inside an initial Calabi-Yau space, as in Figure 11.2. (A beach ball, like all famil-

Figure 11.2 The highlighted region inside a Calabi-Yau shape contains a sphere.

(a)   (b)   (c)   (d)

Figure 11.3 A sphere inside a Calabi-Yau space shrinks down to a point, pinching the fabric of space. We simplify this and subsequent figures by showing only part of the full Calabi-Yau shape.

iar objects, is three-dimensional. Here, however, we are referring solely to its surface; we are ignoring the thickness of the material from which it is made as well as the interior space it encloses. Points on the beach ball's surface can be located by giving two numbers—"latitude" and "longitude"— much as we locate points on the earth's surface. This is why the *surface* of the beach ball, like the surface of the garden hose discussed in preceding chapters, is *two*-dimensional.) They then considered shrinking the sphere until it is pinched down to a single point, as we illustrate with the sequence of shapes in Figure 11.3. This figure, and subsequent ones in this chapter, have been simplified by focusing in on the most relevant "piece" of the Calabi-Yau shape, but in the back of your mind you should note that these shape transformations are occuring within a somewhat larger Calabi-Yau space, as in Figure 11.2. And finally, Tian and Yau imagined slightly tearing the Calabi-Yau space at the pinch (Figure 11.4(a)), opening it up and gluing in another beach ball–like shape (Figure 11.4(b)), which they could then reinflate to a nice plump form (Figures 11.4(c) and 11.4(d)).

(a)   (b)   (c)   (d)

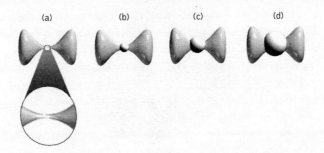

Figure 11.4 A pinched Calabi-Yau space tears open and grows a sphere that smoothes out its surface. The original sphere of Figure 11.3 is "flopped."

Mathematicians call this sequence of manipulations a *flop-transition*. It's as if the original beach ball shape is "flopped" over into a new orientation within the overall Calabi-Yau shape. Yau, Tian, and others noted that under certain circumstances, the new Calabi-Yau shape produced by a flop, as in Figure 11.4(d), is *topologically distinct* from the initial Calabi-Yau shape in Figure 11.3(a). This is a fancy way of saying that there is absolutely no way to deform the initial Calabi-Yau space in Figure 11.3(a) into the final Calabi-Yau space shown in Figure 11.4(d) without tearing the fabric of the Calabi-Yau space at some intermediate stage.

From a mathematical standpoint, this procedure of Yau and Tian is of interest because it provides a way to produce new Calabi-Yau spaces from ones that are known. But its real potential lies in the realm of physics, where it raises a tantalizing question: Could it be that, beyond its being an abstract mathematical procedure, the sequence displayed from Figure 11.3(a) through Figure 11.4(d) might actually occur in nature? Might it be that, contrary to Einstein's expectations, the fabric of space *can tear apart and subsequently be repaired* in the manner described?

The Mirror Perspective

For a couple of years after their 1987 observation, Yau would, every so often, encourage me to think about the possible physical incarnation of these flop transitions. I didn't. To me it seemed that flop transitions were merely a piece of abstract mathematics without any bearing on the physics of string theory. In fact, based on the discussion in Chapter 10 in which we found that circular dimensions have a minimum radius, one might be tempted to say that string theory does not allow the sphere in Figure 11.3 to shrink all the way down to a pinched point. But remember, as also noted in Chapter 10, that if a chunk of space collapses—in this case, a spherical piece of a Calabi-Yau shape—as opposed to the collapse of a complete spatial dimension, the argument identifying small and large radii is not directly applicable. Nevertheless, even though this idea for ruling out flop transitions does not stand up to scrutiny, the possibility that the fabric of space could tear still seemed rather unlikely.

But then, in 1991 the Norwegian physicist Andy Lütken together with Paul Aspinwall, a graduate-school classmate of mine from Oxford and now a professor at Duke University, asked themselves what proved to be a very interesting question: If the spatial fabric of the Calabi-Yau portion of our universe were to undergo a space-tearing flop transition, what would it look like from the perspective of the mirror Calabi-Yau space? To understand the motivation for this question, you must recall that the physics emerging from either member of a mirror pair of Calabi-Yau shapes (if selected for the extra dimensions) is identical, but the complexity of the mathematics that a physicist must employ to extract the physics can differ significantly between the two. Aspinwall and Lütken speculated that the mathematically complicated flop transition of Figures 11.3 and 11.4 might have a far simpler mirror description—one that might give a more transparent view on the associated physics.

At the time of their work, mirror symmetry was not understood at the depth required to answer the question they posed. However, Aspinwall and Lütken noted that there did not seem to be anything in the mirror description that would indicate a disastrous physical consequence associated with the spatial tears of flop transitions. Around the same time, the work Plesser and I had done in finding mirror pairs of Calabi-Yau shapes (see Chapter 10) unexpectedly led us to think about flop transitions as well. It is a well-known mathematical fact that gluing various points together as in Figure 10.4—the procedure we had used to construct mirror pairs— leads to geometrical situations that are identical to the pinch and puncture in Figures 11.3 and 11.4. Physically, though, Plesser and I could find no associated calamity. Moreover, inspired by the observations of Aspinwall and Lütken (as well as a previous paper of theirs with Graham Ross), Plesser and I realized that we could repair the pinch mathematically in two different ways. One way led to the Calabi-Yau shape in Figure 11.3(a) while the other led to that in Figure 11.4(d). This suggested to us that the evolution from Figure 11.3(a) through Figure 11.4(d) was something that could actually occur in nature.

By late 1991, then, at least a few string theorists had a strong feeling that the fabric of space *can* tear. But no one had the technical facility to definitively establish or refute this striking possibility.

Inching Forward

Off and on during 1992, Plesser and I tried to show that the fabric of space can undergo space-tearing flop transitions. Our calculations yielded bits and pieces of supporting circumstantial evidence, but we could not find definitive proof. Sometime during the spring, Plesser visited the Institute for Advanced Study in Princeton to give a talk, and privately told Witten about our recent attempts to realize the mathematics of space-tearing flop transitions within the physics of string theory. After summarizing our ideas, Plesser waited for Witten's response. Witten turned from the blackboard and stared out of his office window. After a minute of silence, maybe two, he turned back to Plesser and told him that if our ideas worked out, "it would be spectacular." This rekindled our efforts. But after a while, with our progress stalled, each of us turned to working on other string theory projects.

Even so, I found myself mulling over the possibility of space-tearing flop transitions. As the months went by, I felt increasingly sure that they had to be part and parcel of string theory. The preliminary calculations Plesser and I had done, together with insightful discussions with David Morrison, a mathematician from Duke University, made it seem that this was the only conclusion that mirror symmetry naturally supported. In fact, during a visit to Duke, Morrison and I, together with some helpful observations from Sheldon Katz of Oklahoma State University, who was also visiting Duke at the time, outlined a strategy for proving that flop transitions can occur in string theory. But when we sat down to do the required calculations, we found that they were extraordinarily intensive. Even on the world's fastest computer, they would take more than a century to complete. We had made progress, but we clearly needed a new idea, one that could greatly enhance the efficiency of our calculational method. Unwittingly, Victor Batyrev, a mathematician from the University of Essen, revealed such an idea through a pair of papers released in the spring and summer of 1992.

Batyrev had become very interested in mirror symmetry, especially in the wake of the success of Candelas and his collaborators in using it to solve the sphere-counting problem described at the end of Chapter 10.

With a mathematician's perspective, though, Batyrev was unsettled by the methods Plesser and I had invoked to find mirror pairs of Calabi-Yau spaces. Although our approach used tools familiar to string theorists, Batyrev later told me that our paper seemed to him to be "black magic." This reflects the large cultural divide between the disciplines of physics and mathematics, and as string theory blurs their borders, the vast differences in language, methods, and styles of each field become increasingly apparent. Physicists are more like avant-garde composers, willing to bend traditional rules and brush the edge of acceptability in the search for solutions. Mathematicians are more like classical composers, typically working within a much tighter framework, reluctant to go to the next step until all previous ones have been established with due rigor. Each approach has its advantages as well as drawbacks; each provides a unique outlet for creative discovery. Like modern and classical music, it's not that one approach is right and the other wrong—the methods one chooses to use are largely a matter of taste and training.

Batyrev set out to recast the construction of mirror manifolds in a more conventional mathematical framework, and he succeeded. Inspired by earlier work of Shi-Shyr Roan, a mathematician from Taiwan, he found a systematic mathematical procedure for producing pairs of Calabi-Yau spaces that are mirrors of one another. His construction reduces to the procedure Plesser and I had found in the examples we had considered, but offers a more general framework that is phrased in a manner more familiar to mathematicians.

The flip side is that Batyrev's papers invoked areas of mathematics that most physicists had never previously encountered. I, for example, could extract the gist of his arguments, but had significant difficulty in understanding many crucial details. One thing, however, was clear: The methods of his paper, if properly understood and applied, could very well open a new line of attack on the issue of space-tearing flop transitions.

By late summer, energized by these developments, I decided that I wanted to return to the problem of flops with full and undistracted intensity. I had learned from Morrison that he was going on leave from Duke to spend a year at the Institute for Advanced Study, and I knew that Aspinwall would also be there, as a postdoctoral fellow. After a few phone calls and e-mails, I arranged to take leave from Cornell University and spend the fall of 1992 at the Institute as well.

A Strategy Emerges

One would be hard pressed to think of a more ideal place for long hours of intense concentration than the Institute for Advanced Study. Founded in 1930, it is set within gently rolling fields on the border of an idyllic forest a few miles from the campus of Princeton University. It is said that you can't get distracted from your work at the Institute, because, well, there aren't any distractions.

After leaving Germany in 1933, Einstein joined the Institute and remained there for the duration of his life. It takes little imagination to picture him pondering unified field theory in the Institute's quiet, lonely, almost ascetic surroundings. The legacy of deep thought infuses the atmosphere, which, depending on your own immediate state of progress, can be either exciting or oppressive.

Shortly after arriving at the Institute, Aspinwall and I were walking down Nassau Street (the main commercial street in the town of Princeton) trying to agree on a place to have dinner. This was no small task since Paul is as devout a meat eater as I am a vegetarian. In the midst of catching up on each other's lives as we were walking along, he asked me if I had any ideas about new things to work on. I told him I did, and recounted my take on the importance of establishing that the universe, if truly described by string theory, can undergo space-tearing flop transitions. I also outlined the strategy I had been pursuing, as well as my newfound hope that Batyrev's work might allow us to fill in the missing pieces. I thought that I was preaching to the converted, and that Paul would be excited by this prospect. He wasn't. In retrospect, his reticence was due largely to our good-natured and long-standing intellectual joust in which we each play devil's advocate to the other's ideas. Within days, he came around and we turned our full attention to flops.

By then, Morrison had also arrived, and the three of us met in the Institute's tea-room to formulate a strategy. We agreed that the central goal was to determine whether the evolution from Figure 11.3(a) to Figure 11.4(d) can actually occur in our universe. But a direct attack on the question was forbidding, because the equations describing this evolution are extremely difficult, especially when the spatial tear occurs. Instead, we chose to rephrase the issue using the mirror description, hoping that the equations involved might be more manageable. This is schematically illustrated in Figure 11.5, in which the top row is the original evolution from Figure 11.3(a) to Figure 11.4(d), and the bottom row is the same evolution from the perspective of the mirror Calabi-Yau shapes. As a number of us had already realized, it turns out that in the mirror rephrasing it appears that string physics is perfectly well behaved and encounters no catastrophes. As you can see, there does not seem to be any pinching or tearing in the bottom row in Figure 11.5. However, the real question this observation raised for us was this: Were we pushing mirror symmetry beyond the bounds of its applicability? Although the upper and lower Calabi-Yau shapes drawn on the far left-hand side of Figure 11.5 yield identical physics, is it true that at every step in the evolution to the right-hand side of Figure 11.5—necessarily passing through the pinch-tear-repair stage in the middle—the physical properties of the original and mirror perspective are identical?

Although we had solid reason to believe that the powerful mirror relationship holds for the shape progression leading to the tear in the upper Calabi-Yau shape in Figure 11.5, we realized that no one knew whether the upper and lower Calabi-Yau shapes in Figure 11.5 continue to be mirrors after the tear has occurred. This is a crucial question, because if they are, then the absence of a catastrophe in the mirror perspective would mean an absence in the original, and we would have demonstrated that space can tear in string theory. We realized that this question could be reduced to a calculation: Extract the physical properties of the universe for the upper Calabi-Yau shape after the tear (using, say, the upper-right Calabi-Yau shape in Figure 11.5) and for its supposed mirror (the lower-right Calabi-Yau shape in Figure 11.5), and see if they are identical.

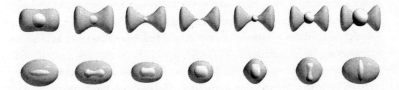

Figure 11.5 A space-tearing flop transition (top row) and its purported mirror rephrasing (bottom row).

It was this calculation to which Aspinwall, Morrison, and I devoted ourselves in the fall of 1992.

Late Nights at Einstein's Final Stomping Ground

Edward Witten's razor-sharp intellect is clothed in a soft-spoken demeanor that often has a wry, almost ironic, edge. He is widely regarded as Einstein's successor in the role of the world's greatest living physicist. Some would go even further and describe him as the greatest physicist of all time. He has an insatiable appetite for cutting-edge physics problems and he wields tremendous influence in setting the direction of research in string theory.

The breadth and depth of Witten's productivity is legendary. His wife, Chiara Nappi, who is also a physicist at the Institute, paints a picture of Witten sitting at their kitchen table, mentally probing the edge of string theory knowledge, and only now and then returning to pick up pen and paper to verify an elusive detail or two.[3] Another story is told by a postdoctoral fellow who, one summer, had an office next to Witten's. He describes the unsettling juxtaposition of laboriously struggling with complex string theory calculations at his desk while hearing the incessant rhythmic patter of Witten's keyboard, as paper after groundbreaking paper poured forth directly from mind to computer file.

A week or so after I arrived, Witten and I were chatting in the Institute's courtyard, and he asked about my research plans. I told him about the space-tearing flops and the strategy we were planning to pursue. He lit up upon hearing the ideas, but cautioned that he thought the calculations would be horrendously difficult. He also pointed out a potential weak link in the strategy I described, having to do with some work I had done a few years earlier with Vafa and Warner. The issue he raised turned out to be only tangential to our approach for understanding flops, but it started him thinking about what ultimately turned out to be related and complementary issues.

Aspinwall, Morrison, and I decided to split our calculation in two pieces. At first a natural division might have seemed to involve first extracting the physics associated with the final Calabi-Yau shape from the upper row of Figure 11.5, and then doing the same for the final Calabi-

Yau shape from the lower row of Figure 11.5. If the mirror relationship is not shattered by the tear in the upper Calabi-Yau, then these two final Calabi-Yau shapes should yield identical physics, just like the two initial Calabi-Yau shapes from which they evolved. (This way of phrasing the problem avoids doing any of the very difficult calculations involving the upper Calabi-Yau shape just when it tears.) It turns out, though, that calculating the physics associated with the final Calabi-Yau shape in the upper row is pretty straightforward. The real difficulty in carrying out this program lies in first figuring out the *precise shape* of the final Calabi-Yau space in the lower row of Figure 11.5—the putative mirror of the upper Calabi-Yau—and then in extracting the associated physics.

A procedure for accomplishing the second task—extracting the physical features of the final Calabi-Yau space in the lower row, once its shape was precisely known—had been worked out a few years earlier by Candelas. His approach, however, was calculationally intensive and we realized that it would require a clever computer program to carry it out in our explicit example. Aspinwall, who in addition to being a renowned physicist is a crackerjack programmer, took on this task. Morrison and I set out to accomplish the first task, namely, to identify the precise shape of the candidate mirror Calabi-Yau space.

It was here that we felt Batyrev's work could provide us some important clues. Once again, though, the cultural divide between mathematics and physics—in this case, between Morrison and me—started to impede progress. We needed to join the power of the two fields to find the *mathematical* form of the lower Calabi-Yau shapes that should correspond to the same *physical* universe as the upper Calabi-Yau shapes, if flop tears are within nature's repertoire. But neither of us was sufficiently conversant in the other's language to see clear to reaching this end. It became obvious to both of us that we needed to bite the bullet: Each of us needed to take a crash course in the other's field of expertise. And so, we decided to spend our days pushing forward as best we could on the calculation, while spending evenings being both professor and student in a class of one: I would lecture to Morrison for an hour or two on the relevant physics; he would then lecture to me for an hour or two on the relevant mathematics. School would typically let out at about 11 P.M.

We stuck to the program, day in and day out. Progress was slow, but we could sense that things were starting to fall into place. Meanwhile, Witten was making significant headway on reformulating the weak link he had earlier identified. His work was establishing a new and more powerful method of translation between the physics of string theory and the mathematics of the Calabi-Yau spaces. Aspinwall, Morrison, and I had almost daily impromptu meetings with Witten at which he would show us new insights following from his approach. As the weeks went by, it gradually became clear that unexpectedly, his work, from a vantage point completely different from our own, was converging on the issue of flop transitions. Aspinwall, Morrison, and I realized that if we didn't complete our calculation soon, Witten would beat us to the punch.

Of Six-Packs and Working Weekends

Nothing focuses the mind of a physicist like a healthy dose of competition. Aspinwall, Morrison, and I went into high gear. It's important to note that this meant one thing to Morrison and me, and quite another to Aspinwall. Aspinwall is a curious mixture of upper-class British sensibility, largely a reflection of the decade he spent at Oxford as both an undergraduate and a graduate student, infused ever so slightly with a prankster's roguishness. As far as work habits go, he is perhaps the most civilized physicist I know. While many of us work deep into the evening, he never works past 5 P.M. While many of us work weekends, Aspinwall does not. He gets away with this because he is both sharp and efficient. Going into high gear for him merely amounts to notching up his efficiency level to even greater heights.

By this time, it was early December. Morrison and I had been lecturing to one another for several months and it was starting to pay off. We were very close to being able to identify the precise shape of the Calabi-Yau space we were seeking. Moreover, Aspinwall had just about finished his computer code, and he now awaited our result, which would be the required input for his program. It was a Thursday night when Morrison and I finally had confidence that we knew how to identify the sought-after Calabi-Yau shape. That, too, boiled down to a procedure that required its own, fairly simple, computer code. By Friday afternoon we had written the program and debugged it; by late Friday night we had our result.

But it was after 5 P.M. and it was Friday. Aspinwall had gone home and would not return until Monday. There was nothing we could do without his computer code. Neither Morrison nor I could imagine waiting out the whole weekend. We were on the verge of answering the long-pondered question of spatial tears in the fabric of the cosmos, and the suspense was too much to bear. We called Aspinwall at home. At first he refused to come to work the next morning as we asked. But then, after much groaning, he consented to join us, as long as we bought him a six-pack of beer. We agreed.

A Moment of Truth

We all met at the Institute Saturday morning as planned. It was a bright sunny morning, and the atmosphere was jokingly relaxed. I, for one, half expected that Aspinwall would not show up; once he did, I spent 15 minutes extolling the import of this first weekend he had come into the office. He assured me it wouldn't happen again.

We all huddled around Morrison's computer in the office he and I shared. Aspinwall told Morrison how to bring his program up on the screen and showed us the precise form for the required input. Morrison appropriately formatted the results we had generated the previous night, and we were set to go.

The particular calculation we were performing amounts, roughly speaking, to determining the mass of a certain particle species—a specific vibrational pattern of a string—when moving through a universe whose Calabi-Yau component we had spent all fall identifying. We hoped, in line with the strategy discussed earlier, that this mass would agree identically with a similar calculation done on the Calabi-Yau shape emerging from the space-tearing flop transition. The latter was the relatively easy calculation, and we had completed it weeks before; the answer turned out to be 3, in the particular units we were using. Since we were now doing the purported mirror calculation numerically on a computer, we expected to get something extremely close to but not exactly 3, something like 3.000001 or 2.999999, with the tiny difference arising from rounding errors.

Morrison sat at the computer with his finger hovering over the enter button. With the tension mounting he said, "Here goes," and set the calculation in motion. In a couple of seconds the computer returned its answer: 8.999999. My heart sank. Could it be that space-tearing flop transitions shatter the mirror relation, likely indicating that they cannot actually occur? Almost immediately, though, we all realized that something funny must be going on. If there was a real mismatch in the physics following from the two shapes, it was extremely unlikely that the computer calculation should yield an answer so close to a whole number. If our ideas were wrong, there was no reason in the world to expect anything but a random collection of digits. We had gotten a wrong answer, but one that suggested, perhaps, that we had just made some simple arithmetic error. Aspinwall and I went to the blackboard, and in a moment we found our mistake: we had dropped a factor of 3 in the "simpler" calculation we had done weeks before; the true result was 9. The computer answer was therefore just what we wanted.

Of course, the after-the-fact agreement was only marginally convincing. When you know the answer you want, it is often all too easy to figure out a way of getting it. We needed to do another example. Having already written all of the necessary computer code, this was not hard to do. We calculated another particle mass on the upper Calabi-Yau shape, being careful this time to make no errors. We found the answer: 12. Once again, we huddled around the computer and set it on its way. Seconds later it returned 11.999999. *Agreement.* We had shown that the supposed mirror *is* the mirror, and hence space-tearing flop transitions are part of the physics of string theory.

At this I jumped out of my chair and ran an unrestrained victory lap around the office. Morrison beamed from behind the computer. Aspinwall's reaction, though, was rather different. "That's great, but I knew it would work," he calmly said. "And where's my beer?"

Witten's Approach

That Monday, we triumphantly went to Witten and told him of our success. He was very pleased with our result. And, as it turned out, he too had just found a way of establishing that flop transitions occur in string theory. His argument was quite different from ours, and it significantly illuminates the microscopic understanding of why the spatial tears do not have any catastrophic consequences.

His approach highlights the difference between a point-particle theory and string theory when such tears occur. The key distinction is that there are two types of string motion near the tear, but only one kind of point-particle motion. Namely, a string can travel adjacent to the tear, like a point particle does, but it can also encircle the tear as it moves forward, as illustrated in Figure 11.6. In essence, Witten's analysis reveals that strings which encircle the tear, something that cannot happen in a point-particle theory, shield the surrounding universe from the catastrophic effects that would otherwise be encountered. It's as if the world-sheet of the string—recall from Chapter 6 that this is a two-dimensional surface that a string sweeps out as it moves through space—provides a protective barrier that precisely cancels out the calamitous aspects of the geometrical degeneration of the spatial fabric.

You might well ask, What if such a tear should occur, and it just so happens that there are no strings in the vicinity to shield it? Moreover, you might also be concerned that at the instant in time that a tear occurs, a string—an infinitely thin loop—would provide as effective a barrier as shielding yourself from a cluster bomb by hiding behind a hula hoop. The resolution to both of these issues relies on a central feature of quantum mechanics that we discussed in Chapter 4. There we saw that in Feynman's formulation of quantum mechanics, an object, be it a particle or a string, travels from one location to another by "sniffing out" all possible trajectories. The resulting motion that is observed is a combination of *all* pos-

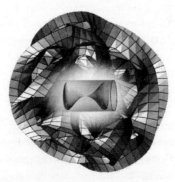

Figure 11.6 The world-sheet swept out by a string provides a shield that cancels the potentially cataclysmic effects associated with a tear in the fabric of space.

sibilities, with the relative contributions of each possible trajectory precisely determined by the mathematics of quantum mechanics. Should a tear in the fabric of space occur, then among the possible trajectories of travelling strings are those that encircle the tear—trajectories such as those in Figure 11.6. Even if no strings seem to be near the tear when it occurs, quantum mechanics takes account of physical effects from all possible string trajectories and among these are numerous (infinite, in fact) protective paths that encircle the tear. It is these contributions that Witten showed precisely to cancel out the cosmic calamity that the tear would otherwise create.

In January 1993, Witten and the three of us released our papers simultaneously to the electronic Internet archive through which physics papers are immediately made available worldwide. The two papers described, from our widely different perspectives, the first examples of *topology-changing transitions*—the technical name for the space-tearing processes we had found. The long-standing question about whether the fabric of space can tear had been settled quantitatively by string theory.

Consequences

We have made much of the realization that spatial tears can occur without physical calamity. But what *does* happen when the spatial fabric rips? What are the observable consequences? We have seen that many properties of the world around us depend upon the detailed structure of the curled-up dimensions. And so, you would think that the fairly drastic transformation from one Calabi-Yau to another as shown in Figure 11.5, would have a significant physical impact. In fact, though, the lower-dimensional drawings that we use to visualize the spaces make the transformation appear to be somewhat more complicated than it actually is. If we could visualize six-dimensional geometry, we would see that, yes, the fabric is tearing, but it does so in a fairly mild way. It's more like the handiwork of a moth on wool than that of a deep knee bend on shrunken trousers.

Our work and that of Witten show that physical characteristics such as the number of families of string vibrations and the types of particles within each family are unaffected by these processes. As the Calabi-Yau space evolves through a tear, what can be affected are the precise values of the masses of the individual particles—the energies of the possible patterns of string vibrations. Our papers showed that these masses will vary continuously in response to the changing geometrical form of the Calabi-Yau component of space, some going up while others go down. Of primary importance, though, is the fact that there is no catastrophic jump, spike, or any unusual feature of these varying masses as the tear actually occurs. From the point of view of physics, the moment of tearing has no distinguishing characteristics.

This point raises two issues. First, we have focused on tears in the spatial fabric that occur in the extra six-dimensional Calabi-Yau component of the universe. Can such tears also occur in the more familiar three extended spatial dimensions? The answer, almost certainly, is yes. After all, space is space—regardless of whether it is tightly curled up into a Calabi-Yau shape or is unfurled into the grand expanse of the universe we perceive on a clear, starry night. In fact, we have seen earlier that the familiar spatial dimensions might themselves actually be curled up into the form of a giant shape that curves back on itself, way on the other side of the universe, and that therefore even the distinction between which dimensions are curled up and which are unfurled is somewhat artificial. Although our and Witten's analyses did rely on special mathematical features of Calabi-Yau shapes, the result—that the fabric of space can tear—is certainly of wider applicability.

Second, could such a topology-changing tear happen today or tomorrow? Could it have happened in the past? Yes. Experimental measurements of elementary particle masses show their values to be quite stable over time. But if we head back to the earliest epochs following the big bang, even non-string-based theories invoke important periods during which elementary particle masses do change over time. These periods, from a string-theoretic perspective, could certainly have involved the topology-changing tears discussed in this chapter. Closer to the present, the observed stability of elementary particle masses implies that if the universe is currently undergoing a topology-changing spatial tear, it must be doing it exceedingly slowly—so slowly that its effect on elementary particle masses is smaller than our present experimental sensitivity. Remarkably, so long as this condition is met, the universe could currently be in the midst of a spatial rupture. If it were occurring slowly enough, we would not even know it was happening. This is one of those rare instances in physics in which the lack of a striking observable phenomenon is cause for great excitement. The absence of an observable calamitous consequence from such an exotic geometrical evolution is testament to how far beyond Einstein's expectations string theory has gone.

Chapter 12

Beyond Strings:
In Search of M-Theory

In his long search for a unified theory, Einstein reflected on whether "God could have made the Universe in a different way; that is, whether the necessity of logical simplicity leaves any freedom at all."[1] With this remark, Einstein articulated the nascent form of a view that is currently shared by many physicists: If there is a final theory of nature, one of the most convincing arguments in support of its particular form would be that the theory couldn't be otherwise. The ultimate theory should take the form that it does because it is the unique explanatory framework capable of describing the universe without running up against any internal inconsistencies or logical absurdities. Such a theory would declare that things are the way they are because they *have* to be that way. Any and all variations, no matter how small, lead to a theory that—like the phrase "This sentence is a lie"—sows the seeds of its own destruction.

Establishing such inevitability in the structure of the universe would take us a long way toward coming to grips with some of the deepest questions of the ages. These questions emphasize the mystery surrounding who or what made the seemingly innumerable choices apparently required to design our universe. Inevitability answers these questions by erasing the options. Inevitability means that, in actuality, there are no choices. Inevitability declares that the universe could not have been different. As we will discuss in Chapter 14, nothing ensures that the universe is so tightly constructed. Nevertheless, the pursuit of such rigidity in the laws of nature lies at the heart of the unification program in modern physics.

By the late 1980s, it appeared to physicists that although string theory came close to providing a unique picture of the universe, it did not quite make the grade. There were two reasons for this. First, as briefly noted in Chapter 7, physicists found that there were actually *five* different versions of string theory. You may recall that they are called the Type I, Type IIA, Type IIB, Heterotic O(32) (Heterotic-O, for short), and Heterotic $E_8 \times E_8$ (Heterotic-E, for short) theories. They all share many basic features—their vibrational patterns determine the possible mass and force charges, they require a total of 10 spacetime dimensions, their curled-up dimensions must be in one of the Calabi-Yau shapes, etc.—and for this reason we have not emphasized their differences in previous chapters. Nevertheless, analyses in the 1980s showed that they do differ. You can read more about their properties in the endnotes, but it's enough to know that they differ in how they incorporate supersymmetry as well as in significant details of the vibrational patterns they support.[2] (Type I string theory, for example, has open strings with two loose ends in addition to the closed loops we have focused on.) This has been an embarrassment for string theorists because although it's impressive to have a serious proposal for the final unified theory, having five proposals takes significant wind from the sails of each.

The second deviation from inevitability is more subtle. To fully appreciate it, you must recognize that all physical theories consist of two parts. The first part is the collection of fundamental ideas of the theory, which are usually expressed by mathematical equations. The second part of a theory comprises the solutions to its equations. Generally speaking, some equations have one and only one solution while others have more than one solution (possibly many more). (For a simple example, the equation "2 times a particular number equals 10" has one solution: 5. But the equation "0 times a particular number equals 0" has infinitely many solutions, since 0 times *any* number is 0.) And so, even if research leads to a unique theory with unique equations, it might be that inevitability is compromised because the equations have many different possible solutions. By the late 1980s, it appeared that this was the case with string theory. When physicists studied the equations of any one of the five string theories, they found that they *do* have many solutions—for example, many different possible ways to curl up the extra dimensions—with each solution corresponding to a universe with different properties. Most of these universes, although emerging as valid solutions to the equations of string theory, appear to be irrelevant to the world as we know it.

These deviations from inevitability might seem to be unfortunate fundamental characteristics of string theory. But research since the mid-1990s has given us dramatic new hope that these features may be merely reflections of the way string theorists have been analyzing the theory. Briefly put, the equations of string theory are so complicated that no one knows their exact form. Physicists have managed to write down only approximate versions of the equations. It is these approximate equations that differ significantly from one string theory to the next. And it is these approximate equations, within the context of any one of the five string theories, that give rise to an abundance of solutions, a cornucopia of unwanted universes.

Since 1995 (the start of the second superstring revolution), there has been a growing body of evidence that the exact equations, whose precise form is still beyond our reach, may resolve these problems, thereby helping to give string theory the stamp of inevitability. In fact, it has already been established to the satisfaction of most string theorists that, when the exact equations are understood, they will show that all five string theories are actually intimately related. Like the appendages on a starfish, they are all part of one connected entity whose detailed properties are currently under intense investigation. Rather than having five distinct string theories, physicists are now convinced that there is *one* theory that sews all five into a unique theoretical framework. And like the clarity that emerges when hitherto hidden relationships are revealed, this union is providing a powerful new vantage point for understanding the universe according to string theory.

To explain these insights we must engage some of the most difficult, cutting-edge developments in string theory. We must understand the nature of the approximations used in studying string theory and their inherent limitations. We must gain some familiarity with the clever techniques—collectively called *dualities*—that physicists have invoked to circumvent some of these approximations. And then we must follow the subtle reasoning that makes use of these techniques to find the remarkable insights alluded to above. But don't worry. The really hard work has already been done by string theorists and we will content ourselves here with explaining their results.

Nevertheless, as there are many seemingly separate pieces that we must develop and assemble, in this chapter it is especially easy to lose the forest for the trees. And so, if at any time in this chapter the discussion gets a little too involved and you feel compelled to rush on to black holes (Chapter 13) or cosmology (Chapter 14), take a quick glance back at the following section, which summarizes the key insights of the second superstring revolution.

A Summary of the Second Superstring Revolution

The primary insight of the second superstring revolution is summarized by Figures 12.1 and 12.2. In Figure 12.1 we see the situation prior to the recent ability to go (partially) beyond the approximation methods physicists have traditionally used to analyze string theory. We see that the five string theories were thought of as being completely separate. But, with the newfound insights emerging from recent research, as indicated in Figure 12.2, we see that, like the starfish's five arms, all of the string theories are now viewed as a single, all-encompassing framework. (In fact, by the end of this chapter we will see that even a sixth theory—a sixth arm—will be

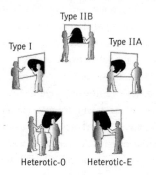

Type IIB

Type I   Type IIA

Heterotic-0   Heterotic-E

Figure 12.1 For many years, physicists working on the five string theories thought they were working on completely separate theories.

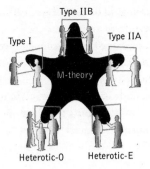

Type IIB

Type I   Type IIA

M-theory

Heterotic-O   Heterotic-E

Figure 12.2 Results from the second superstring revolution have shown that all five string theories are actually part of a single, unified framework, tentatively called M-theory.

merged into this union.) This overarching framework has provisionally been called M-theory, for reasons that will become clear as we proceed. Figure 12.2 represents a landmark achievement in the quest for the ultimate theory. Seemingly disconnected threads of research in string theory have now been woven together into a single tapestry—a unique, all-encompassing theory that may well be the long-sought theory of everything.

Although much work remains to be done, there are two essential features of M-theory that physicists have already uncovered. First, M-theory has *eleven* dimensions (ten space and one time). Somewhat as Kaluza found that one additional spatial dimension allowed for an unexpected merger of general relativity and electromagnetism, string theorists have realized that one additional spatial dimension in string theory—beyond the nine space and one time dimensions discussed in preceding chapters—allows for a deeply satisfying synthesis of all five versions of the theory. Moreover, this extra spatial dimension is not pulled out of thin air; rather, string theorists have realized that the reasoning of the 1970s and 1980s that led to one time and nine space dimensions was *approximate,* and that exact calculations, which can now be completed, show that one spatial dimension had hitherto been overlooked.

The second feature of M-theory that has been discovered is that it contains vibrating strings, but it also includes other objects: vibrating *two-*

dimensional membranes, undulating *three*-dimensional blobs (called "three-branes"), and a host of other ingredients as well. As with the eleventh dimension, this feature of M-theory emerges when calculations are freed from reliance on the approximations used prior to the mid-1990s.

Beyond these and a variety of other insights attained over the last few years, much of the true nature of M-theory remains mysterious—one suggested meaning for the "M." Physicists worldwide are working with great vigor to acquire a full understanding of M-theory, and this may well constitute the central problem of twenty-first-century physics.

An Approximation Method

The limitations of the methods physicists have been using to analyze string theory are bound up with something called *perturbation theory*. Perturbation theory is an elaborate name for making an approximation to try to give a rough answer to a question, and then systematically improving this approximation by paying closer attention to fine details initially ignored. It plays an important part in many areas of scientific research, has been an essential element in understanding string theory, and, as we now illustrate, is also something we encounter frequently in our day-to-day lives.

Imagine that one day your car is acting up, so you go see a mechanic to have it checked out. After giving your car a once-over, he gives you the bad news. The car needs a new engine block, for which parts and labor typically run in the $900 range. This is a ballpark approximation that you expect to be refined as the finer details of the work required become apparent. A few days later, having had time to run additional tests on the car, the mechanic gives you a more precise estimate, $950. He explains that you also need a new regulator, which with parts and labor costs about $50. Finally, when you go to pick up the car, he has added together all of the detailed contributions and presents you with a bill of $987.93. This, he explains, includes the $950 for the engine block and regulator, an additional $27 covering a fan belt, $10 for a battery cable, and $.93 for an insulated bolt. The initial approximate figure of $900 has been refined by including more and more details. In physics terms, these details are referred to as *perturbations* to the initial estimate.

When perturbation theory is properly and effectively applied, the initial estimate will be reasonably close to the final answer; when incorporated, the fine details ignored in the initial estimate make small differences in the final result. But sometimes when you go to pay a final bill it is shockingly different from the initial estimate. Although you might use other, more emotive terms, technically this is called a *failure of perturbation theory*. This means that the initial approximation was not a good guide to the final answer because the "refinements," rather than causing relatively small deviations, resulted in large changes to the ballpark estimate.

As indicated briefly in earlier chapters, our discussion of string theory to this point has relied on a perturbative approach somewhat analogous to that used by the mechanic. The "incomplete understanding" of string theory that we have referred to from time to time has its roots, in one way or another, in this approximation method. Let's build up to an understanding of this important remark by discussing perturbation theory in a context that is less abstract than string theory but closer to its string theory application than the example of the mechanic.

A Classical Example of Perturbation Theory

Understanding the motion of the earth through the solar system provides a classic example of using a perturbative approach. On such large distance scales, we need consider only the gravitational force, but unless further approximations are made, the equations encountered are extremely complicated. Remember that according to both Newton and Einstein, everything exerts a gravitational influence on everything else, and this quickly leads to a complex and mathematically intractable gravitational tug-of-war involving the earth, the sun, the moon, the other planets, and, in principle, all other heavenly bodies as well. As you can imagine, it is impossible to take all of these influences into account and determine the exact motion of the earth. In fact, even if there were only three heavenly participants, the equations become so complicated that no one has been able to solve them in full.[3]

Nevertheless, we *can* predict the motion of the earth through the solar system with great accuracy by making use of a perturbative approach. The enormous mass of the sun, in comparison to that of every other member of our solar system, and its proximity to the earth, in comparison to that of every other star, makes it by far the dominant influence on the earth's motion. And so, we can get a ballpark estimate by considering only the sun's gravitational influence. For many purposes this is perfectly adequate. If necessary, we can refine this approximation by sequentially including the gravitational effects of the next-most-relevant bodies, such as the moon and whichever planets are passing closest by at the moment. The calculations can start to become difficult as the emerging web of gravitational influences gets complicated, but don't let this obscure the perturbative philosophy: The sun-earth gravitational interaction gives us an approximate explanation of the earth's motion, while the remaining complex of other gravitational influences offers a sequence of ever smaller refinements.

A perturbative approach works in this example because there is a dominant physical influence that admits a relatively simple theoretical description. This is not always the case. For example, if we are interested in the motion of three comparable-mass stars orbiting one another in a trinary system, there is no single gravitational relationship whose influence dwarfs the others. Correspondingly, there is no single dominant interaction that provides a ballpark estimate, with the other effects yielding small refinements. If we tried to use a perturbative approach by, say, singling out the gravitational attraction between two stars and using it to determine our ballpark approximation, we would quickly find that our approach had failed. Our calculations would reveal that the "refinement" to the predicted motion arising from the inclusion of the third star is *not* small, but in fact is as significant as the supposed ballpark approximation. This is familiar: The motion of three people dancing the hora bears little resemblance to two people dancing the tango. A large refinement means that the initial approximation was way off the mark and the whole scheme was built on a house of cards. You should note that it is not simply a matter of including the large refinement due to the third star. There is a domino effect: The large refinement has a significant impact on the motion of the other two stars, which in turn has a large impact on the motion of the third star, which then has a substantial impact on the other two, and so on. All strands in the gravitational web are equally important and must be dealt with simultaneously. Oftentimes, in such cases, our only recourse is to make use of the brute power of computers to simulate the resulting motion.

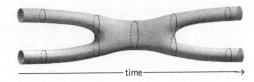

$\longleftarrow$————————————— time ——————————————$\longrightarrow$

Figure 12.3 Strings interact by joining and splitting.

This example highlights the importance, when using a perturbative approach, of determining whether the supposedly ballpark estimate really *is* in the ballpark, and if it is, which and how many of the finer details must be included in order to achieve a desired level of accuracy. As we now discuss, these issues are particularly crucial for applying perturbative tools to physical processes in the microworld.

A Perturbative Approach to String Theory

Physical processes in string theory are built up from the basic interactions between vibrating strings. As we discussed toward the end of Chapter 6,* these interactions involve the splitting apart and joining together of string loops, such as in Figure 6.7, which we reproduce in Figure 12.3 for convenience. String theorists have shown how a precise mathematical formula can be associated with the schematic portrayal of Figure 12.3—a formula that expresses the influence that each incoming string has on the resulting motion of the other. (The details of the formula differ among the five string theories, but for the time being we will ignore such subtle features.) If it weren't for quantum mechanics, this formula would be the end of the story of how the strings interact. But the microscopic frenzy dictated by the uncertainty principle implies that string/antistring pairs (two strings executing opposite vibrational patterns) can momentarily erupt into existence, borrowing energy from the universe, so long as they annihilate one another with sufficient haste, thereby repaying the energy loan. Such pairs of strings, born of the quantum frenzy but which live on borrowed energy

*Those readers who skipped over the "More Precise Answer" section of Chapter 6 might find it helpful to skim the beginning part of that section.

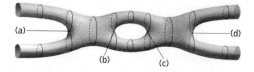

Figure 12.4 The quantum frenzy can cause a string/antistring pair to erupt (b) and annihilate (c), yielding a more complicated interaction.

and hence must shortly recombine into a single loop, are known as *virtual string pairs*. And even though it is only momentary, the transient presence of these additional virtual string pairs affects the detailed properties of the interaction.

This is schematically depicted in Figure 12.4. The two initial strings slam together at the point marked (a), where they merge together into a single loop. This loop travels a bit, but at (b) frenzied quantum fluctuations result in the creation of a virtual string pair that travels along and then subsequently annihilates at (c), producing, once again, a single string. Finally, at (d), this string gives up its energy by dissociating into a pair of strings that head off in new directions. Because of the single loop in the center of Figure 12.4, physicists call this a "one-loop" process. As with the interaction depicted in Figure 12.3, a precise mathematical formula can be associated with this diagram to summarize the effect the virtual string pair has on the motion of the two original strings.

But that's not the end of the story either, because quantum jitters can cause momentary virtual string eruptions to occur any number of times, producing a sequence of virtual string pairs. This gives rise to diagrams with more and more loops, as illustrated in Figure 12.5. Each of these diagrams provides a handy and simple way of depicting the physical processes involved: The incoming strings merge together, quantum jitters cause the resulting loop to split apart into a virtual string pair, these travel along and then annihilate one another by merging together into a single loop, which travels along and produces another virtual string pair, and on and on. As with the other diagrams, there is a corresponding mathematical formula for each of these processes that summarizes the effect on the motion of the original pair of strings.[4]

Moreover, just as the mechanic determined your final car-repair bill through a refinement of his original estimate of $900 by adding to it $50,

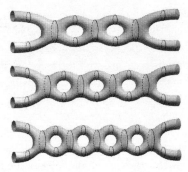

Figure 12.5 The quantum frenzy can cause numerous sequences of string/antistring pairs to erupt and annihilate.

$27, $10, and $.93, and just as we arrived at an ever more precise understanding of the motion of the earth through a refinement of the sun's influence by including the smaller effects of the moon and other planets, string theorists have shown that we can understand the interaction between two strings by adding together the mathematical expressions for diagrams with no loops (no virtual string pairs), with one loop (one pair of virtual strings), with two loops (two pairs of virtual strings), and so forth, as illustrated in Figure 12.6.

An exact calculation requires that we add together the mathematical expressions associated with each of these diagrams, with an increasingly large number of loops. But, since there are an infinite number of such diagrams and the mathematical calculations associated with each get in-

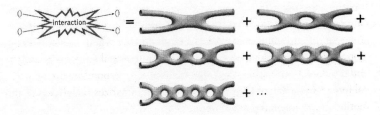

Figure 12.6 The net influence each incoming string has on the other comes from adding together the influences involving diagrams with ever more loops.

creasingly difficult as the number of loops grows, this is an impossible task. Instead, string theorists have cast these calculations into a perturbative framework based on the expectation that a reasonable ballpark estimate is given by the zero-loop processes, with the loop diagrams resulting in refinements that get smaller as the number of loops increases.

In fact, almost everything we know about string theory—including much of the material covered in previous chapters—has been discovered by physicists performing detailed and elaborate calculations using this perturbative approach. But to trust the accuracy of the results found, one must determine whether the supposedly ballpark approximations that ignore all but the first few diagrams in Figure 12.6 are really in the ballpark. This leads us to ask the crucial question: Are we in the ballpark?

Is the Ballpark in the Ballpark?

It depends. Although the mathematical formula associated with each diagram becomes very complicated as the number of loops grows, string theorists have recognized one basic and essential feature. Somewhat as the strength of a rope determines the likelihood that vigorous pulling and shaking will cause it to tear into two pieces, there is a number that determines the likelihood that quantum fluctuations will cause a single string to split into two strings, momentarily yielding a virtual pair. This number is known as the *string coupling constant* (more precisely, each of the five string theories has its own string coupling constant, as we will discuss shortly). The name is quite descriptive: The size of the string coupling constant describes how strongly the quantum jitters of three strings (the initial loop and the two virtual loops into which it splits) are related—how tightly, so to speak, they are *coupled* to one another. The calculational formalism shows that the larger the string coupling constant, the more likely it is that quantum jitters will cause an initial string to split apart (and subsequently rejoin); the smaller the string coupling constant, the less likely it is for such virtual strings to erupt momentarily into existence.

We will shortly take up the question of determining the value of the string coupling constant within any of the five string theories, but first, what do we really mean by "small" or "large" when assessing its size? Well, the mathematics underlying string theory shows that the dividing line between "small" and "large" is the number 1, in the following sense. If the string coupling constant has a value less than 1, then—like multiple strikes of lightning—larger numbers of virtual string pairs are increasingly *unlikely* to erupt momentarily into existence. If the coupling constant is 1 or greater, however, it is increasingly *likely* that ever-larger numbers of such virtual pairs will momentarily burst on the scene.[5] The upshot is that if the string coupling constant is less than 1, the loop diagram contributions become ever smaller as the number of loops grows. This is just what is needed for the perturbative framework, since it indicates that we will get reasonably accurate results even if we ignore all processes except for those with just a few loops. But if the string coupling constant is not less than 1, the loop diagram contributions become more important as the number of loops increases. As in the case of a trinary star system, this invalidates a perturbative approach. The supposed ballpark approximation—the process with no loops—is *not* in the ballpark. (This discussion applies equally well to each of the five string theories—with the value of the string coupling constant in any given theory determining the efficacy of the perturbative approximation scheme.)

This realization leads us to the next crucial question: What is the value of the string coupling constant (or, more precisely, what are the values of the string coupling constants in each of the five string theories)? *At present, no one has been able to answer this question.* It is one of the most important unresolved issues in string theory. We can be sure that conclusions based on a perturbative framework are justified only if the string coupling constant is less than 1. Moreover, the precise value of the string coupling constant has a direct impact on the masses and charges carried by the various string vibrational patterns. Thus, we see that much physics hinges on the value of the string coupling constant. And so, let's take a closer look at why the important question of its value—in any of the five string theories—remains unanswered.

The Equations of String Theory

The perturbative approach for determining how strings interact with one another can also be used to determine the fundamental equations of string theory. In essence, the equations of string theory determine how strings interact and, conversely, the way strings interact directly determines the equations of the theory.

As a prime example, in each of the five string theories there is an equation that is meant to determine the value of the theory's coupling constant. Currently, however, physicists have been able to find only an approximation to this equation, in each of the five string theories, by mathematically evaluating a small number of relevant string diagrams using a perturbative approach. Here is what the approximate equations say: In any of the five string theories, the string coupling constant takes on a value such that if it is multiplied by zero the result is zero. This is a terribly disappointing equation; since any number times zero yields zero, the equation can be solved with any value of the string coupling constant. Thus, in any of the five string theories, the approximate equation for its string coupling constant gives us no information about its value.

While we are at it, in each of the five string theories there is another equation that is supposed to determine the precise form of both the extended and the curled-up spacetime dimensions. The approximate version of this equation that we currently have is far more restrictive than the one dealing with the string coupling constant, but it still admits many solutions. For instance, four extended spacetime dimensions together with any curled-up, six-dimensional Calabi-Yau space provide a whole class of solutions, but even this does not exhaust the possibilities, which also allow for a different split between the number of extended and curled-up dimensions.[6]

What can we make of these results? There are three possibilities. First, starting with the most pessimistic possibility, although each string theory comes equipped with equations to determine the value of its coupling constant as well as the dimensionality and precise geometrical form of spacetime—something no other theory can claim—even the as-yet-unknown exact form of these equations may admit a vast spectrum of solutions, substantially weakening their predictive power. If true, this would be a setback, since the promise of string theory is that it will be able to *explain* these features of the cosmos, rather than require us to determine them from experimental observation and, more or less arbitrarily, insert them into the theory. We will return to this possibility in Chapter 15. Second, the unwanted flexibility in the approximate string equations may be an indication of a subtle flaw in our reasoning. We are attempting to use a perturbative approach to determine the value of the string coupling constant itself. But, as discussed, perturbative methods are sensible only if the coupling constant is less than 1, and hence our calculation may be making an unjustified assumption about its own answer—namely, that the result will be smaller than 1. Our failure could well indicate that this assumption is wrong and that, perhaps, the coupling in any one of the five string theories is greater than 1. Third, the unwanted flexibility may merely be due to our use of approximate rather than exact equations. For instance, even though the coupling constant in a given string theory might be less than 1, the equations of the theory may still depend sensitively on the contributions from *all* diagrams. That is, the accumulated small refinements from diagrams with ever more loops might be essential for modifying the approximate equations—which admit many solutions—into exact equations that are far more restrictive.

By the early 1990s, the latter two possibilities made it clear to most string theorists that complete reliance on the perturbative framework was standing squarely in the way of progress. The next breakthrough, most everyone in the field agreed, would require a *nonperturbative* approach—an approach that was not shackled to approximate calculational techniques and could therefore reach well beyond the limitations of the perturbative framework. As of 1994, finding such a means seemed like a pipe dream. Sometimes, though, pipe dreams spill over into reality.

Duality

Hundreds of string theorists from around the world gather together annually for a conference devoted to recapping the past year's results and assessing the relative merit of various possible research directions. Depending on the state of progress in a given year, one can usually predict the level of interest and excitement among the participants. In the mid-1980s, the heyday of the first superstring revolution, the meetings were filled with unrestrained euphoria. Physicists had widespread hope that they would shortly understand string theory completely, and that they would reveal it to be the ultimate theory of the universe. In retrospect this was naive. The intervening years have shown that there are many deep and subtle aspects of string theory that will undoubtedly take prolonged and dedicated effort to understand. The early, unrealistic expectations resulted in a backlash; when everything did not immediately fall into place, many researchers were crestfallen. The string conferences of the late 1980s reflected the low-level disillusionment—physicists presented interesting results, but the atmosphere lacked inspiration. Some even suggested that the community stop holding an annual strings conference. But things picked up in the early 1990s. Through various breakthroughs, some of which we have discussed in previous chapters, string theory was rebuilding its momentum and researchers were regaining their excitement and optimism. But very little presaged what happened at the strings conference in March 1995 at the University of Southern California.

When his appointed hour to speak had arrived, Edward Witten strode to the podium and delivered a lecture that ignited the second superstring revolution. Inspired by earlier works of Duff, Hull, Townsend, and building on insights of Schwarz, the Indian physicist Ashoke Sen, and others, Witten announced a strategy for transcending the perturbative understanding of string theory. A central part of the plan involves the concept of *duality*.

Physicists use the term duality to describe theoretical models that appear to be different but nevertheless can be shown to describe exactly the same physics. There are "trivial" examples of dualities in which ostensibly different theories are actually identical and appear to be different only because of the way in which they happen to be presented. To someone who knows only English, general relativity might not immediately be recognized as Einstein's theory if presented in Chinese. A physicist fluent in both languages, though, can easily perform a translation from one to the other, establishing their equivalence. We call this example "trivial" because nothing is gained, from the point of view of physics, by such a translation. If someone who is fluent in English and Chinese were studying a difficult problem in general relativity, it would be equally challenging regardless of the language used to expressed it. A switch from English to Chinese, or vice versa, brings no new physical insight.

Nontrivial examples of duality are those in which distinct descriptions of the same physical situation *do* yield different and complementary physical insights and mathematical methods of analysis. In fact, we have already encountered two examples of duality. In Chapter 10, we discussed how string theory in a universe that has a circular dimension of radius R

can equally well be described as a universe with a circular dimension of radius $1/R$. These are distinct geometrical situations that, through the properties of string theory, are actually physically identical. Mirror symmetry is a second example. Here two different Calabi-Yau shapes of the extra six spatial dimensions—universes that at first sight would appear to be completely distinct—yield exactly the same physical properties. They give dual descriptions of a single universe. Of crucial importance, unlike the case of English versus Chinese, there *are* important physical insights that follow from using these dual descriptions, such as a minimum size for circular dimensions and topology-changing processes in string theory.

In his lecture at Strings '95, Witten gave evidence for a new, profound kind of duality. As briefly outlined at the beginning of this chapter, he suggested that the five string theories, although apparently different in their basic construction, are all just different ways of describing the same underlying physics. Rather than having five different string theories, then, we would simply have five different windows onto this single underlying theoretical framework.

Before the developments of the mid-1990s, the possibility of such a grand version of duality was one of those wishful ideas that physicists might harbor, but about which they would rarely if ever speak, since it seems so outlandish. If two string theories differ with regard to significant details of their construction, it's hard to imagine how they could merely be different descriptions of the same underlying physics. Nonetheless, through the subtle power of string theory, there is mounting evidence that all five string theories *are* dual. And furthermore, as we will discuss, Witten gave evidence that even a sixth theory gets mixed into the stew.

These developments are intimately entwined with the issues regarding the applicability of perturbative methods we encountered at the end of the preceding section. The reason is that the five string theories are manifestly different when each is *weakly coupled*—a term of the trade meaning that the coupling constant of a theory is less than 1. Because of their reliance on perturbative methods, physicists have been unable for some time to address the question of what properties any one of the string theories would have if its coupling constant should be larger than 1—the so-called *strongly coupled* behavior. The claim of Witten and others, as we now discuss, is that this crucial question can now be answered. Their results convincingly suggest that, together with a sixth theory we have yet to describe, the strong coupling behavior of any of these theories has a dual description in terms of the weak coupling behavior of another, and vice versa.

To gain a more tangible sense of what this means, you might want to keep the following analogy in mind. Imagine two rather sheltered individuals. One loves ice but, strangely enough, has never seen water (in its liquid form). The other loves water but, equally strangely, has never seen ice. Through a chance meeting, they decide to team up for a camping trip in the desert. When they set out to leave, each is fascinated by the other's gear. The ice-lover is captivated by the water-lover's silky smooth transparent liquid, and the water-lover is strangely drawn to the remarkable solid crystal cubes brought by the ice-lover. Neither has any inkling that there is actually a deep relationship between water and ice; to them, they are two completely different substances. But as they head out into the scorching heat of the desert, they are shocked to find that the ice slowly begins to turn into water. And, in the frigid cold of the desert night, they are equally shocked to find that the liquid water slowly begins to turn into solid ice. They realize that these two substances—which they initially thought to be completely unrelated—are intimately connected.

The duality among the five string theories is somewhat similar: Roughly speaking, the string coupling constants play a role analogous to temperature in our desert analogy. Like ice and water, any two of the five string theories, at first sight, appear to be completely distinct. But as we vary their respective coupling constants, the theories transmute among themselves. Just as ice transmutes into water as we raise its temperature, one string theory can transmute into another as we increase the value of its coupling constant. This takes us a long way toward showing that all of the string theories are dual descriptions of one single underlying structure—the analog of H_2O for water and ice.

The reasoning underlying these results relies almost entirely on the use of arguments rooted in principles of symmetry. Let's discuss this.

The Power of Symmetry

Over the years, no one even attempted to study the properties of any of the five string theories for large values of their string coupling constants because no one had any idea how to proceed without the perturbative framework. However, during the late 1980s and early 1990s, physicists made slow but steady progress in identifying certain special properties— including certain masses and force charges—that are part of the strong-coupling physics of a given string theory, and yet are still within our ability to calculate. The calculation of these properties, which necessarily transcends the perturbative framework, has played a central role in driving the progress of the second superstring revolution and is firmly rooted in the power of symmetry.

Symmetry principles provide insightful tools for understanding a great many things about the physical world. We have discussed, for instance, that the well-supported belief that the laws of physics do not treat any place in the universe or moment in time as special allows us to argue that the laws governing the here and now are the same ones at work everywhere and everywhen. This is a grandiose example, but symmetry principles can be equally important in less all-encompassing circumstances. For instance, if you witness a crime but were able to catch only a glimpse of the right side of the perpetrator's face, a police artist can nonetheless use your information to sketch the whole face. Symmetry is why. Although there are differences between the left and right sides of a person's face, most are symmetric enough that an image of one side can be flipped over to get a good approximation of the other.

In each of these widely different applications, the power of symmetry is its ability to nail down properties in an *indirect* manner—something that is often far easier than a more direct approach. We could learn about fundamental physics in the Andromeda galaxy by going there, finding a planet around some star, building accelerators, and performing the kinds of experiments carried out on earth. But the indirect approach of invoking symmetry under changes of locale is far easier. We could also learn about features on the left side of the perpetrator's face by tracking him down and examining it. But it is often far easier to invoke the left-right symmetry of faces.[7]

Supersymmetry is a more abstract symmetry principle that relates physical properties of elementary constituents that carry different amounts of spin. At best there are only hints from experimental results that the microworld incorporates this symmetry, but, for reasons discussed earlier, there is a strong belief that it does. It is certainly an integral part of string theory. In the 1990s, led by the pioneering work of Nathan Seiberg of the

Institute for Advanced Study, physicists have realized that supersymmetry provides a sharp and incisive tool that can answer some very difficult and important questions by indirect means.

Even without understanding intricate details of a theory, the fact that it has supersymmetry built in allows us to place significant constraints on the properties it can have. Using a linguistic analogy, imagine that we are told that a sequence of letters has been written on a slip of paper, that the sequence has exactly three occurrences, say, of the letter "y," and that the paper has been hidden within a sealed envelope. If we are given no further information, then there is no way that we can guess the sequence—for all we know it might be a random assortment of letters with three y's like *mvcfojziyxidqfqzyycdi* or any one of the infinitely many other possibilities. But imagine that we are subsequently given two further clues: The hidden sequence of letters spells out an English word and it has the minimum number of letters consistent with the first clue of having three y's. From the infinite number of letter sequences at the outset, these clues reduce the possibilities to *one* word—to the shortest English word containing three y's: *syzygy*.

Supersymmetry supplies similar constraining clues for those theories that incorporate its symmetry principles. To get a feel for this, imagine that we are presented with a physics puzzle analogous to the linguistic puzzle just described. Hidden inside a box there is something—its identity is left unspecified—that has a certain force charge. The charge might be electric, magnetic, or any of the other generalizations, but to be concrete let's say it has three units of electric charge. Without further information, the identity of the contents cannot be determined. It might be three particles of charge 1, like positrons or protons; it might be four particles of charge 1 and one particle of charge −1 (like the electron), as this combination still has a net charge of three; it might be nine particles of charge one-third (like the up-quark) or it might be the same nine particles accompanied by any number of chargeless particles (such as photons). As was the case with the hidden sequence of letters when we only had the clue about the three y's, the possibilities for the contents of the box are endless.

But let's now imagine that, as in the case of the linguistic puzzle, we are given two further clues: The theory describing the world—and hence the contents of the box—is supersymmetric, and the content of the box has the *minimum mass* consistent with the first clue of having three units of charge. Based on the insights of E. Bogomol'nyi, Manoj Prasad, and Charles Sommerfield, physicists have shown that this specification of a tight organizational framework (the framework of supersymmetry, the analog of the English language) and a "minimality constraint" (minimum mass for a chosen amount of electric charge, the analog of a minimum word length for a chosen number of y's) implies that the identity of the hidden contents is nailed down *uniquely*. That is, merely by ensuring that the contents of the box is the lightest it could possibly be and still have the specified charge, physicists showed that its identity is fully established. Constituents of minimum mass for a chosen value of charge are known as *BPS states*, in honor of their three discoverers.[8]

The important thing about BPS states is that their properties are uniquely, easily, and exactly determined without resort to a perturbative calculation. This is true regardless of the value of the coupling constants. That is, even if the string coupling constant is large, implying that the perturbative approach is invalid, we are still able to deduce the exact properties of the BPS configurations. The properties are often called *nonperturbative* masses and charges since their values transcend the perturbative approximation scheme. For this reason, you can also think of BPS as standing for "beyond perturbative states."

The BPS properties exhaust only a small part of the full physics of a chosen string theory when its coupling constant is large, but they nonetheless give us a tangible grip on some of its strong coupling characteristics. As the coupling constant in a chosen string theory is increased beyond the realm accessible to perturbation theory, we anchor our limited understanding in the BPS states. Like a few choice words in a foreign tongue, we will find that they will take us quite far.

Duality in String Theory

Following Witten, let's start with one of the five string theories, say the Type I string, and imagine that all of its nine space dimensions are flat and unfurled. This, of course, is not at all realistic, but it makes the discussion simpler; we will return to curled-up dimensions shortly. We begin by assuming that the string coupling constant is much less than 1. In this case, perturbative tools are valid, and hence many of the detailed properties of the theory can and have been worked out with accuracy. If we increase the value of the coupling constant but still keep it a good deal less than 1, perturbative methods can still be used. The detailed properties of the theory will change somewhat—for instance, the numerical values associated with the scattering of one string off another will be a bit different because the multiple loop processes of Figure 12.6 make greater contributions when the coupling constant increases. But beyond these changes in detailed numerical properties, the overall physical content of the theory remains the same, so long as the coupling constant stays in the perturbative realm.

As we increase the Type I string coupling constant beyond the value 1, perturbative methods become invalid and so we focus only on the limited set of nonperturbative masses and charges—the BPS states—that are still within our ability to understand. Here is what Witten argued, and later confirmed through joint work with Joe Polchinski of the University of California at Santa Barbara: *These strong coupling characteristics of Type I string theory exactly agree with known properties of Heterotic-O string theory, when the latter has a small value for its string coupling constant.* That is, when the coupling constant of the Type I string is large, the particular masses and charges that we know how to extract are precisely equal to those of the Heterotic-O string when its coupling constant is small. This gives us a strong indication that these two string theories, which at first sight, like water and ice, seem totally different, are actually dual. It persuasively suggests that the physics of the Type I theory for large values of its coupling constant is *identical* to the physics of the Heterotic-O theory for small values of its coupling constant. Related arguments gave equally persuasive evidence that the reverse is also true: The physics of the Type I theory for small values of its coupling constant is identical to that of the Heterotic-O theory for large values of its coupling constant.[9] Although the two string theories appear to be unrelated when analyzed using the perturbative approximation scheme, we now see that each transforms into the other—somewhat like the transmutation between water and ice—as their coupling constants are varied in value.

This central new kind of result, in which the strong coupling physics of one theory is described by the weak coupling physics of another theory, is known as *strong-weak duality*. As with the other dualities discussed previously, it tells us that the two theories involved are not actually distinct. Rather, they give two dissimilar descriptions of the same underlying theory. Unlike the English-Chinese trivial duality, strong-weak coupling duality is powerful. When the coupling constant of one member of a dual pair of theories is small, we can analyze its physical properties using well-developed perturbative tools. If the coupling constant of the theory is large, however, and thus the perturbative methods break down, we now know that we can use the dual description—a description in which the relevant coupling constant is small—and return to the use of perturbative tools. The translation has resulted in our having quantitative methods to analyze a theory we initially thought to be beyond our theoretical abilities.

Actually proving that the strong coupling physics of the Type I string theory is identical to the weak coupling physics of the Heterotic-O theory, and vice versa, is an extremely difficult task that has not yet been achieved. The reason is simple. One member of the pair of the supposedly dual theories is not amenable to perturbative analysis, as its coupling constant is too big. This prevents direct calculations of many of its physical properties. In fact, it is precisely this point that makes the proposed duality so potent, for, if true, it provides a new tool for analyzing a strongly coupled theory: Use perturbative methods on its weakly coupled dual description.

But even if we cannot prove that the two theories are dual, the perfect alignment between those properties we *can* extract with confidence provides extremely compelling evidence that the conjectured strong-weak coupling relationship between the Type I and Heterotic-O string theories is correct. In fact, increasingly clever calculations that have been performed to test the proposed duality have all resulted in positive results. Most string theorists are convinced that the duality is true.

Following the same approach, one can study the strong coupling properties of another of the remaining string theories, say, the Type IIB string. As originally conjectured by Hull and Townsend and supported by the research of a number of physicists, something equally remarkable appears to occur. As the coupling constant of the Type IIB string gets larger and larger, the physical properties that we are still able to understand appear to match up exactly with that of the weakly coupled Type IIB string itself.

In other words, the Type IIB string is *self-dual*.[10] Specifically, detailed analysis persuasively suggests that if the Type IIB coupling constant were larger than 1, and if we were to change its value to its reciprocal (whose value, therefore, is less than 1), the resulting theory is absolutely identical to the one we started with. Similar to what we found in trying to squeeze a circular dimension to a sub-Planck-scale length, if we try to increase the Type IIB coupling to a value larger than 1, the self-duality shows that the resulting theory is precisely equivalent to the Type IIB string with a coupling smaller than 1.

A Summary, So Far

Let's see where we are. By the mid-1980s, physicists had constructed five different superstring theories. In the approximation scheme of perturbation theory, they all appear to be distinct. But this approximation method is valid only if the string coupling constant in a given string theory is less than 1. The expectation has been that physicists should be able to calculate the precise value of the string coupling constant in any given string theory, but the form of the approximate equations currently available makes this impossible. For this reason, physicists aim to study each of the five string theories for a range of possible values of their respective coupling constants, both less than and greater than 1—i.e., both weak and strong coupling. But traditional perturbative methods give no insight into the strong coupling characteristics of any of the string theories.

Recently, by making use of the power of supersymmetry, physicists have learned how to calculate some of the strong coupling properties of a given string theory. And to the surprise of most everyone in the field, the strong coupling properties of the Heterotic-O string appear to be identical to the weak coupling properties of the Type I string, and vice versa. Moreover, the strong coupling physics of the Type IIB string is identical to its own properties when its coupling is weak. These unexpected links encourage us to follow Witten and press on to the other two string theories, Type IIA and Heterotic-E, to see how they fit into the overall picture. Here we will find even more exotic surprises. To prepare ourselves, we need a brief historical digression.

Supergravity

In the late 1970s and early 1980s, before the surge of interest in string theory, many theoretical physicists sought a unified theory of quantum mechanics, gravity, and the other forces in the framework of point-particle quantum field theory. The hope was that the inconsistencies between point-particle theories involving gravity and quantum mechanics would be overcome by studying theories with a great deal of symmetry. In 1976 Daniel Freedman, Sergio Ferrara, and Peter Van Nieuwenhuizen, all then of the State University of New York at Stony Brook, discovered that the most promising were those involving supersymmetry, since the tendency of bosons and fermions to give cancelling quantum fluctuations helps to calm the violent microscopic frenzy. The authors coined the term *supergravity* to describe supersymmetric quantum field theories that try to incorporate general relativity. Such attempts to merge general relativity with quantum mechanics ultimately met with failure. Nevertheless, as mentioned in Chapter 8, there was a prescient lesson to be learned from these investigations, one that presaged the development of string theory.

The lesson, which perhaps became most clear through the work of Eugene Cremmer, Bernard Julia, and Scherk, all of the Ecole Normale Supérieure in 1978, was that the attempts that came closest to success were supergravity theories formulated not in four dimensions, but in more. Specifically, the most promising were the versions calling for ten or eleven dimensions, with eleven dimensions, it turns out, being the maximal possible.[11] Contact with four observed dimensions was accomplished in the framework, once again, of Kaluza and Klein: The extra dimensions were curled up. In the ten-dimensional theories, as in string theory, six dimensions were curled up, while seven were curled up for the eleven-dimensional theory.

When string theory took physicists by storm in 1984, the perspective on point-particle supergravity theories changed dramatically. As emphasized repeatedly, if we examine a string with the precision available currently and for the foreseeable future, it *looks* like a point particle. We can make this informal remark precise: When studying low-energy processes in string theory—those processes that do not have enough energy to probe the ultramicroscopic, extended nature of the string—we can approximate a string by a structureless point particle, using the framework of point-particle quantum field theory. We cannot use this approximation when dealing with short-distance or high-energy processes because we know that the extended nature of the string is crucial to its ability to resolve the conflicts between general relativity and quantum mechanics that a point-particle theory cannot. But at low enough energies—large enough distances—these problems are not encountered, and such an approximation is often made for the sake of calculational convenience.

The quantum field theory that most closely approximates string theory in this manner is none other than ten-dimensional supergravity. The special properties of ten-dimensional supergravity discovered in the 1970s and 1980s are now understood to be low-energy relics of the underlying power of string theory. Researchers studying ten-dimensional supergravity had uncovered the tip of a very deep iceberg—the rich structure of superstring theory. In fact, it turns out that there are four different ten-dimensional supergravity theories that differ in details regarding the precise way in which supersymmetry is incorporated. Three of these turn out to be the low-energy point-particle approximations to the Type IIA string, the Type IIB string, and the Heterotic-E string. The fourth gives the the low-energy point-particle approximation to both the Type I string and the Heterotic-O string; in retrospect, this was the first indication of the close connection between these two string theories.

This is a very tidy story except that eleven-dimensional supergravity seems to have been left out in the cold. String theory, formulated in ten dimensions, appears to have no room for an eleven-dimensional theory. For a number of years, the general view held by most but not all string theorists was that eleven-dimensional supergravity was a mathematical oddity without any connection to the physics of string theory.[12]

Glimmers of M-Theory

The view now is very different. At Strings '95, Witten argued that if we start with the Type IIA string and increase its coupling constant from a value much less than 1 to a value much greater than 1, the physics we are still able to analyze (essentially that of the BPS saturated configurations) has a low-energy approximation that *is* eleven-dimensional supergravity.

When Witten announced this discovery, it stunned the audience and it has since rocked the string theory community. For almost everyone in the field, it was a completely unexpected development. Your first reaction to this result may echo that of most experts in the field: *How can a theory specific to eleven dimensions be relevant to a different theory in ten?*

The answer is of deep significance. To understand it, we must describe Witten's result more precisely. Actually, it's easier first to illustrate a closely related result discovered later by Witten and a postdoctoral fellow at Princeton University, Petr Hořava, that focuses on the Heterotic-E string. They found that the strongly coupled Heterotic-E string also has an eleven-dimensional description, and Figure 12.7 shows why. In the leftmost part of the figure, we take the Heterotic-E string coupling constant to be much smaller than 1. This is the realm that we have been describing in previous chapters and that string theorists have studied for well over a decade. As we move to the right in Figure 12.7, we sequentially increase the size of the coupling constant. Prior to 1995, string theorists knew that this would make the loop processes (see Figure 12.6) increasingly important and, as the coupling constant got larger, would ultimately invalidate the whole perturbative framework. But what no one suspected is that as the coupling constant is made larger, a new dimension becomes visible! This is the "vertical" dimension shown in Figure 12.7. Bear in mind that in this figure the two-dimensional grid with which we begin represents all nine spatial dimensions of the Heterotic-E string. Thus, the new, vertical dimension represents a *tenth* spatial dimension, which, together with time, takes us to a total of eleven spacetime dimensions.

Moreover, Figure 12.7 illustrates a profound consequence of this new

Figure 12.7 As the Heterotic-E string coupling constant is increased, a new space dimension appears and the string itself gets stretched into a cylindrical membrane shape.

dimension. The *structure* of the Heterotic-E string changes as this dimension grows. It is stretched from a one-dimensional loop into a ribbon and then a deformed cylinder as we increase the size of the coupling constant! In other words, the Heterotic-E string is *actually a two-dimensional membrane* whose width (the vertical extent in Figure 12.7) is controlled by the size of the coupling constant. For over a decade, string theorists have always used perturbative methods that are firmly rooted in the assumption that the coupling constant is very small. As argued by Witten, this assumption has made the fundamental ingredients look and behave like one-dimensional strings even though they actually have a hidden, second spatial dimension. By relaxing the assumption that the coupling constant is very small and considering the physics of the Heterotic-E string when the coupling constant is large, the second dimension becomes manifest.

This realization does not invalidate any of the conclusions we have drawn in previous chapters, but it does force us to see them within a new framework. For instance, how does this all mesh with the one time and nine space dimensions required by string theory? Well, recall from Chapter 8 that this constraint arises from counting the number of independent directions in which a string can vibrate, and requiring that this number be just right to ensure that quantum-mechanical probabilities have sensible values. The new dimension we have just uncovered is *not* one in which a Heterotic-E string can vibrate, since it is a dimension that is locked within the structure of the "strings" themselves. Put another way, the perturbative framework that physicists used in deriving the requirement of a ten-dimensional spacetime assumed from the outset that the Heterotic-E coupling constant is small. Although it was not recognized until much later, this implicitly enforced two mutually consistent approximations: that the width of the membrane in Figure 12.7 is small, making it look like a string, and that the eleventh dimension is *so* small that it is beyond the sensitivity of the perturbative equations. Within this approximation scheme, we are led to envision a ten-dimensional universe filled with one-dimensional strings. Now we see that this is but an approximation to an eleven-dimensional universe containing two-dimensional membranes.

For technical reasons, Witten first came upon the eleventh dimension in his studies of the strong coupling properties of the Type IIA string, and there the story is quite similar. As in the Heterotic-E example, there is an eleventh dimension whose size is controlled by the Type IIA coupling constant. When its value is increased, the new dimension grows. As it does, Witten argued, the Type IIA string, rather than stretching into a ribbon as in the Heterotic-E case, expands into an "inner tube," as illustrated in Figure 12.8. Once again, Witten argued that although theorists have always viewed Type IIA strings as one-dimensional objects, having only length but no thickness, this view is a reflection of the perturbative approximation scheme in which the string coupling constant is assumed to be small. If nature *does* require a small value of this coupling constant then it is a trustworthy approximation. Nevertheless, Witten's arguments and those of other physicists during the second superstring revolution do give strong evidence that the Type IIA and Heterotic-E "strings" are, fundamentally, two-dimensional membranes living in an eleven-dimensional universe.

But what *is* this eleven-dimensional theory? At low energies (low compared to the Planck energy), Witten and others argued, it is approximated by the long-neglected eleven-dimensional supergravity quantum field theory. But for higher energies, how can we describe this theory? This topic is currently under intense scrutiny. We know from Figures 12.7 and 12.8 that the eleven-dimensional theory contains two-dimensional extended objects—two-dimensional membranes. And as we shall soon discuss, extended objects of other dimensions play an important role as well. But beyond a hodgepodge of properties, *no one knows what this eleven-dimensional theory is.* Are membranes its fundamental ingredients? What are its defining properties? How does it purport to make contact with physics as we know it? If the respective coupling constants are small, our best current answers to these questions are described in previous chapters, since at small coupling constants we are led back to the theory of strings.

Figure 12.8 As the Type IIA string coupling constant is increased, strings expand from one-dimensional loops to two-dimensional objects that look like the surface of a bicycle-tire inner tube.

But if the coupling constants are not small, no one currently knows the answers.

Whatever the eleven-dimensional theory is, Witten has provisionally named it *M-theory*. The name stands for as many things as people you poll. Some samples: Mystery Theory, Mother Theory (as in "Mother of all Theories"), Membrane Theory (since, whatever it is, membranes seem to be part of the story), Matrix Theory (after some recent work by Tom Banks of Rutgers University, Willy Fischler of the University of Texas at Austin, Stephen Shenker of Rutgers University, and Susskind that offers a novel interpretation of the theory). But even without having a firm grasp on its name or its properties, it is already clear that M-theory provides a unifying substrate for pulling together all five string theories.

M-Theory and the Web of Interconnections

There is an old proverb about three blind men and an elephant. The first blind man grabs hold of the elephant's ivory tusk and describes the smooth, hard surface that he feels. The second blind man grabs hold of one of the elephant's legs. He describes the tough, muscular girth that he feels. The third blind man grabs hold of the elephant's tail and describes the slender and sinewy appendage that he feels. Since their mutual descriptions are so different, and since none of the men can see the others, each thinks that he has grabbed hold of a different animal. For many years, physicists were as much in the dark as the blind men, thinking that the different string theories *were* very different. But now, through the insights of the second superstring revolution, physicists have realized that M-theory is the unifying pachyderm of the five string theories.

In this chapter we have discussed changes in our understanding of string theory that arise when we venture beyond the domain of the perturbative framework—a framework implicitly in use prior to this chapter. Figure 12.9 summarizes the interrelations we have found so far, with arrows to indicate dual theories. As you can see, we have a web of connections, but it is not yet complete. By also including the dualities of Chapter 10, we can finish the job.

Recall the large/small circular radius duality that interchanges a circular dimension of radius R with one whose radius is $1/R$. Previously, we

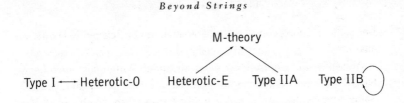

Figure 12.9 The arrows show which theories are dual to others.

glossed over one aspect of this duality, which we now must clarify. In Chapter 10, we discussed the properties of strings in a universe with a circular dimension without carefully specifying which of the five string formulations we were working with. We argued that the interchange of winding and vibration modes of a string allows us to rephrase exactly the string theoretic description of a universe with a circular dimension of radius $1/R$ in terms of one in which the radius is R. The point we glossed over is that the Type IIA and Type IIB string theories actually get exchanged by this duality, as do the Heterotic-O and Heterotic-E strings. That is, the more precise statement of the large/small radius duality is this: The physics of the Type IIA string in a universe with a circular dimension of radius R is absolutely identical to the physics of the Type IIB string in a universe with a circular dimension of radius $1/R$ (a similar statement holds for the Heterotic-E and Heterotic-O strings). This refinement of the large/small radius duality has no significant effect on the conclusions of Chapter 10, but it does have an important impact on the present discussion.

The reason is that by providing a link between the Type IIA and Type IIB string theories, as well as between the Heterotic-O and Heterotic-E theories, the large/small radius duality completes the web of connections, as illustrated by the dotted lines in Figure 12.10. This figure shows that

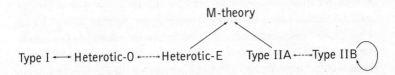

Figure 12.10 By including the dualities involving the geometrical form of spacetime (as in Chapter 10), all five of the string theories and M-Theory are joined together in a web of dualities.

all five string theories, together with M-theory, are dual to one another. They are all sewn together into a single theoretical framework; they provide five different approaches to describing one and the same underlying physics. For some or other application, one phrasing may be far more effective than another. For instance, it's far easier to work with the weakly coupled Heterotic-O theory than it is to work with the strongly coupled Type I string. Nevertheless, they describe exactly the same physics.

The Overall Picture

We can now more fully understand the two figures—Figures 12.1 and 12.2—that we introduced in the beginning of this chapter to summarize the essential points. In Figure 12.1 we see that prior to 1995, without taking any dualities into account, we had five apparently distinct string theories. Various physicists worked on each, but without an understanding of the dualities they appeared to be different theories. Each of the theories had variable features such as the size of their coupling constant and the geometrical form and sizes of curled-up dimensions. The hope was (and still is) that these defining properties would be determined by the theory itself, but without the ability to determine them with the current approximate equations, physicists have naturally studied the physics that follows from a range of possibilities. This is represented in Figure 12.1 by the shaded regions—each point in such a region denotes one specific choice for the coupling constant and the curled-up geometry. Without invoking any dualities, we still have five disjointed (collections of) theories.

But now, if we apply all of the dualities we have discussed, then as we vary the coupling and geometric parameters, we can pass from any one theory to any other, so long as we also include the unifying central region of M-theory; this is shown in Figure 12.2. Even though we have only a scant understanding of M-theory, these indirect arguments lend strong support to the claim that it provides a unifying substrate for our five naively distinct string theories. Moreover, we have learned that M-theory is closely related to yet a sixth theory—eleven-dimensional supergravity—and this is recorded in Figure 12.11, a more precise version of Figure 12.2.[13]

Figure 12.11 illustrates that the fundamental ideas and equations of

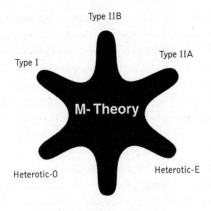

Type IIB

Type IIA

Type I

M-Theory

Heterotic-O

Heterotic-E

11-D Supergravity

Figure 12.11 By incorporating the dualities, all five string theories, eleven-dimensional supergravity, and M-theory are merged together into a unified framework.

M-theory, although only partially understood at the moment, unify those of all of the formulations of string theory. M-theory is the theoretical elephant that has opened the eyes of string theorists to a far grander unifying framework.

A Surprising Feature of M-Theory: Democracy in Extension

When the string coupling constant is small in any of the upper five peninsular regions of the theory map in Figure 12.11, the fundamental ingredient of the theory appears to be a one-dimensional string. We have, however, just gained a new perspective on this observation. If we start in either the Heterotic-E or Type IIA regions and turn the value of the respective string coupling constants up, we migrate toward the center of the map in Figure 12.11, and what appeared to be one-dimensional strings stretch into two-dimensional membranes. Moreover, through a more or less intricate sequence of duality relations involving both the string coupling constants and the detailed form of the curled-up spatial dimensions, we can smoothly and continuously move from any point in Figure 12.11 to any other. Since the two-dimensional membranes we have come upon from the Heterotic-E and Type IIA perspectives can be followed as we migrate to any of the three other string formulations in Figure 12.11, we learn that each of the five string formulations involves two-dimensional membranes as well.

This raises two questions. First, are two-dimensional membranes the true fundamental ingredient of string theory? And second, having made the bold leap in the 1970s and early 1980s from zero-dimensional point particles to one-dimensional strings, and having now seen that string theory actually involves two-dimensional membranes, might it be that there are even higher-dimensional ingredients in the theory as well? As of this writing, the answers to these questions are not fully known, but the situation appears to be the following.

We relied heavily on supersymmetry to give us some understanding of each formulation of string theory beyond the domain of validity of perturbative approximation methods. In particular, the properties of BPS states, their masses and their force charges, are uniquely determined by supersymmetry, and this allowed us to understand some of their strongly coupled characteristics without having to perform direct calculations of unimaginable difficulty. In fact, through the initial efforts of Horowitz and Strominger, and through subsequent groundbreaking work of Polchinski, we now know even more about these BPS states. In particular, not only do we know their masses and the force charges they carry, but we also have a clear understanding of what they *look* like. And the picture is, perhaps, the most surprising development of all. Some of the BPS states are one-dimensional strings. Others are two-dimensional membranes. By now, these shapes are familiar. But, the surprise is that yet others are *three*-dimensional, *four*-dimensional—in fact, the range of possibilities encompasses every spatial dimension up to and including *nine*. String theory or M-theory, or whatever it is finally called, actually contains extended objects of a whole slew of different spatial dimensions. Physicists have coined the term three-brane to describe extended objects with three spatial dimensions, four-brane for those with four spatial dimensions, and so on up to nine-branes (and, more generally, for an object with p space dimensions, where p represents a whole number, physicists have coined the far from euphonious terminology *p-brane*). Sometimes, using this terminology, strings are described as one-branes, and membranes as two-branes. The fact that all of these extended objects are actually part of the theory has led Paul Townsend to declare a "democracy of branes."

Notwithstanding brane democracy, strings—one-dimensional extended objects—are special for the following reason. Physicists have shown that the mass of the extended objects of every dimension except for one-dimensional strings is *inversely* proportional to the value of the associated string coupling constant when we are in any of the five string regions of Figure 12.11. This means that with weak string coupling, in any of the five formulations, all but the strings will be enormously massive—orders of magnitude heavier than the Planck mass. Because they are so heavy and, therefore, from $E = mc^2$, require such unimaginably high energy to be produced, branes have only a small effect on much of physics (but not on all, as we shall see in the next chapter). However, when we venture out-side the peninsular regions of Figure 12.11, the higher-dimensional branes become lighter and hence increasingly important.[14]

And so, the image you should have in mind is the following. In the cen-tral region of Figure 12.11, we have a theory whose fundamental ingredi-ents are not just strings or membranes, but rather "branes" of a variety of dimensions, all more or less on equal footing. Currently, we do not have a firm grasp on many essential features of this full theory. But one thing we do know is that as we move from the central region to any of the penin-sular regions, only the strings (or membranes curled up to look ever more like strings, as in Figures 12.7 and 12.8) are light enough to make contact with physics as we know it—the particles of Table 1.1 and the four forces through which they interact. The perturbative analyses string theorists have made use of for close to two decades have not been refined enough to discover even the existence of the super-massive extended objects of other dimensions; strings dominated the analyses and the theory was given the far-from-democratic name of string theory. Again, in these regions of Figure 12.11 we are justified, for most considerations, in ignoring all but the strings. In essence, this is what we have done so far in this book. We see now, though, that in actuality the theory is more rich than anyone previously imagined.

Does Any of This Solve the Unanswered Questions in String Theory?

Yes and no. We have managed to deepen our understanding by breaking free of certain conclusions that, in retrospect, were a consequence of perturbative approximate analyses rather than true string physics. But the current scope of our nonperturbative tools is quite limited. The discovery of the remarkable web of duality relations affords us far greater insight into string theory, but many issues remain unresolved. At present, for example, we do not know how to go beyond the approximate equations for the value of the string coupling constant—equations that, as we have seen, are too coarse to give us any useful information. Nor do we have any greater insight into why there are precisely three extended spatial dimensions, or how to choose the detailed form for the curled-up dimensions. These questions require more sharply honed nonperturbative methods than we currently possess.

What we do have is a far deeper understanding of the logical structure and theoretical reach of string theory. Prior to the realizations summarized in Figure 12.11, the strong coupling behavior of each string theory was a black box, a complete mystery. As on maps of old, the realm of strong coupling was uncharted territory, potentially filled with dragons and sea monsters. But now we see that although the journey to strong coupling may take us through unfamiliar regions of M-theory, it ultimately lands us back in the comfortable surrounds of weak coupling—albeit in the dual language of what was once thought to be a different string theory.

Duality and M-theory unite the five string theories and they suggest an important conclusion. It may well be that there aren't other surprises, on par with the ones just discussed, that are awaiting our discovery. Once a cartographer can fill in every region on a spherical globe of the earth, the map is done and geographical knowledge is complete. That's not to say explorations in Antarctica or on an isolated island in Micronesia are without scientific or cultural merit. It only means that the age of geographic discovery is over. The absence of blank spots on the globe ensures this. The "theory map" of Figure 12.11 plays a similar role for string theorists. It covers the entire range of theories that can be reached by setting sail from any one of the five string constructions. Although we are far from a full understanding of the terra incognita of M-theory, there are no blank regions on the map. Like the cartographer, the string theorist can now claim with guarded optimism that the spectrum of logically sound theories incorporating the essential discoveries of the past century—special and general relativity; quantum mechanics; gauge theories of the strong, weak, and electromagnetic forces; supersymmetry; extra dimensions of Kaluza and Klein—is fully mapped out by Figure 12.11.

The challenge to the string theorist—or perhaps we should say the M-theorist—is to show that *some* point on the theory map of Figure 12.11 actually describes our universe. To do this requires finding the full and exact equations whose solution will pick out this elusive point on the map, and then understanding the corresponding physics with sufficient precision to allow comparisons with experiment. As Witten has said, "Understanding what M-theory really is—the physics it embodies—would transform our understanding of nature at least as radically as occurred in any of the major scientific upheavals of the past."[15] This is the program for unification in the twenty-first century.

Chapter 13

Black Holes: A String/
M-Theory Perspective

The pre–string theory conflict between general relativity and quantum mechanics was an affront to our visceral sense that the laws of nature should fit together in a seamless, coherent whole. But this antagonism was more than a towering abstract disjunction. The extreme physical conditions that occurred at the moment of the big bang and that prevail within black holes *cannot* be understood without a quantum mechanical formulation of the gravitational force. With the discovery of string theory, we now have a hope of solving these deep mysteries. In this and the next chapter, we describe how far string theorists have gone toward understanding black holes and the origin of the universe.

Black Holes and Elementary Particles

At first sight it's hard to imagine any two things more radically different than black holes and elementary particles. We usually picture black holes as the most gargantuan of heavenly bodies, whereas elementary particles are the most minute specks of matter. But the research of a number of physicists during the late 1960s and early 1970s, including Demetrios Christodoulou, Werner Israel, Richard Price, Brandon Carter, Roy Kerr, David Robinson, Hawking, and Penrose, showed that black holes and elementary particles are perhaps not as different as one might think. These physicists found increasingly persuasive evidence for what John Wheeler has summarized by the statement "black holes have no hair." By this, Wheeler meant that except for a small number of distinguishing features, all black holes appear to be alike. The distinguishing features? One, of course, is the black hole's mass. What are the others? Research has revealed that they are the electric and certain other force charges a black hole can carry, as well as the rate at which it spins. And that's it. Any two black holes with the same mass, force charges, and spin are completely identical. Black holes do not have fancy "hairdos"—that is, other intrinsic traits—that distinguish one from another. This should ring a loud bell. Recall that it is precisely such properties—mass, force charges, and spin—that distinguish one elementary particle from another. The similarity of the defining traits has led a number of physicists over the years to the strange speculation that black holes might actually be gigantic elementary particles.

In fact, according to Einstein's theory, there is no minimum mass for a black hole. If we crush a chunk of matter of any mass to a small enough size, a straightforward application of general relativity shows that it will become a black hole. (The lighter the mass, the smaller we must crush it.) And so, we can imagine a thought experiment in which we start with ever-lighter blobs of matter, crush them into ever-smaller black holes, and compare the properties of the resulting black holes with the properties of elementary particles. Wheeler's no-hair statement leads us to conclude that for small enough masses the black holes we form in this manner will look very much like elementary particles. Both will look like tiny bundles characterized completely by their mass, force charges, and spin.

But there is a catch. Astrophysical black holes, with masses many times that of the sun, are so large and heavy that quantum mechanics is largely irrelevant and only the equations of general relativity need be used to understand their properties. (We are here discussing the overall structure of the black hole, not the singular central point of collapse within a black hole, whose tiny size most certainly requires a quantum-mechanical description.) As we try to make ever less massive black holes, however, there comes a point when they are so light and small that quantum mechanics *does* comes into play. This happens if the total mass of the black hole is about the Planck mass or less. (From the point of view of elementary particle physics, the Planck mass is huge—some ten billion billion times the mass of a proton. From the point of view of black holes, though, the Planck mass, being equal to that of an average grain of dust, is quite tiny.) And so, physicists who speculated that tiny black holes and elementary particles might be closely related immediately ran up against the incompatibility between general relativity—the theoretical heart of black holes—and quantum mechanics. In the past, the incompatibility stymied all progress in this intriguing direction.

Does String Theory Allow Us to Go Forward?

It does. Through a fairly unexpected and sophisticated realization of black holes, string theory provides the first theoretically sound connection between black holes and elementary particles. The road to this connection is a bit circuitous, but it takes us through some of the most interesting developments in string theory, making it a journey well worth taking.

It begins with a seemingly unrelated question that string theorists have kicked around since the late 1980s. Mathematicians and physicists have long known that when six spatial dimensions are curled up into a Calabi-Yau shape, there are generally two kinds of spheres that are embedded within the shape's fabric. One kind are the two-dimensional spheres, like the surface of a beach ball, that played a vital role in the space-tearing flop transitions of Chapter 11. The other kind are harder to picture but they are equally prevalent. They are *three*-dimensional spheres—like the surfaces of beach balls adorning the sandy ocean shores of a universe with *four* extended space dimensions. Of course, as we discussed in Chapter 11, an ordinary beach ball in our world is itself a three-dimensional object, but its *surface*, just like the surface of a garden hose, is *two*-dimensional: You need only two numbers—latitude and longitude, for instance—to locate any position on its surface. But we are now imagining having one more space dimension: a four-dimensional beach ball whose surface is *three*-dimensional. As it's pretty close to impossible to picture such a beach ball in your mind's eye, for the most part we will appeal to lower-dimensional analogs that are more easily visualized. But, as we shall now see, one aspect of the three-dimensional nature of the spherical surfaces is of prime importance.

By studying the equations of string theory, physicists realized that it is possible, and even likely, that as time evolves, these three-dimensional spheres will shrink—collapse—to vanishingly small volume. But what would happen, string theorists asked, if the fabric of space were to collapse in this manner? Will there be some catastrophic effect from this kind of pinching of the spatial fabric? This is much like the question we posed and resolved in Chapter 11, but here we are focusing on collapsing three-dimensional spheres, whereas in Chapter 11 we focused solely on collapsing two-dimensional spheres. (As in Chapter 11, since we are envisioning that a piece of a Calabi-Yau shape is shrinking, as opposed to the whole Calabi-Yau shape itself, the small radius/large radius identification of Chapter 10 does not apply.) Here is the essential qualitative difference arising from the change in dimension.[1] We recall from Chapter 11 that a pivotal realization is that strings, as they move through space, can lasso a two-dimensional sphere. That is, their two-dimensional worldsheet can fully surround a two-dimensional sphere, as in Figure 11.6. This proves to be just enough protection to keep a collapsing, pinching two-dimensional sphere from causing physical catastrophes. But now we are looking at the other kind of sphere inside a Calabi-Yau space, and it has too many dimensions for it to be surrounded by a moving string. If you have trouble seeing this, it is perfectly okay to think of the analogy obtained by lowering all dimensions by one. You can picture three-dimensional spheres as if they are two-dimensional surfaces of ordinary beach balls, so long as you also picture one-dimensional strings as if they are zero-dimensional point particles. Then, in analogy with the fact that a zero-dimensional point-particle cannot lasso anything, let alone a two-dimensional sphere, a one-dimensional string cannot lasso a three-dimensional sphere.

Such reasoning led string theorists to speculate that if a three-dimensional sphere inside a Calabi-Yau space were to collapse, something that the approximate equations showed to be a perfectly possible if not commonplace evolution in string theory, it might yield a cataclysmic result. In fact, the approximate equations of string theory developed prior to the mid-1990s seemed to indicate that the workings of the universe would grind to a halt if such a collapse were to occur; they indicated that certain of the infinities tamed by string theory would be unleashed by such a pinching of the spatial fabric. For a number of years, string theorists had to live with this disturbing, albeit inconclusive, state of understanding. But in 1995, Andrew Strominger showed that these doomsaying speculations were wrong.

Strominger, following earlier groundbreaking work of Witten and Seiberg, made use of the realization that string theory, when analyzed with the newfound precision of the second superstring revolution, is not just a theory of one-dimensional strings. He reasoned as follows. A one-dimensional string—a one-brane in the newer language of the field—can completely surround a one-dimensional piece of space, like a circle, as we illustrate in Figure 13.1. (Notice that this is different from Figure 11.6, in which a one-dimensional string, as it moves through time, lassos a two-dimensional sphere. Figure 13.1 should be viewed as a snapshot taken at one instant in time.) Similarly, in Figure 13.1 we see that a two-dimensional membrane—a two-brane—can wrap around and completely cover a two-dimensional sphere, much as a piece of plastic wrap can be tightly wrapped around the surface of an orange. Although it's harder to visualize, Strominger followed the pattern and realized that the newly discovered three-dimensional ingredients in string theory—the three-branes—can wrap around and completely cover a three-dimensional sphere. Having seen clear to this insight, Strominger then showed, with a simple and standard physics calculation, that the wrapped three-brane provides a tailor-made shield that exactly cancels all of the potentially cataclysmic effects that string theorists had previously feared would occur if a three-dimensional sphere were to collapse.

This was a wonderful and important insight. But its full power was not revealed until a short time later.

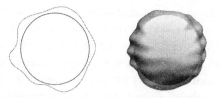

Figure 13.1 A string can encircle a one-dimensional curled-up piece of the spatial fabric; a two-dimensional membrane can wrap around a two-dimensional piece.

Tearing the Fabric of Space—with Conviction

One of the most exciting things about physics is how the state of knowledge can change literally overnight. The morning after Strominger posted his paper on the electronic Internet archive, I read it in my office at Cornell after having retrieved it from the World Wide Web. In one stroke, Strominger had made use of the exciting new insights of string theory to resolve one of the thorniest issues surrounding the curling up of extra dimensions into a Calabi-Yau space. But as I pondered his paper, it struck me that he might have worked out only half of the story.

In the earlier space-tearing flop-transition work described in Chapter 11, we had studied a two-part process in which a two-dimensional sphere pinches down to a point, causing the fabric of space to tear, and then the two-dimensional sphere reinflates in a new way, thereby repairing the tear. In Strominger's paper, he had studied what happens when a three-dimensional sphere pinches down to a point, and had shown that the newfound extended objects in string theory ensure that physics continues to be perfectly well behaved. But that's where his paper stopped. Might it be that there was another half to the story, involving, once again, the tearing of space and its subsequent repair through the reinflation of spheres?

Dave Morrison was visiting me at Cornell during the spring term of 1995, and that afternoon we got together to discuss Strominger's paper. Within a couple of hours we had an outline of what the "second half of the story" might look like. Drawing on some insights from the late 1980s of the mathematicians Herb Clemens of the University of Utah, Robert Friedman of Columbia University, and Miles Reid of the University of Warwick, as applied by Candelas, Green, and Tristan Hübsch, then of the University of Texas at Austin, we realized that when a three-dimensional sphere collapses, it may be possible for the Calabi-Yau space to tear and subsequently repair itself by reinflating the sphere. But there is an important surprise. Whereas the sphere that collapsed had three dimensions, the one that reinflates has only *two*. It's hard to picture what this looks like, but we can get an idea by focusing on a lower-dimensional analogy. Rather than the hard-to-picture case of a three-dimensional sphere collapsing

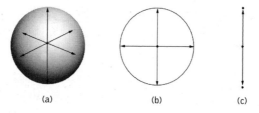

and being replaced by a two-dimensional sphere, let's imagine a *one*-dimensional sphere collapsing and being replaced by a *zero*-dimensional sphere.

First of all, what are one- and zero-dimensional spheres? Well, let's reason by analogy. A two-dimensional sphere is the collection of points in three-dimensional space that are the same distance from a chosen center, as shown in Figure 13.2(a). By following the same idea, a one-dimensional sphere is the collection of points in two-dimensional space (the surface of this page, for example) that are the same distance from a chosen center. As shown in Figure 13.2(b), this is nothing but a circle. Finally, following the pattern, a zero-dimensional sphere is the collection of points in a one-dimensional space (a line) that are the same distance from a chosen center. As shown in Figure 13.2(c), this amounts to *two points*, with the "radius" of the zero-dimensional sphere equal to the distance each point is from their common center. And so, the lower-dimensional analogy alluded to in the preceding paragraph involves a circle (a one-dimensional sphere) pinching down, followed by space tearing, and then being replaced by a zero-dimensional sphere (two points). Figure 13.3 puts this abstract idea into practice.

We imagine beginning with the surface of a doughnut, in which a one-dimensional sphere (a circle) is embedded, as highlighted in Figure 13.3. Now, let's imagine that as time goes by, the highlighted circle collapses, causing the fabric of space to pinch. We can repair the pinch by allowing the fabric to momentarily tear, and then replacing the pinched one-dimensional sphere—the collapsed circle—with a zero-dimensional

Figure 13.3 A circular piece of a doughnut (a torus) collapses to a point. The surface tears open, yielding two puncture holes. A zero-dimensional sphere (two points) is "glued in," replacing the original one-dimensional sphere (the circle) and repairing the torn surface. This allows a transformation to a completely different shape—a beach ball.

sphere—two points—plugging the holes in the upper and lower portions of the shape arising from the tear. As shown in Figure 13.3, the resulting shape looks like a warped banana, which through gentle deformation (non–space tearing) can be reshaped smoothly into the surface of a beach ball. We see, therefore, that when a one-dimensional sphere collapses and is replaced by a zero-dimensional sphere, the topology of the original doughnut, that is, its fundamental shape, is drastically altered. In the context of the curled-up spatial dimensions, the space-tearing progression of Figure 13.3 would result in the universe depicted in Figure 8.8 evolving into that depicted in Figure 8.7.

Although this is a lower-dimensional analogy, it captures the essential features of what Morrison and I foresaw for the second half of Strominger's story. After the collapse of a three-dimensional sphere inside a Calabi-Yau space, it seemed to us that space could tear and subsequently repair itself by growing a two-dimensional sphere, leading to far more drastic changes in topology than Witten and we had found in our earlier work (discussed in Chapter 11). In this way, one Calabi-Yau shape could, in essence, transform itself into a completely different Calabi-Yau shape—much like the doughnut transforming into the beach ball in Figure 13.3—while string physics remained perfectly well behaved. Although a picture was starting to emerge, we knew that there were significant aspects that we would need to work out before we could establish that our second half of the story did not introduce any singularities—that is, pernicious and physically unacceptable consequences. We each went home that evening with the tentative elation that we were sitting on a major new insight.

A Flurry of E-Mail

The next morning I received an e-mail from Strominger asking me for any comments or reactions to his paper. He mentioned that "it should tie in somehow with your work with Aspinwall and Morrison," because, as it turned out, he too had been exploring a possible connection to the phenomenon of topology change. I immediately sent him an e-mail describing the rough outline Morrison and I had come up with. When he responded, it was clear that his level of excitement matched what Morrison and I had been riding since the preceding day.

During the next few days a continuous stream of e-mail messages circulated between the three of us as we sought feverishly to put quantitative rigor behind our idea of drastic space-tearing topology change. Slowly but surely, all the details fell into place. By the following Wednesday, a week after Strominger posted his initial insight, we had a draft of a joint paper spelling out the dramatic new transformation of the spatial fabric that can follow the collapse of a three-dimensional sphere.

Strominger was scheduled to give a seminar at Harvard the next day, and so left Santa Barbara in the early morning. We agreed that Morrison and I would continue to fine-tune the paper and then submit it to the electronic archive that evening. By 11:45 P.M., we had checked and rechecked our calculations and everything seemed to hang together perfectly. And so, we electronically submitted our paper and headed out of the physics building. As Morrison and I walked toward my car (I was going to drive him to the house he had rented for the term) our discussion turned to one of devil's advocacy, in which we imagined the harshest criticisms that someone determined not to accept our results might level. As we drove out of the parking lot and left the campus, we realized that although our arguments were strong and convincing, they were not thoroughly airtight. Neither of us felt that there was any real chance that our work was wrong, but we did recognize that the strength of our claims and the particular wording we had chosen at a few points in the paper might leave the ideas open to rancorous debate, potentially obscuring the importance of the results. We agreed that it might have been better had we written the paper in a somewhat lower key, underplaying the depth of the claims, and allowing the physics community to judge the paper on its merits, rather than possibly reacting to the form of its presentation.

As we drove on, Morrison reminded me that under the rules of the electronic archive we could revise our paper until 2 A.M., when it would then be released for public Internet access. I immediately turned the car around and we drove back to the physics building, retrieved our initial submission, and set to work on toning down the prose. Thankfully, it was quite easy to do. A few word changes in critical paragraphs softened the edge of our claims without compromising the technical content. Within an hour, we resubmitted the paper, and agreed not to talk about it at all during the drive to Morrison's house.

By early the next afternoon it was evident that the response to our paper was enthusiastic. Among the many e-mail responses was one from Plesser, who gave us one of the highest compliments one physicist can give another by declaring, "I wish that I had thought of that!" Notwithstanding our fears the previous night, we had convinced the string theory community that not only can the fabric of space undergo the mild tears discovered earlier (Chapter 11), but that far more drastic rips, roughly illustrated by Figure 13.3, can occur as well.

Returning to Black Holes and Elementary Particles

What does this have to do with black holes and elementary particles? A lot. To see this, we must ask ourselves the same question we posed in Chapter 11. What are the observable physical consequences of such tears in the fabric of space? For flop transitions, as we have seen, the surprising answer to this question was that not much happens at all. For *conifold transitions*—the technical name for the drastic space-tearing transitions we had now found—there is, once again, no physical catastrophe (as there would be in conventional general relativity), but there are more pronounced observable consequences.

Two related notions underlie these observable consequences; we will explain each in turn. First, as we have discussed, Strominger's initial breakthrough was his realization that a three-dimensional sphere inside a Calabi-Yau space can collapse without an ensuing disaster, because a three-brane wrapped around it provides a perfect protective shield. But what does such a wrapped-brane configuration look like? The answer comes from earlier work of Horowitz and Strominger, which showed that to persons such as ourselves who are directly cognizant only of the three extended spatial dimensions, the three-brane "smeared" around the three-dimensional sphere will set up a gravitational field that looks like that of a black hole.[2] This is *not* obvious and becomes clear only from a detailed study of the equations governing the branes. Again, it's hard to draw such higher-dimensional configurations accurately on a page, but Figure 13.4 conveys the rough idea with a lower-dimensional analogy involving two-dimensional spheres. We see that a two-dimensional membrane can smear itself around a two-dimensional sphere (which itself is sitting inside a Calabi-Yau space positioned at some location in the extended dimensions). Someone looking through the extended dimensions toward this location will sense the wrapped brane by its mass and the force charges it carries, properties that Horowitz and Strominger had shown would look just like those of a black hole. Moreover, in Strominger's 1995 breakthrough paper, he argued that the mass of the three-brane—the mass of the black hole, that is—is proportional to the volume of the three-dimensional sphere it wraps: The bigger the volume of the sphere, the bigger the three-brane must be in order to wrap around it, and the more massive it becomes. Similarly, the smaller the volume of the sphere, the smaller the mass of the three-brane that wraps it. As this sphere collapses, then, a three-brane that

Figure 13.4 When a brane wraps around a sphere that is within the curled-up dimensions, it appears as a black hole in the familiar extended dimensions.

wraps around the sphere, which is perceived as a black hole, appears to become ever lighter. When the three-dimensional sphere has collapsed to a pinched point, the corresponding black hole—brace yourself—is massless. Although it sounds completely mysterious—what in the world is a *massless* black hole?—we will soon connect this enigma with more familiar string physics.

The second ingredient we need to recall is that the number of holes in a Calabi-Yau shape, as discussed in Chapter 9, determines the number of low-energy, and hence low-mass, vibrational string patterns, the patterns that can possibly account for the particles in Table 1.1 as well as the force carriers. Since the space-tearing conifold transitions change the number of holes (as, for example in Figure 13.3, in which the hole of the doughnut is eliminated by the tearing/repairing process), we expect a change in the number of low-mass vibrational patterns. Indeed, when Morrison, Strominger, and I studied this in detail, we found that as a new two-dimensional sphere replaces the pinched three-dimensional sphere in the curled-up Calabi-Yau dimensions, the number of massless string vibrational patterns increases by exactly one. (The example of the doughnut turning into a beach ball in Figure 13.3 would lead you to believe that the number of holes—and thus the number of patterns—decreases, but this proves to be a misleading property of the lower-dimensional analogy.)

To combine the observations of the preceding two paragraphs, imagine a sequence of snapshots of a Calabi-Yau space in which the size of a particular three-dimensional sphere gets smaller and smaller. The first observation implies that a three-brane wrapping around this three-dimensional sphere—which appears to us as a black hole—will have ever smaller mass until, at the final point of collapse, it will be massless. But, as we asked above, what does this mean? The answer became clear to us by invoking the second observation. Our work showed that the new massless pattern of string vibration arising from the space-tearing conifold transition *is the microscopic description of a massless particle into which the black hole has transmuted.* We concluded that as a Calabi-Yau shape goes through a space-tearing conifold transition, an initially massive black hole becomes ever lighter until it is massless and then it transmutes into a massless particle—such as a massless photon—which in string theory is nothing but a single string executing a particular vibrational pattern. In this way, for the first time, string theory explicitly establishes a direct, concrete, and quantitatively unassailable connection between black holes and elementary particles.

"Melting" Black Holes

The connection between black holes and elementary particles which we found is closely akin to something we are all familiar with from day-to-day life, known technically as a phase transition. A simple example of a phase transition is the one we mentioned in the last chapter: water can exist as a solid (ice), as a liquid (liquid water), and a gas (steam). These are known as the *phases* of water, and the transformation from one form to another is called a *phase transition*. Morrison, Strominger, and I showed that there is a tight mathematical and physical analogy between such phase transitions and the space-tearing conifold transitions from one Calabi-Yau shape to another. Again, just as someone who has never before encountered liquid water or solid ice would not immediately recognize that they are two phases of the same underlying substance, physicists had not realized previously that the kinds of black holes we were studying and elementary particles are actually two phases of the same underlying stringy material. Whereas the surrounding temperature determines the phase in which water will exist, the topological form—the shape—of the extra Calabi-Yau dimensions determines whether certain physical configurations within string theory appear as black holes or elementary particles. That is, in the first phase, the initial Calabi-Yau shape (the analog of the ice phase, say), we find that there are certain black holes present. In the second phase, the second Calabi-Yau shape (the analog of the liquid water phase), these black holes have gone through a phase transition—they have "melted" so to speak—into fundamental vibrational string patterns. The tearing of space through conifold transitions takes us from one Calabi-Yau phase to the other. In so doing, we see that black holes and elementary particles, like water and ice, are two sides of the same coin. We see that black holes snugly fit within the framework of string theory.

We have purposely used the same water analogy for these drastic space-tearing transmutations and for the transmutations from one of the five formulations of string theory to another (Chapter 12) because they are deeply connected. Recall that we expressed through Figure 12.11 that the five string theories are dual to one another and thereby are unified under the rubric of a single overarching theory. But does the ability to move continuously from one description to another—to set sail from any point on the map of Figure 12.11 and reach any other—persist even after we allow the extra dimensions to curl up into some Calabi-Yau shape or another? Prior to the discovery of the drastic topology-changing results, the anticipated answer was no, since there was no known way to continuously deform one Calabi-Yau shape into any other. But now we see that the answer is yes: Through these physically sensible space-tearing conifold transitions, we can continuously change any given Calabi-Yau space into any other. By varying coupling constants and curled-up Calabi-Yau geometry, we see that all string constructions are, once again, different phases of a single theory. Even after curling up all extra dimensions, the unity of Figure 12.11 firmly holds.

Black Hole Entropy

For many years, some of the most accomplished theoretical physicists speculated about the possibility of space-tearing processes and of a connection between black holes and elementary particles. Although such speculations might have sounded like science fiction at first, the discovery of string theory, with its ability to merge general relativity and quantum mechanics, has allowed us now to plant these possibilities firmly at the forefront of cutting-edge science. This success emboldens us to ask whether any of the other mysterious properties of our universe that have stubbornly resisted resolution for decades might also succumb to the powers of string theory. Foremost among these is the notion of *black hole entropy*. This is the arena in which string theory has most impressively flexed its muscles, successfully solving a quarter-century-old problem of profound significance.

Entropy is a measure of disorder or randomness. For instance, if your desk is cluttered high with layer upon layer of open books, half-read articles, old newspapers, and junk mail, it is in a state of high disorder, or *high entropy*. On the other hand, if it is fully organized with articles in alphabetized folders, newspapers neatly stacked in chronological order, books arranged in alphabetical order by author, and pens placed in their designated holders, your desk is in state of high order or, equivalently, *low entropy*. This example illustrates the essential idea, but physicists have given a fully quantitative definition to entropy that allows one to describe something's entropy by using a definite numerical value: Larger numbers mean greater entropy, smaller numbers mean less entropy. Although the details are a little complicated, this number, roughly speaking, counts the possible rearrangements of the ingredients in a given physical system that leave its overall appearance intact. When your desk is neat and clean, almost any rearrangement—changing the order of the newspapers, books, or articles, moving the pens from their holders—will disturb its highly ordered organization. This accounts for its having low entropy. On the contrary, when your desk is a mess, numerous rearrangements of the newspapers, articles, and junk mail will leave it a mess and therefore will not disturb its overall appearance. This accounts for its having high entropy.

Of course, a description of rearranging books, articles, and newspapers on a desktop—and deciding which rearrangements "leave its overall appearance intact"—lacks scientific precision. The rigorous definition of entropy actually involves counting or calculating the number of possible rearrangements of the microscopic quantum-mechanical properties of the elementary constituents of a physical system that do not affect its gross macroscopic properties (such as its energy or pressure). The details are not essential so long as you realize that entropy is a fully quantitative quantum-mechanical concept that precisely measures the overall disorder of a physical system.

In 1970, Jacob Bekenstein, then a graduate student of John Wheeler's at Princeton, made an audacious suggestion. He put forward the remarkable idea that black holes might have entropy—and a huge amount of it. Bekenstein was motivated by the venerable and well-tested *second law of thermodynamics*, which declares that the entropy of a system always increases: Everything tends toward greater disorder. Even if you clean your cluttered desk, decreasing its entropy, the total entropy, including that of your body and the air in the room, actually increases. You see, to clean your desk you have to expend energy; you have to disrupt some of the orderly molecules of fat in your body to create this energy for your muscles, and as you clean, your body gives off heat, which jostles the surrounding air molecules into a higher state of agitation and disorder. When all of these effects are accounted for, they more than compensate for your desk's decrease in entropy, and thus the total entropy increases.

But what happens, Bekenstein in effect asked, if you clean your desk near the event horizon of a black hole and you set up a vacuum pump to suck all of the newly agitated air molecules from the room into the hidden depths of the black hole's interior? We can be even more extreme: What if the vacuum pumps all the air, and all the contents on the desk, and even the desk itself into the black hole, leaving you in a cold, airless, thoroughly ordered room? Since the entropy in your room has certainly decreased, Bekenstein reasoned that the only way to satisfy the second law of thermodynamics would be for the black hole to have entropy, and for this entropy to sufficiently increase as matter is pumped into it to offset the observed entropic decrease outside the black hole's exterior.

In fact, Bekenstein was able to draw on a famous result of Stephen Hawking's to strengthen his case. Hawking had shown that the area of the event horizon of a black hole—recall, this is the surface of no return that enshrouds every black hole—always increases in any physical interaction. Hawking demonstrated that if an asteroid falls into a black hole, or if some of the surface gas of a nearby star accretes onto the black hole, or if two black holes collide and combine—in any of these processes and all others as well, the total area of the event horizon of a black hole always increases. To Bekenstein, the inexorable evolution to greater total area suggested a link with the inexorable evolution to greater total entropy embodied in the second law of thermodynamics. He proposed that the area of the event horizon of a black hole provides a precise measure of its entropy.

On closer inspection, though, there are two reasons why most physicists thought that Bekenstein's idea could not be right. First, black holes would seem to be among the most ordered and organized objects in the whole universe. Once one measures the black hole's mass, the force charges it carries, and its spin, its identity has been nailed down precisely. With so few defining features, a black hole appears to lack sufficient structure to allow for disorder. Just as there is little havoc one can wreak on a desktop that holds solely a book and a pencil, black holes seem too simple to support disorder. The second reason that Bekenstein's proposal was hard to swallow is that entropy, as we have discussed it here, is a quantum-mechanical concept, whereas black holes, until recently, were firmly entrenched in the antagonistic camp of classical general relativity. In the early 1970s, without a way to merge general relativity and quantum mechanics, it seemed awkward, at best, to discuss the possible entropy of a black hole.

How Black Is Black?

As it turns out, Hawking too had thought of the analogy between his black hole area-increase law and the law of inevitable increase of entropy, but he dismissed it as nothing more than a coincidence. After all, Hawking argued, based upon his area-increase law and other results he had found with James Bardeen and Brandon Carter, if one did take the analogy between the laws of black holes and the laws of thermodynamics seriously, not only would one be forced to identify the area of the black hole's event horizon with entropy, but it turns out that one would also have to assign a *temperature* to the black hole (with the precise value determined by the strength of the black hole's gravitational field at its event horizon). But if a black hole has a nonzero temperature—no matter how small—the most basic and well-established physical principles would *require* it to emit radiation, much like a glowing poker. But black holes, as everyone knows, are black; they supposedly do not emit anything. Hawking and most everyone else agreed that this definitively ruled out Bekenstein's suggestion. Instead, Hawking was willing to accept that if matter carrying entropy is dropped into a black hole, this entropy is lost, plain and simple. So much for the second law of thermodynamics.

This was the case until Hawking, in 1974, discovered something truly amazing. Black holes, Hawking announced, are *not* completely black. If one ignores quantum mechanics and invokes only the laws of classical general relativity, then as originally found some six decades previously, black holes certainly do not allow anything—not even light—to escape their gravitational grip. But the inclusion of quantum mechanics modifies this conclusion in a profound way. Even though he was not in possession of a quantum-mechanical version of general relativity, Hawking was able to finesse a partial union of these two theoretical tools that yielded some limited yet reliable results. And the most important result he found was that black holes *do* emit radiation, quantum mechanically.

The calculations are long and arduous, but Hawking's basic idea is simple. We have seen that the uncertainty principle ensures that even the vacuum of empty space is a teeming, roiling frenzy of virtual particles momentarily erupting into existence and subsequently annihilating one another. This jittery quantum behavior also occurs in the region of space just outside the event horizon of a black hole. Hawking realized, however, that the gravitational might of the black hole can inject energy into a pair of virtual photons, say, that tears them just far enough apart so that one gets sucked into the hole. With its partner having disappeared into the abyss of the hole, the other photon of the pair no longer has a partner with which to annihilate. Instead, Hawking showed that the remaining photon gets an energy boost from the gravitational force of the black hole and, as its partner falls inward, it gets shot outward, away from the black hole. Hawking realized that to someone looking at the black hole from the safety of afar, the combined effect of this tearing apart of virtual photon pairs, happening over and over again all around the horizon of the black hole, will appear as a steady stream of outgoing radiation. Black holes *glow*.

Moreover, Hawking was able to calculate the temperature that a far-off observer would associate with the emitted radiation and found that it is given by the strength of the gravitational field at the black hole's horizon, exactly as the analogy between the laws of black hole physics and the laws of thermodynamics suggested.[3] Bekenstein was right: Hawking's results showed that the analogy should be taken seriously. In fact, these results showed that it is much more than an analogy—it is an *identity*. A black hole has entropy. A black hole has temperature. And the gravitational laws of black hole physics are nothing but a rewriting of the laws of thermodynamics in an extremely exotic gravitational context. This was Hawking's 1974 bombshell.

To give you a sense of the scales involved, it turns out that when one carefully accounts for all of the details, a black hole whose mass is about three times that of the sun has a temperature of about a hundred-millionth of a degree above absolute zero. It's not zero, but only just. Black holes are not black, but only barely. Unfortunately, this makes a black hole's emitted radiation meager, and impossible to detect experimentally. There is, however, an exception. Hawking's calculations also showed that the less massive a black hole is, the higher its temperature and the greater the radiation it emits. For instance, a black hole as light as a small asteroid would emit about as much radiation as a million-megaton hydrogen bomb, with radiation concentrated in the gamma-ray part of the electromagnetic spectrum. Astronomers have searched the night sky for such radiation, but except for a few long-shot possibilities they have come up empty-handed, a likely indication that such low-mass black holes, if they exist, are very rare.[4] As Hawking often jokingly points out, this is too bad, for if the black hole radiation that his work predicts were to be detected, he would undoubtedly win a Nobel Prize.[5]

By contrast with its tiny, sub-millionth of a degree temperature, when one calculates the entropy of, say, a three-solar-mass black hole, the result is an absolutely enormous number: a one followed by about 78 zeros! And the more massive the hole, the greater the entropy. The success of Hawking's calculations unequivocally established that this truly reflects the enormous amount of disorder embodied by a black hole.

But disorder of what? As we have seen, black holes appear to be terribly simple objects, so what is the source of this overwhelming disorder? On this question, Hawking's calculations were completely silent. His partial merger of general relativity and quantum mechanics could be used to find the numerical value of a black hole's entropy, but offered no insight into its microscopic meaning. For nearly a quarter of a century, some of the greatest physicists tried to understand what possible microscopic properties of black holes could account for their entropy. But without a fully trustworthy amalgam of quantum mechanics and general relativity, glimpses of an answer may have been uncovered, but the mystery remained unsolved.

Enter String Theory

Or, it did until January 1996, when Strominger and Vafa—building on earlier insights of Susskind and Sen—released a paper to the electronic physics archive entitled "Microscopic Origin of the Beckenstein-Hawking Entropy." In this work, Strominger and Vafa were able to use string theory to identify the microscopic constituents of a certain class of black holes and to calculate precisely their associated entropy. Their work relied on the newfound ability to partially circumvent the perturbative approximations in use during the 1980s and early 1990s, and the result they found agreed exactly with that predicted by Bekenstein and Hawking, finally completing a picture partially painted more than two decades previously.

Strominger and Vafa focused on the class of so-called *extremal* black holes. These are black holes that are imbued with charge—you can think of it as electric charge—and that, moreover, have the minimal possible mass consistent with the charge they carry. As can be seen from this definition, they are closely related to the BPS states discussed in Chapter 12. In fact, Strominger and Vafa exploited this similarity to the hilt. They showed that they could build—theoretically, of course—certain extremal black holes by starting with a particular collection of BPS branes (of certain specified dimensions) and binding them together according to a precise mathematical blueprint. In much the same way that an atom can be built—theoretically, again—by starting with a bunch of quarks and electrons and then precisely arranging them into protons and neutrons, surrounded by orbiting electrons, Strominger and Vafa showed how some of the newfound ingredients in string theory could similarly be molded together to yield particular black holes.

In actuality, black holes are one possible end product of stellar evolution. After a star has burned all its nuclear fuel through billions of years of atomic fusion, it no longer has the strength—the outward-directed pressure—to withstand the enormous inward force of gravity. Under a broad spectrum of conditions, this results in a cataclysmic implosion of the star's enormous mass; it violently collapses under its own tremendous weight, forming a black hole. Contrary to this realistic means of formation, Strominger and Vafa advocated "designer" black holes. They turned the tables on black hole formation by showing how they could be systematically constructed—in a theorist's imagination—by carefully, slowly, and meticulously weaving together a precise combination of the branes that had emerged from the second superstring revolution.

The power of this approach became immediately clear. By maintaining full theoretical control over the microscopic construction of their black holes, Strominger and Vafa could easily and directly count the number of rearrangements of the black hole's microscopic constituents that would leave its overall observable properties, its mass and force charges, unchanged. They could then compare this number with the area of the black hole's horizon—the entropy predicted by Bekenstein and Hawking. When Strominger and Vafa did so, they found perfect agreement. At least for the class of extremal black holes, they had succeeded in using string theory to account for the microscopic constituents and the associated entropy precisely. A quarter-century-old puzzle had been solved.[6]

Many string theorists view this success as an important and convincing piece of evidence in support of the theory. Our understanding of string theory is still too coarse to be able to make direct and precise contact with experimental observations of, say, the mass of a quark or an electron. But we now see that string theory has provided the first fundamental explanation of a long-established property of black holes that has stumped physicists using more conventional theories for many years. And this property of black holes is intimately tied up with Hawking's prediction that they should radiate, a prediction that, in principle, should be experimentally measurable. Of course, this requires that we definitively find a black hole in the heavens and then construct equipment sensitive enough to detect the radiation that it emits. If the black hole were light enough, the latter step is well within the reach of current technology. Even though this experimental program has not as yet met with success, it does re-emphasize that the chasm between string theory and definitive physical statements about the natural world can be bridged. Even Sheldon Glashow—the archrival of string theory through the 1980s—has said recently, "when string theorists talk about black holes they are almost talking about observable phenomena—and that is impressive."[7]

The Remaining Mysteries of Black Holes

Even with these impressive developments, there are still two central mysteries surrounding black holes. The first concerns the impact black holes have on the concept of determinism. In the beginning of the nineteenth century the French mathematician Pierre-Simon de Laplace enunciated the strictest and most far-reaching consequence of the clockwork universe that followed from Newton's laws of motion:

> An intelligence that, at a given instant, could comprehend all the forces
> by which nature is animated and the respective situation of the beings
> that make it up, if moreover it were vast enough to submit these data
> to analysis, would encompass in the same formula the movements of
> the greatest bodies of the universe and those of the lightest atoms. For
> such an intelligence nothing would be uncertain, and the future, like
> the past, would be open to its eyes.[8]

In other words, if at some instant you know the positions and velocities of
every particle in the universe, you can use Newton's laws of motion to
determine—at least in principle—their positions and velocities at any
other prior or future time. From this perspective, any and all occurrences,
from the formation of the sun to the crucifixion of Christ, to the motion
of your eyes across this word, strictly follow from the precise positions and
velocities of the particulate ingredients of the universe a moment after the
big bang. This rigid lock-step view of the unfolding of the universe raises
all sorts of perplexing philosophical dilemmas surrounding the question of
free will, but its import was substantially diminished by the discovery of
quantum mechanics. We have seen that Heisenberg's uncertainty principle undercuts Laplacian determinism because we fundamentally cannot
know the precise positions and velocities of the constituents of the universe. Instead, these classical properties are replaced by quantum wave
functions, which tell us only the probability that any given particle is here
or there, or that it has this or that velocity.

The downfall of Laplace's vision, however, does not leave the concept
of determinism in total ruins. Wave functions—the probability waves of
quantum mechanics—evolve in time according to precise mathematical
rules, such as the Schrödinger equation (or its more precise relativistic
counterparts, such as the Dirac equation and the Klein-Gordon equation). This informs us that *quantum determinism* replaces Laplace's classical determinism: Knowledge of the wave functions of all of the
fundamental ingredients of the universe at some moment in time allows
a "vast enough" intelligence to determine the wave functions at any prior
or future time. Quantum determinism tells us that the *probability* that
any particular event will occur at some chosen time in the future is fully
determined by knowledge of the wave functions at any prior time. The
probabilistic aspect of quantum mechanics significantly softens Lapla-

cian determinism by shifting inevitability from outcomes to outcome-likelihoods, but the latter are fully determined within the conventional framework of quantum theory.

In 1976, Hawking declared that even this softer form of determinism is violated by the presence of black holes. Once again, the calculations behind this declaration are formidable, but the essential idea is fairly straightforward. When anything falls into a black hole, its wave function gets sucked in as well. But this means that in the quest to work out wave functions at all future times, our "vast enough" intelligence will be irreparably shortchanged. To predict the future fully we need to know all wave functions fully today. But, if some have escaped down the abyss of black holes, the information they contain is lost.

At first sight, this complication arising from black holes may not seem worth worrying about. Since everything behind the event horizon of a black hole is cut off from the rest of the universe, can't we just completely ignore anything that is unfortunate enough to have fallen in? Philosophically, moreover, can't we tell ourselves that the universe has not really lost the information carried by the stuff that has fallen into the black hole; it is simply locked within a region of space that we rational beings choose to avoid at all costs? Prior to Hawking's realization that black holes are not completely black, the answer to these questions was yes. But once Hawking informed the world that black holes radiate, the story changed. Radiation carries energy and so, as a black hole radiates, its mass slowly decreases—it slowly evaporates. As it does so, the distance from the center of the hole to the event horizon slowly shrinks, and as this shroud recedes, regions of space that were previously cut off re-enter the cosmic arena. Now our philosophical musings must face the music: Does the information contained in the things swallowed by the black hole—the data we imagined existing within the black hole's interior—re-emerge as the black hole evaporates? This is the information required for quantum determinism to hold, and so this question goes to the heart of whether black holes imbue the evolution of our universe with an even deeper element of happenstance.

As of this writing, there is no consensus among physicists regarding the answer to this question. For many years, Hawking has strongly claimed that the information does not re-emerge—that black holes destroy information thereby "introducing a new level of uncertainty into physics, over and above the usual uncertainty associated with quantum theory."[9] In fact, Hawking, together with Kip Thorne of the California Institute of Technology, has a bet with John Preskill, also of the California Institute of Technology, regarding what happens to the information captured by a black hole: Hawking and Thorne bet that the information is forever lost, while Preskill has taken the opposite position and bet that the information re-emerges as the black hole radiates and shrinks. The wager? Information itself: "The loser(s) will reward the winner(s) with an encyclopedia of the winner's choice."

The bet remains unsettled, but Hawking has recently acknowledged that the newfound understanding of black holes from string theory, as discussed above, shows that there might be a way for the information to re-emerge.[10] The new idea is that for the kind of black holes studied by Strominger and Vafa, and by many other physicists since their initial paper, information can be stored and recovered from the constituent branes. This insight, Strominger recently said, "has led some string theorists to want to claim victory—to claim that the information is recovered as black holes evaporate. In my opinion this conclusion is premature; there is still much work to be done in order to see if this is true."[11] Vafa concurs, saying that he "is agnostic on this question—it could still turn out either way."[12] Answering this question is a central goal of current research. As Hawking has put it,

> Most physicists want to believe that information is not lost, as this would make the world safe and predictable. But I believe that if one takes Einstein's general relativity seriously, one must allow for the possibility that spacetime ties itself in knots and that information gets lost in the folds. Determining whether or not information actually does get lost is one of the major questions in theoretical physics today.[13]

The second unresolved black hole mystery concerns the nature of spacetime at the central point of the hole.[14] A straightforward application of general relativity, going all the way back to Schwarzschild in 1916, shows that the enormous mass and energy crushed together at the black hole's center causes the fabric of spacetime to suffer a devastating rift, to be radically warped into a state of infinite curvature—to be punctured by a spacetime singularity. One conclusion that physicists drew from this is that since all of the matter that has crossed the event horizon is inexorably drawn to the center of the black hole, and since once there the matter has no future, time itself comes to an end at the heart of a black hole. Other physicists, who over the years have explored the properties of the black hole's core using Einstein's equations, revealed the wild possibility that it might be a gateway to another universe that tenuously attaches to ours only at a black hole's center. Roughly speaking, where time in our universe comes to an end, time in the attached universe just begins.

We will take up some of the implications of this mind-boggling possibility in the next chapter, but for now we want to stress one important point. We must recall the central lesson: Extremes of huge mass and small size leading to unimaginably large density invalidate the sole use of Einstein's classical theory and require that quantum mechanics be brought to bear as well. This leads us to ask, What does string theory have to say about the spacetime singularity at the center of a black hole? This is a topic of intense current research, but as with the question of information loss, it has not yet been settled. String theory deftly deals with a variety of other singularities—the rips and tears in space discussed in Chapter 11 and in the first part of this chapter.[15] But if you have seen one singularity you have *not* seen them all. The fabric of our universe can be ripped, punctured, and torn in many different ways. String theory has given us profound insights into some of these singularities, but others, the black hole singularity among them, have so far eluded the string theorists' reach. The essential reason for this, once again, is the reliance on perturbative tools in string theory whose approximations, in this case, cloud our ability to analyze reliably and fully what happens at the deep interior point of a black hole.

However, given the recent tremendous progress in nonperturbative methods and their successful application to other aspects of black holes, string theorists have high hopes that it won't be long before the mysteries residing at the center of black holes start to unravel.

Chapter 14

Reflections on Cosmology

Humans throughout history have had a passionate drive to understand the origin of the universe. There is, perhaps, no single question that so transcends cultural and temporal divides, inspiring the imagination of our ancient forebears as well as the research of the modern cosmologist. At a deep level, there is a collective longing for an explanation of why there is a universe, how it has come to take the form we witness, and for the rationale—the principle—that drives its evolution. The astounding thing is that humanity has now come to a point where a framework is emerging for answering some of these questions scientifically.

The currently accepted scientific theory of creation declares that the universe experienced the most extreme of conditions—enormous energy, temperature, and density—during its earliest moments. These conditions, as is by now familiar, require that both quantum mechanics and gravity be taken into account, and hence the birth of the universe provides a profound arena for exercising the insights of superstring theory. We will discuss these nascent insights shortly, but first, we briefly recount the pre–string theory cosmological story, which is often referred to as the *standard model of cosmology*.

The Standard Model of Cosmology

The modern theory of cosmic origins dates from the decade and a half after Einstein's completion of general relativity. Although Einstein refused to take his own theory at face value and accept that it implies that the universe is neither eternal nor static, Alexander Friedmann did. And as we discussed in Chapter 3, Friedmann found what is now known as the big bang solution to Einstein's equations—a solution that declares that the universe violently emerged from a state of infinite compression, and is currently in the expanding aftermath of that primeval explosion. So certain was Einstein that such time-varying solutions were not a result of his theory that he published a short article claiming to have found a fatal flaw in Friedmann's work. Some eight months later, however, Friedmann succeeded in convincing Einstein that there was, in fact, no flaw; Einstein publicly but curtly retracted his objection. Nevertheless, it is clear that Einstein did not think Friedmann's results had any relevance to the universe. But about five years later, Hubble's detailed observations of a few dozen galaxies with the hundred-inch telescope at Mount Wilson Observatory confirmed that, indeed, the universe is expanding. Friedmann's work, refashioned in a more systematic and efficient form by the physicists Howard Robertson and Arthur Walker, still forms the foundation of modern cosmology.

In a little more detail, the modern theory of cosmic origins goes like this. Some 15 billion or so years ago, the universe erupted from an enormously energetic, singular event, which spewed forth all of space and all of matter. (You don't have to search far to locate where the big bang occurred, for it took place where you are now as well as everywhere else; in the beginning, all locations we now see as separate were the *same* location.) The temperature of the universe a mere 10^{-43} seconds after the bang, the so-called *Planck time,* is calculated to have been about 10^{32} Kelvin, some 10 trillion trillion times hotter than the deep interior of the sun. As time passed, the universe expanded and cooled, and as it did, the initial homogeneous, roiling hot, primordial cosmic plasma began to form eddies and clumps. At about a hundred-thousandth of a second after the bang, things had cooled sufficiently (to about 10 trillion Kelvin—about a million times hotter than the sun's interior) for quarks to clump together in groups of three, forming protons and neutrons. About a hundredth of a second later, conditions were right for the nuclei of some of the lightest elements in the periodic table to start congealing out of the cooling plasma of particles. For the next three minutes, as the simmering universe cooled to about a billion degrees, the predominant nuclei that emerged were those of hydrogen and helium, along with trace amounts of deuterium ("heavy" hydrogen) and lithium. This is known as the period of *primordial nucleosynthesis*.

Not a whole lot happened for the next few hundred thousand years, other than further expansion and cooling. But then, when the temperature had dropped to a few thousand degrees, wildly streaming electrons slowed down to the point where atomic nuclei, mostly hydrogen and helium, could capture them, forming the first electrically neutral atoms. This was a pivotal moment: from this point forward the universe, by and large, became transparent. Prior to the era of electron capture, the universe was filled with a dense plasma of electrically charged particles—some with positive charges like nuclei and others with negative charges, like electrons. Photons, which interact only with electrically charged objects, were bumped and jostled incessantly by the thick bath of charged particles, traversing hardly any distance before being deflected or absorbed. The charged-particle barrier to the free motion of photons would have made the universe appear almost completely opaque, much like what you may have experienced in a dense morning fog or a blinding, gusty snowstorm. But when negatively charged electrons were brought into orbit around positively charged nuclei, yielding electrically neutral atoms, the charged obstructions disappeared and the dense fog lifted. From that time onward, photons from the big bang have traveled unhindered and the full expanse of the universe gradually came into view.

About a billion years later, with the universe having substantially calmed down from its frenetic beginnings, galaxies, stars, and ultimately planets began to emerge as gravitationally bound clumps of the primordial elements. Today, some 15 billion or so years after the bang, we can marvel at both the magnificence of the cosmos and at our collective ability to have pieced together a reasonable and experimentally testable theory of cosmic origin.

But how much faith should we *really* have in the big bang theory?

Putting the Big Bang to the Test

By looking out into the universe with their most powerful telescopes, astronomers can see light that was emitted from galaxies and quasars just a few billion years after the big bang. This allows them to verify the expansion of the universe predicted by the big bang theory back to this early phase of the universe, and everything checks out to a "T." To test the theory to yet earlier times, physicists and astronomers must make use of more indirect methods. One of the most refined approaches involves something known as *cosmic background radiation*.

If you've ever felt a bicycle tire after vigorously pumping it full of air, you know that it is warm to the touch. Some of the energy you expend in the repeated pumping motion is transferred to an increase in temperature of the air in the tire. This reflects a general principle: Under a wide variety of conditions, when things are compressed they heat up. Reasoning in reverse, when things are allowed to decompress—to expand—they cool down. Air conditioners and refrigerators rely on these principles, subjecting substances like freon to repeated cycles of compression and expansion (as well as evaporation and condensation) to cause heat flow in the desired direction. Although these are simple facts of terrestrial physics, it turns out that they have a profound incarnation in the cosmos as a whole.

We saw above that after electrons and nuclei join together to form atoms, photons are free to travel unimpeded through the universe. This means that the universe is filled with a "gas" of photons travelling this way and that, uniformly distributed throughout the cosmos. As the universe expands, this gas of freely streaming photons expands as well since, in essence, the universe is its container. And just as the temperature of a more conventional gas (like the air in a bicycle tire) decreases as it expands, the temperature of this photon gas decreases as the universe expands. In fact, physicists as far back as George Gamow and his students Ralph Alpher and Robert Hermann in the 1950s, and Robert Dicke and Jim Peebles in the mid-1960s, realized that the present-day universe should be permeated by an almost uniform bath of these primordial photons, which, through the last 15 billion years of cosmic expansion, have cooled to a mere handful of degrees above absolute zero.[1] In 1965, Arno Penzias and Robert Wilson of Bell Laboratories in New Jersey accidentally made one of the most important discoveries of our age when they detected this afterglow of the big bang while working on an antenna intended for use with communication satellites. Subsequent research has refined both theory and experiment, culminating in measurements taken by NASA's COBE (Cosmic Background Explorer) satellite in the early 1990s. With these data, physicists and astronomers have confirmed to high precision that the universe *is* filled with microwave radiation (if our eyes were sensitive to microwaves, we would see a diffuse glow in the world around us) whose temperature is about 2.7 degrees above absolute zero, exactly in keeping with the expectation of the big bang theory. In concrete terms, in *every* cubic meter of the universe—including the one you now occupy—there are, on average, about 400 million photons that collectively compose the vast cosmic sea of microwave radiation, an echo of creation. A percentage of the "snow" you see on your television screen when you disconnect the cable feed and tune to a station that has ceased its scheduled broadcasts is due to this dim aftermath of the big bang. This match between theory and experiment confirms the big bang picture of cosmology as far back as the time that photons first moved freely through the universe, about a few hundred thousand years after the bang (ATB).

Can we push further in our tests of the big bang theory to even earlier times? We can. By using standard principles of nuclear theory and thermodynamics, physicists can make definite predictions about the relative abundance of the light elements produced during the period of primordial nucleosynthesis, between a hundredth of a second and a few minutes ATB. According to theory, for example, about 23 percent of the universe should be composed of helium. By measuring the helium abundance in stars and nebulae, astronomers have amassed impressive support that, indeed, this prediction is right on the mark. Perhaps even more impressive is the prediction and confirmation regarding deuterium abundance, since there is essentially no astrophysical process, other than the big bang, that can account for its small but definite presence throughout the cosmos. The confirmation of these abundances, and more recently that of lithium, is a sensitive test of our understanding of early universe physics back to the time of their primordial synthesis.

This is impressive almost to the point of hubris. All the data we possess confirm a theory of cosmology capable of describing the universe from about a hundredth of a second ATB to the present, some 15 billion years later. Nevertheless, one should not lose sight of the fact that the newborn universe evolved with phenomenal haste. Tiny fractions of a second—fractions *much* smaller than a hundredth of a second—form cosmic epochs during which long-lasting features of the world were first imprinted. And so, physicists have continued to push onward, trying to explain the universe at ever earlier times. Since the universe gets ever smaller, hotter, and denser as we push back, an accurate quantum-mechanical description of matter and the forces becomes increasingly important. As we have seen from other viewpoints in earlier chapters, point-particle quantum field theory works until typical particle energies are around the Planck energy. In a cosmological context, this occurred when the whole of the known universe fit within a Planck-sized nugget, yielding a density so great that it strains one's ability to find a fitting metaphor or an enlightening analogy: the density of the universe at the Planck time was simply *colossal*. At such energies and densities gravity and quantum mechanics can no longer be treated as two separate entities as they are in point-particle quantum field theory. Instead, the central message of this book is that at and beyond these enormous energies we must invoke string theory. In temporal terms, we encounter these energies and densities when we probe earlier than the Planck time of 10^{-43} seconds ATB, and hence this earliest epoch is the cosmological arena of string theory.

Let's head toward this era by first seeing what the standard cosmological theory tells us about the universe before a hundredth of a second ATB, but after the Planck time.

From the Planck Time to a Hundredth of a Second ATB

Recall from Chapter 7 (especially Figure 7.1) that the three nongravitational forces appear to merge together in the intensely hot environment of the early universe. Physicists' calculations of how the strengths of these forces vary with energy and temperature show that prior to about 10^{-35} seconds ATB, the strong, weak, and electromagnetic forces were all one "grand unified" or "super" force. In this state the universe was far more symmetric than it is today. Like the homogeneity that follows when a collection of disparate metals is heated to a smooth molten liquid, the significant differences between the forces as we now observe them were all erased by the extremes of energy and temperature encountered in the very early universe. But as time went by and the universe expanded and cooled, the formalism of quantum field theory shows that this symmetry would have been sharply reduced through a number of rather abrupt steps, ultimately leading to the comparatively asymmetric form with which we are familiar.

It's not hard to understand the physics behind such reduction of symmetry, or *symmetry breaking*, as it is more precisely called. Picture a large container filled with water. The molecules of H_2O are uniformly spread throughout the container and regardless of the angle from which you view it, the water looks the same. Now watch the container as you lower the temperature. At first not much happens. On microscopic scales, the average speed of the water molecules decreases, but that's about all. When you decrease the temperature to 0 degrees Celsius, however, you suddenly see that something drastic occurs. The liquid water begins to freeze and turn into solid ice. As discussed in the preceding chapter, this is a simple example of a phase transition. For our present purpose, the important thing to note is that the phase transition results in a decrease in the amount of symmetry displayed by the H_2O molecules. Whereas liquid water looks the same regardless of the angle from which it is viewed—it appears to be rotationally symmetric—solid ice is different. It has a crystalline block structure, which means that if you examine it with adequate precision, it will, like any crystal, look different from different angles. The phase transition has resulted in a decrease in the amount of rotational symmetry that is manifest.

Although we have discussed only one familiar example, the point is true more generally: as we lower the temperature of many physical systems, at some point they undergo a phase transition that typically results in a decrease or a "breaking" of some of their previous symmetries. In fact, a system can go through a series of phase transitions if its temperature is varied over a wide enough range. Water, again, provides a simple example. If we start with H_2O above 100 degrees Celsius, it is a gas: steam. In this form, the system has even more symmetry than in the liquid phase since now the individual H_2O molecules have been liberated from their congested, stuck-together liquid form. Instead, they all zip around the container on completely equal footing, without forming any clumps or "cliques" in which groups of molecules single each other out for a close association at the expense of others. Molecular democracy prevails at high enough temperatures. As we lower the temperature below 100 degrees, of course, water droplets do form as we pass through a gas-liquid phase transition, and the symmetry is reduced. Continuing on to yet lower temperatures, nothing too dramatic happens until we pass through 0 degrees Celsius, when, as above, the liquid-water/solid-ice phase transition results in another abrupt decrease in symmetry.

Physicists believe that between the Planck time and a hundredth of a second ATB, the universe behaved in a very similar way, passing through at least two analogous phase transitions. At temperatures above 10^{28} Kelvin, the three nongravitational forces appeared as one, as symmetric as they could possibly be. (At the end of this chapter we will discuss string theory's inclusion of the gravitational force into this high-temperature merger.) But as the temperature dropped below 10^{28} Kelvin, the universe underwent a phase transition in which the three forces crystallized out from their common union in different ways. Their relative strengths and the details of how they act on matter began to diverge. And so, the symmetry among the forces evident at higher temperatures was broken as the universe cooled. Nevertheless, the work of Glashow, Salam, and Weinberg (see Chapter 5) shows that not all of the high-temperature symmetry was erased: The weak and electromagnetic forces were still deeply interwoven. As the universe further expanded and cooled, nothing much happened until things simmered down to 10^{15} Kelvin—about 100 million times the sun's core temperature—when the universe went through another phase transition that affected the electromagnetic and weak forces. At this temperature, they too crystallized out from their previous, more symmetric union, and as the universe continued to cool, their differences became magnified. The two phase transitions are responsible for the three apparently distinct nongravitational forces at work in the world, even though this review of cosmic history shows that the forces, in fact, are deeply related.

A Cosmological Puzzle

This post–Planck era cosmology provides an elegant, consistent, and calculationally tractable framework for understanding the universe as far back as the briefest moments after the bang. But, as with most success-ful theories, our new insights raise yet more detailed questions. And it turns out that some of these questions, while not invalidating the standard cosmological scenario as presented, do highlight awkward aspects that point toward the need for a deeper theory. Let's focus on one. It is called the *horizon problem,* and it is one of the most important issues in modern cosmology.

Detailed studies of the cosmic background radiation have shown that regardless of which direction in the sky one points the measuring antenna, the temperature of the radiation is the same, to about one part in 100,000. If you think about it for a moment, you will realize that this is quite strange. Why should different locations in the universe, separated by enor-mous distances, have temperatures that are so finely matched? A seem-ingly natural resolution to this puzzle is to note that, yes, two diametrically opposite places in the heavens are far apart today, but like twins separated at birth, during the earliest moments of the universe they (and everything else) were very close together. Since they emerged from a common start-ing point, you might suggest that it's not at all surprising that they share common physical traits such as their temperature.

In the standard big bang cosmology this suggestion fails. Here's why. A bowl of hot soup gradually cools to room temperature because it is in con-tact with the colder surrounding air. If you wait long enough, the temper-ature of the soup and the air will, through their mutual contact, become the same. But if the soup is in a thermos, of course, it retains its heat for much longer, since there is far less communication with the outside en-vironment. This reflects that the homogenization of temperature between two bodies relies on their having prolonged and unimpaired communica-tion. To test the suggestion that positions in space that are currently sep-arated by vast distances share the same temperature because of their initial contact, we must therefore examine the efficacy of information ex-change between them in the early universe. At first you might think that since the positions were closer together at earlier times, communication was ever easier. But spatial proximity is only one part of the story. The other part is temporal duration.

To examine this more fully, let's imagine studying a "film" of the cosmic expansion, but let's review it in reverse, running the film backward in time from today toward the moment of the big bang. Since the speed of light sets a limit to how fast any signal or information of any kind can travel, matter in two regions of space can exchange heat energy and thereby have a chance of coming to a common temperature only if the distance between them at a given moment is less than the distance light can have traveled since the time of the big bang. And so, as we roll the film backward in time we see that there is a competition between how close together our spatial regions become versus how far back we have to turn the clock for them to get there. For instance, if in order for the separation of our two spatial locations to be 186,000 miles, we have to run the film back to less than a second ATB, then even though they are much closer, there is still no way for them to have any influence on each other since light would require a whole second to travel the distance between them.[2] If in order for their separation to be much less, say 186 miles, we have to run the film back to less than a thousandth of a second ATB, then, again, the same conclusion follows: They can't influence each other since in less than a thousandth of a second light can't travel the 186 miles separating them. Carrying on in the same vein, if we have to run the film back to less than a billionth of a second ATB in order for these regions to be within one foot of each other, they still cannot influence each other since there is just not enough time since the bang for light to have traveled the 12 inches between them. This shows that just because two points in the universe get closer and closer as we head back to the bang, it is not necessarily the case that they can have had the thermal contact—like that between soup and air—necessary to bring them to the same temperature.

Physicists have shown that precisely this problem arises in the standard big bang model. Detailed calculations show that there is no way for regions of space that are currently widely separated to have had the exchange of heat energy that would explain their having the same temperature. As the word *horizon* refers to how far we can see—how far light can travel, so to speak—physicists call the unexplained uniformity of temperature throughout the vast expanse of the cosmos the "horizon problem." The puzzle does not mean the standard cosmological theory is wrong. But the uniformity of temperature does strongly suggest that we are missing an important part of the cosmological story. In 1979, the physicist Alan Guth, now of the Massachusetts Institute of Technology, wrote the missing chapter.

Inflation

The root of the horizon problem is that in order to get two widely separated regions of the universe close together, we have to run the cosmic film way back toward the beginning of time. So far back, in fact, that there is not enough time for any physical influence to have traveled from one region to the other. The difficulty, therefore, is that as we run the cosmological film backward and approach the big bang, the universe does not shrink at a fast enough rate.

Well, that's the rough idea, but it's worthwhile sharpening the description a bit. The horizon problem stems from the fact that like a ball tossed upward, the dragging pull of gravity causes the expansion rate of the universe to *slow down*. This means that, for example, to halve the separation between two locations in the cosmos we must run the film back more than halfway toward its beginning. In turn, we see that to halve the separation we must more than halve the time since the big bang. Less time since the bang—proportionally speaking—means it is *harder* for the two regions to communicate, even though they get closer.

Guth's resolution of the horizon problem is now simple to state. He found another solution to Einstein's equations in which the very early universe undergoes a brief period of enormously fast expansion—a period during which it "inflates" in size at an unheralded *exponential* expansion rate. Unlike the case of a ball that slows down after being tossed upward, exponential expansion gets *faster* as it proceeds. When we run the cosmic film in reverse, rapid accelerating expansion turns into rapid decelerating contraction. This means that to halve the separation between two locations in the cosmos (during the exponential epoch) we need run the the film back less than halfway—much less, in fact. Running the film back less implies that the two regions will have had more time to communicate thermally and, like hot soup and air, they will have had ample time to come to the same temperature.

Through Guth's discovery and later important refinements made by Andrei Linde, now of Stanford University, Paul Steinhardt and Andreas Albrecht, then of the University of Pennsylvania, and many others, the standard cosmological model was revamped into the *inflationary* cosmological model. In this framework, the standard cosmological model is modified during a tiny window of time—around 10^{-36} to 10^{-34} seconds ATB—in which the universe expanded by a colossal factor of at least 10^{30}, compared with a factor of about a hundred during the same time interval in the standard scenario. This means that in a brief flicker of time, about a trillionth of a trillionth of a trillionth of a second ATB, the size of the universe increased by a greater percentage than it has in the 15 billion years since. Before this expansion, matter that is now in far-flung regions of the cosmos was much closer together than in the standard cosmological model, making it possible for a common temperature to be easily established. Then, through Guth's momentary burst of cosmological inflation—followed by the more usual expansion of the standard cosmological model—these regions of space were able to become separated by the vast distances we witness currently. And so, the brief but profound inflationary modification of the standard cosmological model solves the horizon problem (as well as a number of other important problems we have not discussed) and has gained wide acceptance among cosmologists.[3]

We summarize the history of the universe from just after the Planck time to the present, according to the current theory, in Figure 14.1.

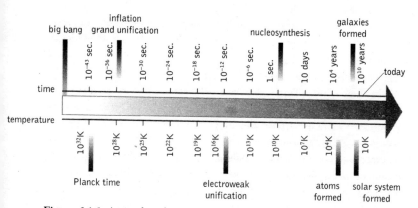

Figure 14.1 A time line denoting a few key moments in the history of the universe.

Cosmology and Superstring Theory

There remains a sliver of Figure 14.1, between the big bang and the Planck time, that we have not yet discussed. By blindly applying the equations of general relativity to that region, physicists have found that the universe continues to get ever smaller, ever hotter, and ever denser, as we move backward in time toward the bang. At time zero, as the size of the universe vanishes, the temperature and density soar to infinity, giving us the most extreme signal that this theoretical model of the universe, firmly rooted in the classical gravitational framework of general relativity, has completely broken down.

Nature is telling us emphatically that under such conditions we must merge general relativity and quantum mechanics—in other words, we must make use of string theory. Currently, research on the implications of string theory for cosmology is at an early stage of development. Perturbative methods can, at best, give skeletal insights, since the extremes of energy, temperature, and density require precision analysis. Although the second superstring revolution has provided some nonperturbative techniques, it will be some time before they are honed for the kinds of calculations required in a cosmological setting. Nevertheless, as we now discuss, during the last decade or so, physicists have taken the first steps toward understanding string cosmology. Here is what they have found.

It appears that there are three essential ways in which string theory modifies the standard cosmological model. First, in a manner that current research continues to clarify, string theory implies that the universe has what amounts to a smallest possible size. This has profound consequences for our understanding of the universe at the moment of the bang itself, when the standard theory claims that its size has shrunk all the way to zero. Second, string theory has a small-radius/large-radius duality (intimately related to its having a smallest possible size), which also has deep cosmological significance, as we will see in a moment. Finally, string theory has more than four spacetime dimensions, and from a cosmological standpoint, we must address the evolution of them all. Let's discuss these points in greater detail.

357

In the Beginning There Was a Planck-Sized Nugget

In the late 1980s, Robert Brandenberger and Cumrun Vafa made the first important strides toward understanding how the application of these string theoretic features modifies the conclusions of the standard cosmological framework. They came to two important realizations. First, as we run the clock backward in time toward the beginning, the temperature continues to rise until the size of the universe is about the Planck length in all directions. But then, the temperature hits a *maximum* and begins to *decrease*. The intuitive reason behind this is not hard to come by. Imagine for simplicity (as Brandenberger and Vafa did) that all of the space dimensions of the universe are circular. As we run the clock backward and the radius of each of these circles shrinks, the temperature of the universe increases. But as each of the radii collapses toward and then through the Planck length, we know that, within string theory, this is physically identical to the radii shrinking to the Planck length and then bouncing back toward increasing size. Since temperature goes down as the universe expands, we would expect that the futile attempt to squeeze the universe to sub-Planck size means that the temperature stops rising, hits a maximum, and then begins to decrease. Through detailed calculations, Brandenberger and Vafa explicitly verified that indeed this is the case.

This led Brandenberger and Vafa to the following cosmological picture. In the beginning, all of the spatial dimensions of string theory are tightly curled up to their smallest possible extent, which is roughly the Planck length. The temperature and energy are high, but not infinite, since string theory has avoided the conundrums of an infinitely compressed zero-size starting point. At this beginning moment of the universe, all the spatial dimensions of string theory are on completely equal footing—they are completely symmetric—all curled up into a multidimensional, Planck-sized nugget. Then, according to Brandenberger and Vafa, the universe goes through its first stage of symmetry reduction when, at about the Planck time, three of the spatial dimensions are singled out for expansion, while all others retain their initial Planck-scale size. These three space dimensions are then identified with those in the inflationary cosmological scenario, the post-Planck-time evolution summarized in Figure 14.1 takes over, and these three dimensions expand to their currently observed form.

Why Three?

An immediate question is, What drives the symmetry reduction that singles out precisely three spatial dimensions for expansion? That is, beyond the experimental fact that only three of the space dimensions have expanded to observably large size, does string theory provide a fundamental reason for why some other number (four, five, six, and so on) or, even more symmetrically, all of the space dimensions don't expand as well? Brandenberger and Vafa came up with a possible explanation. Remember that the small-radius/large-radius duality of string theory rests upon the fact that when a dimension is curled up like a circle, a string can wrap around it. Brandenberger and Vafa realized that, like rubber bands wrapped around a bicycle tire inner tube, such wrapped strings tend to constrict the dimensions they encircle, keeping them from expanding. At first sight, this would seem to mean that each of the dimensions will be constricted, since the strings can and do wrap them all. The loophole is that if a wrapped string and its antistring partner (roughly, a string that wraps the dimension in the opposite direction) should come into contact, they will swiftly annihilate one other, producing an *unwrapped* string. If these processes happen with sufficient rapidity and efficiency, enough of the rubber band–like constriction will be eliminated, allowing the dimensions to expand. Brandenberger and Vafa suggested that this reduction in the choking effect of wrapped strings will happen in only three of the spatial dimensions. Here's why.

Imagine two point particles rolling along a one-dimensional line such as the spatial extent of Lineland. Unless they happen to have identical velocities, sooner or later one will overtake the other, and they will collide. Notice, however, that if these same point particles are randomly rolling around on a two-dimensional plane such as the spatial extent of Flatland, it is likely that they will never collide. The second spatial dimension opens up a new world of trajectories for each particle, most of which do not cross each other at the same point at the same time. In three, four, or any higher number of dimensions, it gets increasingly unlikely that the two particles will ever meet. Brandenberger and Vafa realized that an analogous idea holds if we replace point particles with loops of string, wrapped around spatial dimensions. Although it's significantly harder to see, if there are *three* (or fewer) circular spatial dimensions, two wrapped strings will likely collide with one another—the analog of what happens for two particles moving in one dimension. But in four or more space dimensions, wrapped strings are less and less likely ever to collide—the analog of what happens for point particles in two or more dimensions.[4]

This leads to the following picture. In the first moment of the universe, the tumult from the high, but finite, temperature drives all of the circular dimensions to try to expand. As they do, the wrapped strings constrict the expansion, driving the dimensions back to their original Planck-size radii. But, sooner or later a random thermal fluctuation will drive three dimensions momentarily to grow larger than the others, and our discussion then shows that strings which wrap these dimensions are highly likely to collide. About half of the collisions will involve string/antistring pairs, leading to annihilations that continually lessen the constriction, allowing these three dimensions to continue to expand. The more they expand, the less likely it is for other strings to get entangled around them since it takes more energy for a string to wrap around a larger dimension. Thus, the expansion feeds on itself, becoming ever less constricted as the dimensions get ever larger. We can now imagine that these three spatial dimensions continue to evolve in the manner described in the previous sections, and expand to a size as large as or larger than the currently observable universe.

Cosmology and Calabi-Yau Shapes

For simplicity, Brandenberger and Vafa imagined that all of the spatial dimensions are circular. In fact, as noted in Chapter 8, so long as the circular dimensions are large enough that they curve back on themselves only beyond the range of our current observational capacity, a circular shape is consistent with the universe we observe. But for dimensions that stay small, a more realistic scenario is one in which they are curled up into a more intricate Calabi-Yau space. Of course, the key question is, Which Calabi-Yau space? How is this particular space determined? No one has been able to answer this question. But by combining the drastic topology-changing results described in the preceding chapter with these cosmological insights, we can suggest a framework for doing so. Through the space-tearing conifold transitions, we now know that any Calabi-Yau shape can evolve into any other. So, we can imagine that in the tumultuous, hot moments after the bang, the curled-up Calabi-Yau component of space stays small, but goes through a frenetic dance in which its fabric rips apart and reconnects over and over again, rapidly taking us through a long sequence of different Calabi-Yau shapes. As the universe cools and three of the spatial dimensions get large, the transitions from one Calabi-Yau to another slow down, with the extra dimensions ultimately settling into a Calabi-Yau shape that, optimistically, gives rise to the physical features we observe in the world around us. The challenge facing physicists is to understand, in detail, the evolution of the Calabi-Yau component of space so that its present form can be predicted from theoretical principles. With the newfound ability of one Calabi-Yau to change smoothly into another, we see that the issue of selecting one Calabi-Yau shape from the many may in fact be reduced to a problem of cosmology.[5]

Before the Beginning?

Lacking the exact equations of string theory, Brandenberger and Vafa were forced to make numerous approximations and assumptions in their cosmological studies. As Vafa recently said,

> Our work highlights the new way in which string theory allows us to start addressing persistent problems in the standard approach to cosmology. We see, for example, that the whole notion of an initial singularity may be completely avoided by string theory. But, because of difficulties in performing fully trustworthy calculations in such extreme situations with our present understanding of string theory, our work only provides a first look into string cosmology, and is very far from the final word.[6]

Since their work, physicists have made steady progress in furthering the understanding of string cosmology, spearheaded by, among others,

Gabriele Veneziano and his collaborator Maurizio Gasperini of the University of Torino. Gasperini and Veneziano have come up with their own intriguing version of string cosmology that shares certain features with the scenario described above, but also differs in significant ways. As in the Brandenberger and Vafa work, they too rely on string theory's having a minimal length in order to avoid the infinite temperature and energy density that arises in the standard and inflationary cosmological theories. But rather than concluding that this means the universe begins as an extremely hot Planck-size nugget, Gasperini and Veneziano suggest that there may be a whole *prehistory* to the universe—starting long before what we have so far been calling time zero—that leads up to the Planckian cosmic embryo.

In this so-called *pre–big bang* scenario, the universe began in a vastly different state than it does in the big bang framework. Gasperini and Veneziano's work suggests that rather than being enormously hot and tightly curled into a tiny spatial speck, the universe started out as cold and essentially *infinite* in spatial extent. The equations of string theory then indicate that—somewhat as in Guth's inflationary epoch—an instability kicked in, driving every point in the universe to rush rapidly away from every other. Gasperini and Veneziano show that this caused space to become increasingly curved and results in a dramatic increase in temperature and energy density.[7] After some time, a millimeter-sized three-dimensional region *within* this vast expanse could look just like the superhot and dense patch emerging from Guth's inflationary expansion. Then, through the standard expansion of ordinary big bang cosmology, this patch can account for the whole of the universe with which we are familiar. Moreover, because the pre–big bang epoch involves its own inflationary expansion, Guth's solution to the horizon problem is automatically built into the pre–big bang cosmological scenario. As Veneziano has said, "String theory offers us a version of inflationary cosmology on a silver platter."[8]

The study of superstring cosmology is rapidly becoming an active and fertile arena of research. The pre–big bang scenario, for example, has already generated a significant amount of heated, yet fruitful debate, and it is far from clear what role it will have in the cosmological framework that will ultimately emerge from string theory. Achieving these cosmological insights will, no doubt, rely heavily on the ability of physicists to come to grips with all aspects of the second superstring revolution. What, for example, are the cosmological consequences of the existence of fundamental higher-dimensional branes? How do the cosmological properties we have discussed change if string theory happens to have a coupling constant whose value places us more toward the center of Figure 12.11 rather than in one of the peninsular regions? That is, what is the impact of full-fledged M-theory on the earliest moments of the universe? These central questions are now being studied vigorously. Already, one important insight has emerged.

M-Theory and the Merging of All Forces

In Figure 7.1 we showed how the strengths of the three nongravitational couplings merge together when the temperature of the universe is high enough. How does the strength of the gravitational force fit into this picture? Before the emergence of M-theory, string theorists were able to show that with the simplest of choices for the Calabi-Yau component of space, the gravitational force almost, but not quite, merges with the other three, as shown in Figure 14.2. String theorists found that the mismatch could be avoided by carefully molding the shape of the chosen Calabi-Yau, among other tricks of the trade, but such after-the-fact fine tuning always

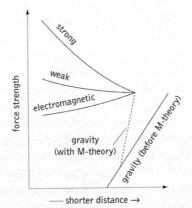

Figure 14.2 Within M-theory, the strengths of all four forces can naturally merge.

makes a physicist uncomfortable. Since no one currently knows how to predict the precise form of the Calabi-Yau dimensions, it seems dangerous to rely upon solutions to problems that hinge so delicately on the fine details of their shape.

Witten has shown, however, that the second superstring revolution provides a far more robust solution. By investigating how the strengths of the forces vary when the string coupling constant is not necessarily small, Witten found that the gravitational force curve can be gently nudged to merge with the other forces, as in Figure 14.2, without any special molding of the Calabi-Yau portion of space. Although it is far too early to tell, this may indicate that cosmological unity is more easily achieved by making use of the larger framework of M-theory.

The developments discussed in this and the previous sections represent the first, somewhat tentative steps toward understanding the cosmological implications of string/M-theory. During the coming years, as the non-perturbative tools of string/M-theory are sharpened, physicists anticipate that some of the most profound insights will emerge from their application to cosmological questions.

But without currently having methods that are sufficiently powerful to understand cosmology according to string theory fully, it is worthwhile to think about some general considerations concerning the possible role of cosmology in the search for the ultimate theory. We caution that some of these ideas are of a more speculative nature than much of what we have discussed previously, but they do raise issues that any purported final theory may one day have to address.

Cosmological Speculation and the Ultimate Theory

Cosmology has the ability to grab hold of us at a deep, visceral level because an understanding of how things began feels—at least to some—like the closest we may ever come to understanding *why* they began. That is not to say that modern science provides a connection between the question of how and the question of why—it doesn't—and it may well be that no such scientific connection is ever found. But the study of cosmology does hold the promise of giving us our most complete understanding of the arena of the why—the birth of the universe—and this at least allows for a scientifically informed view of the frame within which the questions are asked. Sometimes attaining the deepest familiarity with a question is our best substitute for actually having the answer.

In the context of searching for the ultimate theory, these lofty reflections on cosmology give way to far more concrete considerations. The way things in the universe appear to us today—way on the far right-hand side of the time line in Figure 14.1—depends upon the fundamental laws of physics, to be sure, but it may also depend on aspects of cosmological evolution, from the far left-hand side of the time line, that potentially lie outside the scope of even the deepest theory.

It's not hard to imagine how this might be. Think of what happens, for example, when you toss a ball in the air. The laws of gravity govern the ball's subsequent motion, but we can't predict where the ball will land exclusively from those laws. We must also know the velocity of the ball—its speed and direction—as it left your hand. That is, we must know the *initial conditions* of the ball's motion. Similarly, there are features of the universe that also have a historical contingency—the reason why a star formed here or a planet there depends upon a complicated chain of events that, at least in principle, we can imagine tracing back to some feature of how the universe was when it all began. But it is possible that even more basic features of the universe, perhaps even the properties of the fundamental matter and force particles, also have a direct dependence on historical evolution—evolution that itself is contingent upon the initial conditions of the universe.

In fact, we've already noted one possible incarnation of this idea in string theory: As the hot, early universe evolved, the extra dimensions may have transmuted from shape to shape, ultimately settling down to one particular Calabi-Yau space once things had cooled off sufficiently. But, like a ball tossed in the air, the result of that journey through numerous Calabi-Yau shapes may well depend on details of how the journey got started in the first place. And through the influence of the resulting Calabi-Yau shape on particle masses and on properties of forces, we see that cosmological evolution and the state of the universe when it began can have a profound impact on the physics we currently observe.

We don't know what the initial conditions of the universe were, or even the ideas, concepts, and language that should be used to describe them. We believe that the outrageous initial state of *infinite* energy, density, and temperature that arises in the standard and inflationary cosmological models is a signal that these theories have broken down rather than a correct description of the physical conditions that actually existed. String theory offers an improvement by showing how such infinite extremes might be avoided; nevertheless, no one has any insight on the question of how things actually did begin. In fact, our ignorance persists on an even higher plane: We don't know whether the question of determining the initial conditions is one that is even sensible to ask or whether—like asking general relativity to give insight into how hard you happened to toss a ball in the air—it is a question that lies forever beyond the grasp of any theory. Valiant attempts by physicists such as Hawking and James Hartle of the University of California at Santa Barbara have tried to bring the question of cosmological initial conditions within the umbrella of physical theory, but all such attempts remain inconclusive. In the context of string/M-theory, our cosmological understanding is, at present, just too primitive to determine whether our candidate "theory of everything" truly lives up to its name and determines its own cosmological initial conditions, thereby elevating them to the status of physical law. This is a prime question for future research.

But even beyond the issue of initial conditions and their impact on the ensuing historical twists and turns of cosmic evolution, some recent and highly speculative proposals have argued for yet other potential limits on the explanatory power of any final theory. No one knows if these ideas are right or wrong, and certainly they currently lie on the outskirts of mainstream science. But they do highlight—albeit in a rather provocative and speculative manner—an obstacle that any proposed final theory may encounter.

The basic idea rests upon the following possibility. Imagine that what we call *the* universe is actually only one tiny part of a vastly larger cosmological expanse, one of an enormous number of island universes scattered across a grand cosmological archipelago. Although this might sound rather far-fetched—and in the end it may well be—André Linde has suggested a concrete mechanism that might lead to such a gargantuan universe. Linde has found that the brief but crucial burst of inflationary expansion discussed earlier may not have been a unique, one-time event. Instead, he argues, the conditions for inflationary expansion may happen repeatedly in isolated regions peppered throughout the cosmos, which then undergo their own inflationary ballooning in size, evolving into new, separate universes. And in each of these universes, the process continues, with new universes sprouting from far-flung regions in the old, generating a never ending web of ballooning cosmic expanses. The terminology gets a little cumbersome, but let's follow fashion and call this greatly expanded notion of the universe the *multiverse*, with each of the constituent parts being called a universe.

The central observation is that whereas in Chapter 7 we noted that everything we know points toward a consistent and uniform physics throughout our universe, this may have no bearing on the physical attributes in these other universes so long as they are separate from us, or at least so far away that their light has not had time to reach us. And so we can imagine that physics varies from one universe to another. In some, the differences may be subtle: For example, the electron mass or the strength of the strong force might be a thousandth of a percent larger or smaller than in our universe. In others, physics may differ in more pronounced ways: The up-quark might weigh ten times what it weighs in our universe, or the strength of the electromagnetic force might be ten times the value we measure, with all the profound implications that this has on stars and on life as we know it (as indicated in Chapter 1). And in other universes, physics may differ in still more dramatic ways: The list of elementary particles and forces may be completely distinct from ours, or, taking a cue from string theory, even the number of extended dimensions may differ, with some cramped universes having as few as zero or one large spatial dimension, while other expansive universes possess eight, nine, or even ten extended spatial dimensions. If we let our imaginations run free, even the laws themselves can drastically differ from universe to universe. The range of possibilities is endless.

Here's the point. If we scan through this huge maze of universes, the vast majority will not have conditions hospitable to life, or at least to anything remotely akin to life as we know it. For drastic changes in familiar physics, this is clear: If our universe truly looked like the Garden-hose universe, life as we know it would not exist. But even rather conservative changes to physics would interfere with the formation of stars, for example, disrupting their ability to act as cosmic furnaces that synthesize complex life-supporting atoms such as carbon and oxygen that, normally, are spewed throughout the universe by supernova explosions. In light of the sensitive dependence of life on the details of physics, if we now ask, for instance, why the forces and particles of nature have the particular properties we observe, a possible answer emerges: Across the entire multiverse, these features vary widely; their properties *can* be different and *are* different in other universes. What's special about the particular combination of particle and force properties we observe is that, clearly, they allow life to form. And life, intelligent life in particular, is a prerequisite even to ask the question of why our universe has the properties it does. In plain language, things are the way they are in our universe because if they weren't, we wouldn't be here to notice. Like the winners of a mass game of Russian roulette, whose surprise at surviving is tempered by the realization that had they not won, they wouldn't have been able *not* to feel surprised, the multiverse hypothesis has the capacity to lessen our insistence on explaining why our universe appears as it does.

This line of argument is a version of an idea with a long history known as the *anthropic principle*. As presented, it is a perspective that is diametrically opposed to the dream of a rigid, fully predictive, unified theory in which things are the way they are because the universe could not be otherwise. Rather than being the epitome of poetic grace in which everything fits together with inflexible elegance, the multiverse and the anthropic principle paint a picture of a wildly excessive collection of universes with an insatiable appetite for variety. It will be extremely hard, if not impossible, for us ever to know if the multiverse picture is true. Even if there are other universes, we can imagine that we will never come into contact with any of them. But by vastly increasing the scope of "what's out there"—in a manner that dwarfs Hubble's realization that the Milky Way is but one galaxy among many—the concept of the multiverse does at least alert us to the possibility that we may be asking too much of an ultimate theory.

We should require that our ultimate theory give a quantum-mechanically consistent description of all forces and all matter. We should require that our ultimate theory give a cogent cosmology within our universe. However, if the multiverse picture is correct—a huge if—it *may* be asking too much for our theory to explain, as well, the detailed properties of the particle masses, charges, and the force strengths.

But we must emphasize that even if we accept the speculative premise of the multiverse, the conclusion that this compromises our predictive power is far from airtight. The reason, simply put, is that if we unleash our imaginations and allow ourselves to contemplate a multiverse, we should also unleash our theoretical musings and contemplate ways in which the apparent randomness of the multiverse can be tamed. For one relatively conservative musing, we can imagine that—were the multiverse picture true—we would be able to extend our ultimate theory to its full sprawling expanse, and that our "extended ultimate theory" might tell us precisely why and how the values of the fundamental parameters are sprinkled across the constituent universes.

A more radical musing comes from a proposal of Lee Smolin of Penn State University, who, inspired by the similarity between conditions at the big bang and at the centers of black holes—each being characterized by a colossal density of crushed matter—has suggested that every black hole is the seed for a new universe that erupts into existence through a big bang–like explosion, but is forever hidden from our view by the black hole's event horizon. Beyond proposing another mechanism for generating a multiverse, Smolin has injected a new element—a cosmic version of genetic mutation—that does an end run around the scientific limitations associated with the anthropic principle.[9] Imagine, he suggests, that when a universe sprouts from the core of a black hole, its physical attributes, such as particle masses and force strengths, are close, but not identical, to those of its parent universe. Since black holes arise from extinguished stars, and star formation depends upon the precise values of the particle masses and force strengths, the fecundity of any given universe—the number of black hole progeny it can produce—depends sensitively on these parameters. Small variations in the parameters of the progeny universes will therefore lead to some that are even more optimized for black hole production than their parent universe, and have an even greater number of offspring universes of their own.[10] After many "generations," the descendants of universes optimized for producing black holes will thus be so numerous that they will overwhelm the population of the multiverse. And so, rather than invoking the anthropic principle, Smolin's suggestion provides a dynamic mechanism that, on average, drives the parameters of each next-generation universe ever closer to particular values—those that are optimum for black hole production.

This approach gives another method, even in the context of the multiverse, in which the fundamental matter and force parameters can be explained. If Smolin's theory is right, and if we are a typical member of a mature multiverse (these are big "ifs" and can be debated on many fronts, of course), the parameters of the particles and forces that we measure should be optimized for black hole production. That is, any fiddling with these parameters of our universe should make it harder for black holes to form. Physicists have begun to investigate this prediction; at present there is no consensus on its validity. But even if Smolin's specific proposal turns out to be wrong, it does present yet another shape that the ultimate theory might take. The ultimate theory may, at first sight, appear to lack rigidity. We may find that it can describe a wealth of universes, most of which have no relevance to the one we inhabit. And moreover, we can imagine that this wealth of universes may be physically realized, leading to a multiverse—something that, at first sight, forever limits our predictive power. In fact, however, this discussion illustrates that an ultimate explanation can yet be achieved, so long as we grasp not only the ultimate laws but also their implications for cosmological evolution on an unexpectedly grand scale.

Undoubtedly, the cosmological implications of string/M-theory will be a major field of study well into the twenty-first century. Without accelerators capable of producing Planck-scale energies, we will increasingly have to rely on the cosmological accelerator of the big bang, and the relics it has left for us throughout the universe, for our experimental data. With luck and perseverance, we may finally be able to answer questions such as how the universe began, and why it has evolved to the form we behold in the heavens and on earth. There is, of course, much uncharted territory between where we are and where full answers to these fundamental questions lie. But the development of a quantum theory of gravity through superstring theory lends credence to the hope that we now possess theoretical tools for pushing into the vast regions of the unknown, and, no doubt after many a struggle, possibly emerging with answers to some of the deepest questions ever posed.

Part V

Unification in the
Twenty-First Century

Chapter 15

Prospects

Centuries from now, superstring theory, or its evolution within M-theory, may have developed so far beyond our current formulation that it might be unrecognizable even to today's leading researchers. As we continue to seek the ultimate theory, we may well find that string theory is but one of many pivotal steps on a path toward a far grander conception of the cosmos—a conception that involves ideas that differ radically from anything we have previously encountered. The history of science teaches us that each time we think that we have it all figured out, nature has a radical surprise in store for us that requires significant and sometimes drastic changes in how we think the world works. Then again, in a bit of brash posturing, we can also imagine, as others before us have perhaps naively done, that we are living through a landmark period in humanity's history in which the search for the ultimate laws of the universe will finally draw to a close. As Edward Witten has said,

> I feel that we are so close with string theory that—in my moments of greatest optimism—I imagine that any day, the final form of the theory might drop out of the sky and land in someone's lap. But more realistically, I feel that we are now in the process of constructing a much deeper theory than anything we have had before and that well into the twenty-first century, when I am too old to have any useful thoughts on the subject, younger physicists will have to decide whether we have in fact found the final theory.[1]

Although we are still feeling the aftershocks of the second superstring revolution and absorbing the panoply of new insights that it has engendered, most string theorists agree that it will likely take a third and maybe a fourth such theoretical upheaval before the full power of string theory is unleashed and its possible role as the final theory assessed. As we have seen, string theory has already painted a remarkable new picture of how the universe works, but there are significant hurdles and loose ends that will no doubt be the primary focus of string theorists in the twenty-first century. And so, in this last chapter we will not be able to finish telling the story of humanity's search for the deepest laws of the universe, because the search continues. Instead, let's guide our gaze into the future of string theory by discussing five central questions string theorists will face as they continue the pursuit of the ultimate theory.

What Is the Fundamental Principle Underlying String Theory?

One overarching lesson we have learned during the past hundred years is that the known laws of physics are associated with principles of symmetry. Special relativity is based on the symmetry embodied in the principle of relativity—the symmetry between all constant-velocity vantage points. The gravitational force, as embodied in the general theory of relativity, is based on the equivalence principle—the extension of the principle of relativity to embrace all possible vantage points regardless of the complexity of their states of motion. And the strong, weak, and electromagnetic forces are based on the more abstract gauge symmetry principles.

Physicists, as we have discussed, tend to elevate symmetry principles to a place of prominence by putting them squarely on the pedestal of explanation. Gravity, in this view, *exists* in order that all possible observational vantage points are on completely equal footing—i.e., so that the equivalence principle holds. Similarly, the nongravitational forces *exist* in order that nature respect their associated gauge symmetries. Of course, this approach shifts the question of why a certain force exists to why nature respects its associated symmetry principle. But this certainly feels like progress, especially when the symmetry in question is one that seems eminently natural. For example, why should one observer's frame of reference be treated differently from another's? It seems far more natural for the laws of the universe to treat all observational vantage points equally; this is accomplished through the equivalence principle and the introduction of gravity into the structure of the cosmos. Although it requires some mathematical background to appreciate fully, as we indicated in Chapter 5, there is a similar rationale behind the gauge symmetries underlying the three nongravitational forces.

String theory takes us down another notch on the scale of explanatory depth because all of these symmetry principles, as well as another—supersymmetry—emerge from its structure. In fact, had history followed a different course—and had physicists come upon string theory some hundred years earlier—we can imagine that these symmetry principles would have all been discovered by studying its properties. But bear in mind that whereas the equivalence principle gives us some understanding of why gravity exists, and the gauge symmetries give us some sense of why the nongravitational forces exist, in the context of string theory these symmetries are *consequences*; although their importance is in no way diminished, they are part of the end product of a much larger theoretical structure.

This discussion brings the following question into sharp relief: Is string theory itself an inevitable consequence of some broader principle—possibly but not necessarily a symmetry principle—in much the same way that the equivalence principle inexorably leads to general relativity or that gauge symmetries lead to the nongravitational forces? As of this writing, no one has any insight into the answer to this question. To appreciate its importance, we need only imagine Einstein trying to formulate general relativity without having had the happy thought he experienced in the Bern patent office in 1907 that led him to the principle of equivalence. It would not have been impossible to formulate general relativity without first having this key insight, but it certainly would have been extremely difficult. The equivalence principle provides a succinct, systematic, and powerful organizational framework for analyzing the gravitational force. The description of general relativity we gave in Chapter 3, for example, relied centrally on the equivalence principle, and its role in the full mathematical formalism of the theory is even more crucial.

Currently, string theorists are in a position analogous to an Einstein bereft of the equivalence principle. Since Veneziano's insightful guess in 1968, the theory has been pieced together, discovery by discovery, revolution by revolution. But a central organizing principle that embraces these discoveries and all other features of the theory within one overarching and systematic framework—a framework that makes the existence of each individual ingredient absolutely inevitable—is still missing. The discovery of this principle would mark a pivotal moment in the development of string theory, as it would likely expose the theory's inner workings with unforeseen clarity. There is, of course, no guarantee that such a fundamental principle exists, but the evolution of physics during the last hundred years encourages string theorists to have high hopes that it does. As we look to the next stage in the development of string theory, finding its "principle of inevitability"—that underlying idea from which the whole theory necessarily springs forth—is of the highest priority.[2]

What Are Space and Time, Really, and Can We Do without Them?

In many of the preceding chapters, we have freely made use of the concepts of space and of spacetime. In Chapter 2 we described Einstein's realization that space and time are inextricably interwoven by the unexpected fact that an object's motion through space has an influence on its passage through time. In Chapter 3, we deepened our understanding of spacetime's role in the unfolding of the cosmos through general relativity, which shows that the detailed shape of the spacetime fabric communicates the force of gravity from one place to another. The violent quantum undulations in the microscopic structure of the fabric, as discussed in Chapters 4 and 5, established the need for a new theory, leading us to string theory. And finally, in a number of the chapters that followed, we have seen that string theory proclaims that the universe has many more dimensions than we are aware of, some of which are curled up into tiny but complicated shapes that can undergo wondrous transformations in which their fabric punctures, tears, and then repairs itself.

Through graphic representations such as Figures 3.4, 3.6, and 8.10, we have tried to illustrate these ideas by envisioning the fabric of space and spacetime as if it were somewhat like a piece of material out of which the universe is tailored. These images have considerable explanatory power; they are used regularly by physicists as a visual guide in their own technical work. Although staring at figures such as the ones just mentioned gives a gradual impression of meaning, one can still be left asking, What do we *really* mean by the fabric of the universe?

This is a profound question that has, in one form or another, been the subject of debate for hundreds of years. Newton declared space and time to be eternal and immutable ingredients in the makeup of the cosmos, pristine structures lying beyond the bounds of question and explanation. As he wrote in the *Principia,* "Absolute space, in its own nature, without relation to anything external, remains always similar and immovable. Absolute, true, and mathematical time, of itself, and from its own nature, flows equably without relation to anything external."[3] Gottfried Leibniz and others vociferously disagreed, claiming that space and time are merely bookkeeping devices for conveniently summarizing relationships between objects and events within the universe. The location of an object in space and in time has meaning only in comparison with another. Space and time are the vocabulary of these relations, but nothing more. Although Newton's view, supported by his experimentally successful three laws of motion, held sway for more than two hundred years, Leibniz's conception, further developed by the Austrian physicist Ernst Mach, is much closer to our current picture. As we have seen, Einstein's special and general theories of relativity firmly did away with the concept of an absolute and universal notion of space and time. But we can still ask whether the geometrical model of spacetime that plays such a pivotal role in general relativity and in string theory is solely a convenient shorthand for the spatial and temporal relations between various locations, or whether we should view ourselves as truly being embedded in *something* when we refer to our immersion within the spacetime fabric.

Although we are heading into speculative territory, string theory does suggest an answer to this question. The graviton, the smallest bundle of gravitational force, is one particular pattern of string vibration. And just as an electromagnetic field such as visible light is composed of an enormous number of photons, a gravitational field is composed of an enormous number of gravitons—that is, an enormous number of strings executing the graviton vibrational pattern. Gravitational fields, in turn, are encoded in the warping of the spacetime fabric, and hence we are led to identify the fabric of spacetime itself with a colossal number of strings all undergoing the same, orderly, graviton pattern of vibration. In the language of the field, such an enormous, organized array of similarly vibrating strings is known as a *coherent state* of strings. It's a rather poetic image—the strings of string theory as the threads of the spacetime fabric—but we should note that its rigorous meaning has yet to be worked out completely.

Nevertheless, describing the spacetime fabric in this string-stitched form does lead us to contemplate the following question. An ordinary piece of fabric is the end product of someone having carefully woven together individual threads, the raw material of common textiles. Similarly, we can ask ourselves whether there is a raw precursor to the fabric of spacetime—a configuration of the strings of the cosmic fabric in which they have not yet coalesced into the organized form that we recognize as spacetime. Notice that it is somewhat inaccurate to picture this state as a jumbled mass of individual vibrating strings that have yet to stitch themselves together into an ordered whole because, in our usual way of thinking, this presupposes a notion of both space and time—the space in which a string vibrates and the progression of time that allows us to follow its changes in shape from one moment to the next. But in the raw state, before the strings that make up the cosmic fabric engage in the orderly, coherent vibrational dance we are discussing, *there is no realization of space or time.* Even our language is too coarse to handle these ideas, for, in fact, there is even no notion of *before.* In a sense, it's as if individual strings are "shards" of space and time, and only when they appropriately undergo sympathetic vibrations do the conventional notions of space and time emerge.

Imagining such a structureless, primal state of existence, one in which there is no notion of space or time as we know it, pushes most people's powers of comprehension to their limit (it certainly pushes mine). Like the Stephen Wright one-liner about the photographer who is obsessed with getting a close-up shot of the horizon, we run up against a clash of paradigms when we try to envision a universe that *is,* but that somehow does not invoke the concepts of space or time. Nevertheless, it is likely that we will need to come to terms with such ideas and understand their implementation before we can fully assess string theory. The reason is that our present formulation of string theory presupposes the existence of space and time within which strings (and the other ingredients found in M-theory) move about and vibrate. This allows us to deduce the physical properties of string theory in a universe with one time dimension, a certain number of extended space dimensions (usually taken to be three), and additional dimensions that are curled up into one of the shapes allowed by the equations of the theory. But this is somewhat like assessing an artist's creative talent by requiring that she work from a paint-by-number kit. She will, undoubtedly, add a personal flair here or there, but by so tightly constraining the format of her work, we are blinding ourselves to all but a slender view of her abilities. Similarly, since the triumph of string theory is its natural incorporation of quantum mechanics and gravity, and since gravity is bound up with the form of space and time, we should not constrain the theory by forcing it to operate within an already existing spacetime framework. Rather, just as we should allow our artist to work from a blank canvas, we should allow string theory to *create* its own spacetime arena by starting in a spaceless and timeless configuration.

The hope is that from this blank slate starting point—possibly in an era that existed before the big bang or the pre–big bang (if we can use temporal terms, for lack of any other linguistic framework)—the theory will describe a universe that evolves to a form in which a background of coherent string vibrations emerges, yielding the conventional notions of space and time. Such a framework, if realized, would show that space, time, and, by association, dimension are not essential defining elements of the universe. Rather, they are convenient notions that emerge from a more basic, atavistic, and primary state.

Already, cutting-edge research on aspects of M-theory, spearheaded by Stephen Shenker, Edward Witten, Tom Banks, Willy Fischler, Leonard Susskind, and others too numerous to name, has shown that something known as a *zero-brane*—possibly the most fundamental ingredient in M-theory, an object that behaves somewhat like a point particle at large distances but has drastically different properties at short ones—may give us a glimpse of the spaceless and timeless realm. Their work has revealed that whereas strings show us that conventional notions of space cease to have relevance below the Planck scale, the zero-branes give essentially the same conclusion but also provide a tiny window on the new unconventional framework that takes over. Studies with these zero-branes indicate that ordinary geometry is replaced by something known as *noncommutative* geometry, an area of mathematics developed in large part by the French mathematician Alain Connes.[4] In this geometrical framework, the conventional notions of space and of distance between points melt away, leaving us in a vastly different conceptual landscape. Nevertheless, as we focus our attention on scales larger than the Planck length, physicists have shown that our conventional notion of space does re-emerge. It is likely that the framework of noncommutative geometry is still some significant steps away from the blank-slate state anticipated above, but it does give us a hint of what the more complete framework for incorporating space and time may involve.

Finding the correct mathematical apparatus for formulating string theory without recourse to a pre-existing notion of space and time is one of the most important issues facing string theorists. An understanding of how space and time emerge would take us a huge step closer to answering the crucial question of which geometrical form actually *does* emerge.

Will String Theory Lead to a Reformulation of Quantum Mechanics?

The universe is governed by the principles of quantum mechanics to fantastic accuracy. Even so, in formulating theories over the past half century, physicists have followed a strategy that, structurally speaking, places quantum mechanics in a somewhat secondary position. In devising theories, physicists often start by working in a purely classical language that ignores quantum probabilities, wave functions, and so forth—a language that would be perfectly intelligible to physicists in the age of Maxwell and even in the age of Newton—and then, subsequently, overlaying quantum concepts upon the classical framework. This approach is not particularly surprising, since it directly mirrors our experiences. At first blush, the universe appears to be governed by laws rooted in classical concepts such as a particle having a definite position and a definite velocity at any given moment in time. It is only after detailed microscopic scrutiny that we realize that we must modify such familiar classical ideas. Our process of discovery has gone from a classical framework to one that is modified by quantum revelations, and this progression is echoed in the way that physicists, to this day, go about constructing their theories.

This is certainly the case with string theory. The mathematical formalism describing string theory begins with equations that describe the motion of a tiny, infinitely thin piece of *classical* thread—equations that, to a large extent, Newton could have written down some three hundred years ago. These equations are then *quantized*. That is, in a systematic manner developed by physicists over the course of more than 50 years, the classical equations are converted into a quantum-mechanical framework in which probabilities, uncertainty, quantum jitters, and so on are directly incorporated. In fact, in Chapter 12 we have seen this procedure in action: The loop processes (see Figure 12.6) incorporate quantum concepts—in this case, the momentary quantum-mechanical creation of virtual string pairs—with the number of loops determining the precision with which quantum-mechanical effects are accounted for.

The strategy of beginning with a theoretical description that is classical and then subsequently including the features of quantum mechanics has been extremely fruitful for many years. It underlies, for example, the standard model of particle physics. But it is possible, and there is growing evidence that it is likely, that this method is too conservative for dealing with theories that are as far-reaching as string theory and M-theory. The reason is that once we realize that the universe is governed by quantum-mechanical principles, our theories really should be quantum mechanical from the start. We have successfully gotten away with starting from a classical perspective until now because we have not been probing the universe at a deep enough level for this coarse approach to mislead us. But with the depth of string/M-theory, we may well have come to the end of the line for this battle-tested strategy.

We can find specific evidence for this by reconsidering some of the insights emerging from the second superstring revolution (as summarized, for example, by Figure 12.11). As we discussed in Chapter 12, the dualities underlying the unity of the five string theories show us that physical processes that occur in any one string formulation can be reinterpreted in the dual language of any of the others. This rephrasing will at first appear to have little to do with the original description, but, in fact, this is simply the power of duality at work: Through duality, one physical process can be described in a number of vastly different ways. These results are both subtle and remarkable, but we have not yet mentioned what may well be their most important feature.

The duality translations often take a process, described in one of the five string theories, that is *strongly* dependent on quantum mechanics (for example, a process involving string interactions that would not happen if the world were governed by classical, as opposed to quantum, physics) and reformulate it as a process that is *weakly* dependent on quantum mechanics from the perspective of one of the other string theories (for example, a process whose detailed numerical properties are influenced by quantum considerations but whose qualitative form is similar to what it would be in a purely classical world). This means that quantum mechanics is thoroughly intertwined within the duality symmetries underlying string/M-theory: They are *inherently quantum-mechanical symmetries*, since one of the dual descriptions is strongly influenced by quantum considerations. This indicates forcefully that the complete formulation of string/M-theory—a formulation that fundamentally incorporates the new-found duality symmetries—cannot begin classically and then undergo quantization, in the traditional mold. A classical starting point will necessarily omit the duality symmetries, since they hold true only when quantum mechanics is taken into account. Rather, it appears that the complete formulation of string/M-theory must break the traditional mold and spring into existence as a full-fledged quantum-mechanical theory.

Currently, no one knows how to do this. But many string theorists foresee a reformulation of how quantum principles are incorporated into our theoretical description of the universe as the next major upheaval in our understanding. For example, as Cumrun Vafa has said, "I think that a reformulation of quantum mechanics which will resolve many of its puzzles is just around the corner. I think many share the view that the recently uncovered dualities point toward a new, more geometrical framework for quantum mechanics, in which space, time, and quantum properties will be inseparably joined together."[5] And according to Edward Witten, "I believe the logical status of quantum mechanics is going to change in a manner that is similar to the way that the logical status of gravity changed when Einstein discovered the equivalence principle. This process is far from complete with quantum mechanics, but I think that people will one day look back on our epoch as the period when it began."[6]

With guarded optimism, we can envision that a reframing of the principles of quantum mechanics within string theory may yield a more powerful formalism that is capable of giving us an answer to the question of how the universe began and why there are such things as space and time—a formalism that will take us one step closer to answering Leibniz's question of why there is something rather than nothing.

Can String Theory Be Experimentally Tested?

Among the many features of string theory that we have discussed in the preceding chapters, the following three are perhaps the most important ones to keep firmly in mind. First, gravity and quantum mechanics are part and parcel of how the universe works and therefore any purported unified theory must incorporate both. String theory accomplishes this. Second, studies by physicists over the past century have revealed that there are other key ideas—many of which have been experimentally confirmed—that appear central to our understanding of the universe. These include the concepts of spin, the family structure of matter particles, messenger particles, gauge symmetry, the equivalence principle, symmetry breaking, and supersymmetry, to name a few. All of these concepts emerge naturally from string theory. Third, unlike more conventional theories such as the standard model, which has 19 free parameters that can be adjusted to ensure agreement with experimental measurements, string theory has no adjustable parameters. In principle, its implications should be thoroughly definitive—they should provide an unambiguous test of whether the theory is right or wrong.

The road from this "in principle" ratiocination to an "in practice" fact is encumbered by many hurdles. In Chapter 9 we described some of the technical obstacles, such as determining the form of the extra dimensions, that currently stand in our way. In Chapters 12 and 13 we placed these and other obstacles in the broader context of our need for an exact understanding of string theory, which, as we have seen, naturally leads us to the consideration of M-theory. No doubt, achieving a full understanding of string/M-theory will require a great deal of hard work and an equal amount of ingenuity.

At every step of the way, string theorists have sought and will continue to seek experimentally observable consequences of the theory. We must not lose sight of the long-shot possibilities for finding evidence of string theory discussed in Chapter 9. Furthermore, as our understanding deepens there will, no doubt, be other rare processes or features of string theory that will suggest yet other indirect experimental signatures.

But most notably, the confirmation of supersymmetry, through the discovery of superpartner particles as discussed in Chapter 9, would be a major milestone for string theory. We recall that supersymmetry was discovered in the course of theoretical investigations of string theory, and that it is a central part of the theory. Its experimental confirmation would be a compelling, albeit circumstantial, piece of evidence for strings. Moreover, finding the superpartner particles would provide a welcome challenge, since the discovery of supersymmetry would do far more than merely answer the yes-no question of its relevance to our world. The masses and charges of the superpartner particles would reveal the detailed way in which supersymmetry is incorporated into the laws of nature. String theorists would then face the challenge of seeing whether this implementation can be fully realized or explained by string theory. Of course, we can be even more optimistic and hope that within the next decade—before the Large Hadron Collider in Geneva comes on-line—the understanding of string theory will have progressed sufficiently for detailed predictions about the superpartners to be made prior to their hoped-for discovery. Confirmation of such predictions would be a monumental moment in the history of science.

Are There Limits to Explanation?

Explaining everything, even in the circumscribed sense of understanding all aspects of the forces and the elementary constituents of the universe, is one of the greatest challenges science has ever faced. And for the first time, superstring theory gives us a framework that appears to have sufficient depth to meet the challenge. But will we ever realize the promise of the theory fully and, for example, calculate the masses of the quarks or the strength of the electromagnetic force, numbers whose precise values dictate so much about the universe? As in the previous sections, we will have to surmount numerous theoretical hurdles on the way to these goals—currently, the most prominent is achieving a full nonperturbative formulation of string/M-theory.

But is it possible that even if we had an exact understanding of string/-M-theory, framed within a new and far more transparent formulation of quantum mechanics, we could still fail in our quest to calculate particle masses and force strength? Is it possible that we would still have to resort to experimental measurements, rather than theoretical calculations, for their values? And, moreover, might it be that this failing does not mean that we need to look for an even deeper theory, but simply reflects that there *is no* explanation for these observed properties of reality?

One immediate answer to all these questions is yes. As Einstein said some time ago, "The most incomprehensible thing about the universe is that it is comprehensible."[7] The astonishment at our ability to understand the universe at all is easily lost sight of in an age of rapid and impressive progress. However, maybe there is a limit to comprehensibility. Maybe we have to accept that after reaching the deepest possible level of understanding science can offer, there will nevertheless be aspects of the universe that remain unexplained. Maybe we will have to accept that certain features of the universe are the way they are because of happenstance, accident, or divine choice. The success of the scientific method in the past has encouraged us to think that with enough time and effort we *can* unravel nature's mysteries. But hitting the absolute limit of scientific explanation—not a technological obstacle or the current but progressing edge of human understanding—would be a singular event, one for which past experience could not prepare us.

Although of great relevance to our quest for the ultimate theory, this is an issue we cannot yet resolve; indeed, the possibility that there are limits to scientific explanation, in the broad way we have stated it, is an issue that may never be resolved. We have seen, for instance, that even the speculative notion of the multiverse, which at first sight appears to present a definite limit to scientific explanation, can be dealt with by dreaming up equally speculative theories that, at least in principle, can restore predictive power.

One highlight emerging from these considerations is the role of cosmology in determining the implications of an ultimate theory. As we have discussed, superstring cosmology is a young field, even by the youthful standards set by string theory itself. It will, undoubtedly, be an area of primary research focus for years to come, and it is likely to be one of the major growth areas of the field. As we continue to gain new insight into the properties of string/M-theory, our ability to assess the cosmological implications of this rich attempt at a unified theory will become ever sharper. It is possible, of course, that such studies may one day convince us that, indeed, there is a limit to scientific explanation. But it is also possible, to the contrary, that they will usher in a new era—an era in which we can declare that a fundamental explanation of the universe has finally been found.

Reaching for the Stars

Although we are technologically bound to the earth and its immediate neighbors in the solar system, through the power of thought and experiment we have probed the far reaches of both inner and outer space. During the last hundred years in particular, the collective effort of numerous physicists has revealed some of nature's best-kept secrets. And once revealed, these explanatory gems have opened vistas on a world we thought we knew, but whose splendor we had not even come close to imagining. One measure of the depth of a physical theory is the extent to which it poses serious challenges to aspects of our worldview that had previously seemed immutable. By this measure, quantum mechanics and the theories of relativity are deep beyond anyone's wildest expectations: Wave functions, probabilities, quantum tunneling, the ceaseless roiling energy fluctuations of the vacuum, the smearing together of space and time, the relative nature of simultaneity, the warping of the spacetime fabric, black holes, the big bang. Who could have guessed that the intuitive, mechanical, clockwork Newtonian perspective would turn out to be so thoroughly parochial—that there was a whole new mind-boggling world lying just beneath the surface of things as they are ordinarily experienced?

But even these paradigm-shaking discoveries are only part of a larger, all-encompassing story. With solid faith that laws of the large and the small should fit together into a coherent whole, physicists are relentlessly hunting down the elusive unified theory. The search is not over, but through superstring theory and its evolution into M-theory, a cogent framework for merging quantum mechanics, general relativity, and the strong, weak, and electromagnetic forces has finally emerged. And the challenges these developments pose to our previous way of seeing the world are monumental: loops of strings and oscillating globules, uniting all of creation into vibrational patterns that are meticulously executed in a universe with numerous hidden dimensions capable of undergoing extreme contortions in which their spatial fabric tears apart and then repairs itself. Who could have guessed that the merging of gravity and quantum mechanics into a unified theory of all matter and all forces would yield such a revolution in our understanding of how the universe works?

No doubt, there are even grander surprises in store for us as we continue to seek a full and calculationally tractable understanding of superstring theory. Already, through studies in M-theory, we have seen glimpses of a strange new domain of the universe lurking beneath the Planck length, possibly one in which there is no notion of time or space. At the opposite extreme, we have also seen that our universe may merely be one of the innumerable frothing bubbles on the surface of a vast and turbulent cosmic ocean called the multiverse. These ideas are at the current edge of speculation, but they may presage the next leap in our understanding of the universe.

As we fix our sight on the future and anticipate all the wonders yet in store for us, we should also reflect back and marvel at the journey we have taken so far. The search for the fundamental laws of the universe is a distinctly human drama, one that has stretched the mind and enriched the spirit. Einstein's vivid description of his own quest to understand gravity—"the years of anxious searching in the dark, with their intense longing, their alternations of confidence and exhaustion, and final emergence into the light"[8]—encompasses, surely, the whole human struggle. We are all, each in our own way, seekers of the truth and we each long for an answer to why we are here. As we collectively scale the mountain of explanation, each generation stands firmly on the shoulders of the previous, bravely reaching for the peak. Whether any of our descendants will ever take in the view from the summit and gaze out on the vast and elegant universe with a perspective of infinite clarity, we cannot predict. But as each generation climbs a little higher, we realize Jacob Bronowski's pronouncement that "in every age there is a turning point, a new way of seeing and asserting the coherence of the world."[9] And as our generation marvels at our new view of the universe—our new way of asserting the world's coherence—we are fulfilling our part, contributing our rung to the human ladder reaching for the stars.

Notes

Chapter 1

1. The table below is an elaboration of Table 1.1. It records the masses and force charges of the particles of all three families. Each type of quark can carry three possible strong-force charges that are, somewhat fancifully, labeled as colors—they stand for numerical strong-force charges values. The weak charges recorded are, more precisely, the "third-component" of weak isospin. (We have not listed the "right-handed" components of the particles—they differ by having no weak charge.)

Particle	Mass	Family 1 Electric charge	Weak charge	Strong charge
Electron	.00054	−1	−1/2	0
Electron-Neutrino	< 10^{-8}	0	1/2	0
Up Quark	.0047	2/3	1/2	red, green, blue
Down Quark	.0074	−1/3	−1/2	red, green, blue

Particle	Mass	Family 2 Electric charge	Weak charge	Strong charge
Muon	.11	−1	−1/2	0
Muon-Neutrino	< .0003	0	1/2	0
Charm Quark	1.6	2/3	1/2	red, green, blue
Strange Quark	.16	−1/3	−1/2	red, green, blue

Particle	Mass	Family 3 Electric charge	Weak charge	Strong charge
Tau	1.9	−1	−1/2	0
Tau-Neutrino	< .033	0	1/2	0
Top Quark	189	2/3	1/2	red, green, blue
Bottom Quark	5.2	−1/3	−1/2	red, green, blue

2. Strings can also have two freely moving ends (so-called *open strings*) in addition to the loops (*closed strings*) illustrated in Figure 1.1. To ease our presentation, for the most part we will focus on closed strings, although essentially all of what we say applies to both.

3. Albert Einstein, in a 1942 letter to a friend, as quoted in Tony Hey and Patrick Walters, *Einstein's Mirror* (Cambridge, Eng.: Cambridge University Press, 1997).

4. Steven Weinberg, *Dreams of a Final Theory* (New York: Pantheon, 1992), p.52.

5. Interview with Edward Witten, May 11, 1998.

Chapter 2

1. The presence of massive bodies like the earth does complicate matters by introducing gravitational forces. Since we are now focusing on motion in the horizontal direction—not the vertical direction—we can and will ignore the earth's presence. In the next chapter we will undertake a thorough discussion of gravity.

2. More precisely, the speed of light through the *vacuum of empty space* is 670 million miles per hour. When light travels through a substance such as air or glass its speed is decreased in roughly the same way that a rock dropped from a cliff is dragged

to a slower speed when it enters a body of water. This slowing of light relative to its speed through a vacuum is of no consequence for our discussion of relativity and is justifiably ignored throughout the text.

3. For the mathematically inclined reader, we note that these observations can be turned into quantitative statements. For instance, if the moving light clock has speed v and it takes t seconds for its photon to complete one round-trip journey (as measured by our stationary light clock), then the light clock will have traveled a distance vt when its photon has returned to the lower mirror. We can now use the Pythagorean theorem to calculate that the length of each of the diagonal paths in Figure 2.3 is $\sqrt{(vt/2)^2 + h^2}$, where h is the distance between the two mirrors of a light clock (taken to be six inches in the text). The two diagonal paths, taken together, therefore have length $2\sqrt{(vt/2)^2 + h^2}$. Since the speed of light has a constant value, conventionally called c, it takes light $2\sqrt{(vt/2)^2 + h^2}/c$ seconds to complete the double diagonal journey. And so, we have the equality $t = 2\sqrt{(vt/2)^2 + h^2}/c$, which can be solved for t, yielding $t = 2h/\sqrt{c^2 - v^2}$. To avoid confusion, let's write this as $t_{moving} = 2h/\sqrt{c^2 - v^2}$, where the subscript indicates that this is the time we measure for one tick to occur on the moving clock. On the other hand, the time for one tick on our stationary clock is $t_{stationary} = 2h/c$ and as a little algebra reveals, $t_{moving} = t_{stationary}/\sqrt{1 - v^2/c^2}$, directly showing that one tick on the moving clock takes longer than one tick on the stationary clock. This means that between chosen events, fewer total ticks will take place on the moving clock than on the stationary, ensuring that less time has elapsed for the observer in motion.

4. In case you would be more convinced by an experiment carried out in a less esoteric setting than a particle accelerator, consider the following. During October 1971, J. C. Hafele, then of Washington University in St. Louis, and Richard Keating of the United States Naval Observatory flew cesium-beam atomic clocks on commercial airliners for some 40 hours. After taking into account a number of subtle features having to do with gravitational effects (to be discussed in the next chapter), special relativity claims that the total elapsed time on the moving atomic clocks should be less than the elapsed time on stationary earthbound counterparts by a few hundred billionths of a second. This is just what Hafele and Keating found: Time *really does slow down* for a clock in motion.

5. Although Figure 2.4 correctly illustrates the shrinking of an object along its direction of motion, the image does not illustrate what we would actually see if an object were somehow to blaze by at nearly light speed (assuming our eyesight or photographic equipment were sharp enough to see anything at all!). To see something, our eyes—or our camera—must receive light that has reflected off the object's surface. But since the reflected light travels to us from various locations on the object, the light we see at any moment traveled to us along paths of different lengths. This results in

a kind of relativistic visual illusion in which the object will appear both foreshortened and rotated.

6. For the mathematically inclined reader, we note that from the spacetime position 4-vector $x = (ct, x_1, x_2, x_3) = (ct, \vec{x})$ we can produce the velocity 4-vector $u = dx/d\tau$, where τ is the proper time defined by $d\tau^2 = dt^2 - c^{-2}(dx_1^2 + dx_2^2 + dx_3^2)$. Then, the "speed through spacetime" is the magnitude of the 4-vector u, $\sqrt{((c^2 dt^2 - d\vec{x}^2)/(dt^2 - c^{-2}d\vec{x}^2))}$, which is identically the speed of light, c. Now, we can rearrange the equation $c^2(dt/d\tau)^2 - (d\vec{x}/d\tau)^2 = c^2$, to be $c^2(d\tau/dt)^2 + (d\vec{x}/dt)^2 = c^2$. This shows that an increase in an object's speed through space, $\sqrt{(d\vec{x}/dt)^2}$ must be accompanied by a decrease in $d\tau/dt$, the latter being the object's speed through time (the rate at which time elapses on its own clock, $d\tau$, as compared with that on our stationary clock, dt).

Chapter 3

1. Isaac Newton, *Sir Isaac Newton's Mathematical Principle of Natural Philosophy and His System of the World*, trans. A. Motte and Florian Cajori (Berkeley: University of California Press, 1962), Vol. I, p. 634.

2. A bit more precisely, Einstein realized that the equivalence principle holds so long as your observations are confined to a small enough region of space—that is, so long as your "compartment" is small enough. The reason is the following. Gravitational fields can vary in strength (and in direction) from place to place. But we are imagining that your whole compartment accelerates as a single unit and therefore your acceleration simulates a single, uniform gravitational force field. As your compartment gets ever smaller, though, there is ever less room over which a gravitational field can vary, and hence the equivalence principle becomes ever more applicable. Technically, the difference between the uniform gravitational field simulated by an accelerated vantage point and a possibly nonuniform "real" gravitational field created by some collection of massive bodies is known as the "tidal" gravitational field (since it accounts for the moon's gravitational effect on tides on earth). This endnote, therefore, can be summarized by saying that tidal gravitational fields become less noticeable as the size of your compartment gets smaller, making accelerated motion and a "real" gravitational field indistinguishable.

3. Albert Einstein, as quoted in Albrecht Fölsing, *Albert Einstein* (New York: Viking, 1997), p. 315.

4. John Stachel, "Einstein and the Rigidly Rotating Disk," in *General Relativity and Gravitation*, ed. A. Held (New York: Plenum, 1980), p. 1.

5. Analysis of the Tornado ride, or the "rigidly rotating disk," as it is called in more technical language, easily leads to confusion. In fact, to this day there is not universal agreement on a number of subtle aspects of this example. In the text we have followed the spirit of Einstein's own analysis, and in this endnote we continue to take

this viewpoint and seek to clarify a couple of features that you may have found confusing. First, you may be puzzled about why the circumference of the ride is not Lorentz contracted in exactly the same way as the ruler, and hence measured by Slim to have the same length as we originally found. Bear in mind, though, that throughout our discussion the ride was always spinning; we *never* analyzed the ride when it was at rest. Thus, from our perspective as stationary observers, the only difference between our and Slim's measurement of the ride's circumference is that Slim's ruler is Lorentz contracted; the spinning Tornado ride was spinning when we performed our measurement, and it is spinning as we watch Slim carry out his. Since we see that his ruler is contracted, we realize that he will have to lay it out more times to traverse the entire circumference, thereby measuring a longer length than we did. Lorentz contraction of the ride's circumference would have been relevant only if we compared the properties of the ride when spinning and when at rest, but this is a comparison we did not need.

Second, notwithstanding the fact that we did not need to analyze the ride when it was at rest, you may still be wondering about what *would* happen when it does slow down and stop. Now, it would seem, we must take account of the changing circumference with changing speed due to different degrees of Lorentz contraction. But how can this be squared with an unchanging radius? This is a subtle problem whose resolution hinges on the fact that there are no *fully rigid* objects in the real world. Objects can stretch and bend and thereby accommodate the stretching or contracting we have come upon; if not, as Einstein pointed out, a rotating disk that was initially formed by allowing a spinning cast of molten metal to cool while in motion would break apart if its rate of spinning were subsequently changed. For more details on the history of the rigidly rotating disk, see Stachel, "Einstein and the Rigidly Rotating Disk."

6. The expert reader will recognize that in the example of the Tornado ride, that is, in the case of a uniformly rotating frame of reference, the curved three-dimensional spatial sections on which we have focused fit together into a four-dimensional spacetime whose curvature still vanishes.

7. Hermann Minkowski, as quoted in Fölsing, *Albert Einstein,* p. 189.

8. Interview with John Wheeler, January 27, 1998.

9. Even so, existing atomic clocks are sufficiently accurate to detect such tiny—and even tinier—time warps. For instance, in 1976 Robert Vessot and Martin Levine of the Harvard-Smithsonian Astrophysical Observatory, together with collaboraters at the National Aeronautics and Space Administration (NASA), launched a Scout D rocket from Wallops Island, Virginia, that carried an atomic clock accurate to about a trillionth of a second per hour. They hoped to show that as the rocket gained altitude (thereby decreasing the effect of the earth's gravitational pull), an identical earth-

bound atomic clock (still subject to the full force of the earth's gravity) would tick more slowly. Through a two-way stream of microwave signals, the researchers were able to compare the rate of ticking of the two atomic clocks and, indeed, at the rocket's maximum altitude of 6,000 miles, its atomic clock ran fast by about 4 parts per billion relative to its counterpart on earth, agreeing with theoretical predictions to better than a hundredth of a percent.

10. In the mid-1800s, the French scientist Urbain Jean Joseph Le Verrier discovered that the planet Mercury deviates slightly from the orbit around the sun that is predicted by Newton's law of gravity. For more than half a century, explanations for this so-called excess orbital perihelion precession (in plain language, at the end of each orbit, Mercury does not quite wind up where Newton's theory says it should) ran the gamut—the gravitational influence of an undiscovered planet or planetary ring, an undiscovered moon, the effect of interplanetary dust, the oblateness of the sun—but none was sufficiently compelling to win general acceptance. In 1915, Einstein calculated the perihelion precession of Mercury using his newfound equations of general relativity and found an answer that, by his own admission, gave him heart palpitations: The result from general relativity precisely matched observations. This success, certainly, was one significant reason that Einstein had such faith in his theory, but most everyone else awaited confirmation of a *pre*diction, rather than an explanation of a previously known anomaly. For more details, see Abraham Pais, *Subtle Is the Lord* (New York: Oxford University Press, 1982), p. 253.

11. Robert P. Crease and Charles C. Mann, *The Second Creation* (New Brunswick, N.J.: Rutgers University Press, 1996), p. 39.

12. Surprisingly, recent research on the detailed rate of cosmic expansion suggests that the universe may in fact incorporate a very small but nonzero cosmological constant.

Chapter 4

1. Richard Feynman, *The Character of Physical Law* (Cambridge, Mass.: MIT Press, 1965), p. 129.

2. Although Planck's work did solve the infinite energy puzzle, apparently this goal was not what directly motivated his work. Rather, Planck was seeking to understand a closely related issue: the experimental results concerning how energy in a hot oven—a "black body" to be more precise—is distributed over various wavelength ranges. For more details on the history of these developments, the interested reader should consult Thomas S. Kuhn, *Black-Body Theory and the Quantum Discontinuity, 1894–1912* (Oxford, Eng.: Clarendon, 1978).

3. A little more precisely, Planck showed that waves whose minimum energy con-

tent exceeds their purported *average* energy contribution (according to nineteenth-century thermodynamics) are exponentially suppressed. This suppression is increasingly sharp as we examine waves of ever larger frequency.

4. Planck's constant is 1.05×10^{-27} grams-centimeters^2/second.

5. Timothy Ferris, *Coming of Age in the Milky Way* (New York: Anchor, 1989), p. 286.

6. Stephen Hawking, lecture at the Amsterdam Symposium on Gravity, Black Holes, and String Theory, June 21, 1997.

7. It is worthwhile to note that Feynman's approach to quantum mechanics can be used to derive the approach based on wave functions, and vice versa; the two approaches, therefore, are fully equivalent. Nevertheless, the concepts, the language, and the interpretation that each approach emphasizes are rather different, even though the answers each gives are absolutely identical.

8. Richard Feynman, *QED: The Strange Theory of Light and Matter* (Princeton: Princeton University Press, 1988).

Chapter 5

1. Stephen Hawking, *A Brief History of Time* (New York: Bantam Books, 1988), p. 175.

2. Richard Feynman, as quoted in Timothy Ferris, *The Whole Shebang* (New York: Simon & Schuster, 1997), p. 97.

3. In case you are still perplexed about how anything at all can happen within a region of space that is empty, it is important to realize that the uncertainty principle places a limit on how "empty" a region of space can actually be; it modifies what we mean by empty space. For example, when applied to wave disturbances in a field (such as electromagnetic waves traveling in the electromagnetic field) the uncertainty principle shows that the amplitude of a wave and the speed with which its amplitude changes are subject to the same inverse relationship as are the position and speed of a particle: The more precisely the amplitude is specified the less we can possibly know about the speed with which its amplitude changes. Now, when we say that a region of space is empty, we typically mean that, among other things, there are no waves passing through it, and that all fields have value zero. In clumsy but ultimately useful language, we can rephrase this by saying that the amplitudes of all waves that pass through the region are zero, exactly. But if we know the amplitudes exactly, the uncertainty principle implies that the rate of change of the amplitudes is completely uncertain and can take on essentially any value. But if the amplitudes change, this means that in the next moment they will *no longer be zero*, even though the region of space is still "empty." Again, *on average* the field *will* be zero since at some places its

value will be positive while at others negative; on average the net energy in the region has not changed. But this is only on average. Quantum uncertainty implies that the energy in the field—even in an empty region of space—fluctuates up and down, with the size of the fluctuations getting larger as the distance and time scales on which the region is examined get smaller. The energy embodied in such momentary field fluctuations can then, through $E = mc^2$, be converted into the momentary creation of pairs of particles and their antiparticles, which annihilate each other in great haste, to keep the energy from changing, on average.

4. Even though the initial equation that Schrödinger wrote down—the one incorporating special relativity—did not accurately describe the quantum-mechanical properties of electrons in hydrogen atoms, it was soon realized to be a valuable equation when appropriately used in other contexts, and, in fact, is still in use today. However, by the time Schrödinger published his equation he had been scooped by Oskar Klein and Walter Gordon, and hence his relativistic equation is called the "Klein-Gordon equation."

5. For the mathematically inclined reader, we note that the symmetry principles used in elementary particle physics are generally based on groups, most notably, Lie groups. Elementary particles are arranged in representations of various groups and the equations governing their time evolution are required to respect the associated symmetry transformations. For the strong force, this symmetry is called SU(3) (the analog of ordinary three-dimensional rotations, but acting on a complex space), and the three colors of a given quark species transform in a three-dimensional representation. The shifting (from red, green, blue to yellow, indigo, violet) mentioned in the text is, more precisely, an SU(3) transformation acting on the "color coordinates" of a quark. A gauge symmetry is one in which the group transformations can have a spacetime dependence: in this case, "rotating" the quark colors differently at different locations in space and moments in time.

6. During the development of the quantum theories of the three nongravitational forces, physicists also came upon calculations that gave infinite results. In time, though, they gradually realized that these infinities could be done away with through a tool known as *renormalization*. The infinities arising in attempts to merge general relativity and quantum mechanics are far more severe and are not amenable to the renormalization cure. Even more recently, physicists have realized that infinite answers are a signal that a theory is being used to analyze a realm that is beyond the bounds of its applicability. Since the goal of current research is to find a theory whose range of applicability is, in principle, unbounded—the "ultimate" or "final" theory—physicists want to find a theory in which infinite answers do not crop up, regardless of how extreme the physical system being analyzed might be.

7. The size of the Planck length can be understood based upon simple reasoning

rooted in what physicists call *dimensional analysis*. The idea is this. When a theory is formulated as a collection of equations, the abstract symbols must be tied to physical features of the world if the theory is to make contact with reality. In particular, we must introduce a system of units so that if a symbol, say, is meant to refer to a length, we have a scale by which its value can be interpreted. After all, if equations show that the length in question is 5, we need to know if that means 5 centimeters, 5 kilometers, or 5 light years, etc. In a theory that involves general relativity and quantum mechanics, a choice of units emerges naturally, in the following way. There are two constants of nature upon which general relativity depends: the speed of light, c, and Newton's gravitation constant, G. Quantum mechanics depends on one constant of nature $\hbar$. By examining the units of these constants (e.g., c is a velocity, so is expressed as distance divided by time, etc.), one can see that the combination $\sqrt{\hbar G/c^3}$ has the units of a length; in fact, it is 1.616×10^{-33} centimeters. This is the Planck length. Since it involves gravitational and spacetime inputs (G and c) and has a quantum mechanical dependence ($\hbar$) as well, it sets the scale for measurements—the natural unit of length—in any theory that attempts to merge general relativity and quantum mechanics. When we use the term "Planck length" in the text, it is often meant in an approximate sense, indicating a length that is within a few orders of magnitude of 10^{-33} centimeters.

8. Currently, in addition to string theory, two other approaches for merging general relativity and quantum mechanics are being pursued vigorously. One approach is led by Roger Penrose of Oxford University and is known as *twistor theory*. The other approach—inspired in part by Penrose's work—is led by Abhay Ashtekar of Pennsylvania State University and is known as the *new variables* method. Although these other approaches will not be discussed further in this book, there is growing speculation that they may have a deep connection to string theory and that possibly, together with string theory, all three approaches are honing in on the same solution for merging general relativity and quantum mechanics.

Chapter 6

1. The expert reader will recognize that this chapter focuses solely on *perturbative* string theory; nonperturbative aspects are discussed in Chapters 12 and 13.

2. Interview with John Schwarz, December 23, 1997.

3. Similar suggestions were made independently by Tamiaki Yoneya and by Korkut Bardakci and Martin Halpern. The Swedish physicist Lars Brink also contributed significantly to the early development of string theory.

4. Interview with John Schwarz, December 23, 1997.

5. Interview with Michael Green, December 20, 1997.

6. The standard model does suggest a mechanism by which particles acquire

mass—the *Higgs* mechanism, named after the Scottish physicist Peter Higgs. But from the point of view of explaining the particle masses, this merely shifts the burden to explaining properties of a hypothetical "mass-giving particle"—the so-called *Higgs boson.* Experimental searches for this particle are underway, but once again, if it is found and its properties measured, these will be *input* data for the standard model, for which the theory offers no explanation.

7. For the mathematically inclined reader, we note that the association between string vibrational patterns and force charges can be described more precisely as follows. When the motion of a string is quantized, its possible vibrational states are represented by vectors in a Hilbert space, much as for any quantum-mechanical system. These vectors can be labeled by their eigenvalues under a set of commuting hermitian operators. Among these operators are the Hamiltonian, whose eigenvalues give the energy and hence the mass of the vibrational state, as well as operators generating various gauge symmetries that the theory respects. The eigenvalues of these latter operators give the force charges carried by the associated vibrational string state.

8. Based upon insights gleaned from the second superstring revolution (discussed in Chapter 12), Witten and, most notably, Joe Lykken of the Fermi National Accelerator Laboratory have identified a subtle, yet possible, loophole in this conclusion. Lykken, exploiting this realization, has suggested that it might be possible for strings to be under far less tension, and therefore be substantially larger in size, than originally thought. So large, in fact, that they might be observable by the next generation of particle accelerators. If this long-shot possibility turns out to be the case, there is the exciting prospect that many of the remarkable implications of string theory discussed in this and the following chapters will be verifiable experimentally within the next decade. But even in the more "conventional" scenario espoused by string theorists, in which strings are typically on the order of 10^{-33} centimeters in length, there are indirect ways to search for them experimentally, as we will discuss in Chapter 9.

9. The expert reader will recognize that the photon produced in a collision between an electron and a positron is a virtual photon and therefore must shortly relinquish its energy by dissociating into a particle-antiparticle pair.

10. Of course, a camera works by collecting photons that bounce off the object of interest and recording them on a piece of photographic film. Our use of a camera in this example is symbolic, since we are not imagining bouncing photons off of the colliding strings. Rather, we simply want to record in Figure 6.7(c) the whole history of the interaction. Having said that, we should point out one further subtle point that the discussion in the text glosses over. We learned in Chapter 4 that we can formulate quantum mechanics using Feynman's sum-over-paths method, in which we analyze the motion of objects by combining contributions from *all* possible trajectories that lead from some chosen starting point to some chosen destination (with each trajec-

tory contributing with a statistical weight determined by Feynman). In Figures 6.6 and 6.7 we show *one* of the infinite number of possible trajectories followed by point particles (Figure 6.6) or by strings (Figure 6.7) taking them from their initial positions to their final destinations. The discussion in this section, however, applies equally well to any of the other possible trajectories and therefore applies to the whole quantum-mechanical process itself. (Feynman's formulation of point-particle quantum mechanics in the sum-over-paths framework was generalized to string theory through the work of Stanley Mandelstam of the University of California at Berkeley and by the Russian physicist Alexander Polyakov, who is now on the faculty of the physics department of Princeton University.)

Chapter 7

1. Albert Einstein, as quoted in R. Clark, *Einstein: The Life and Times* (New York: Avon Books, 1984), p. 287.

2. More precisely, spin-1/2 means that the *angular momentum* of the electron from its spin is $\hbar/2$.

3. The discovery and development of supersymmetry has a complicated history. In addition to those cited in the text, essential early contributions were made by R. Haag, M. Sohnius, J. T. Lopuszanski, Y. A. Gol'fand, E. P. Lichtman, J. L. Gervais, B. Sakita, V. P. Akulov, D. V. Volkov, and V. A. Soroka, among many others. Some of their work is documented in Rosanne Di Stefano, *Notes on the Conceptual Development of Supersymmetry,* Institute for Theoretical Physics, State University of New York at Stony Brook, preprint ITP-SB-8878.

4. For the mathematically inclined reader we note that this extension involves augmenting the familiar Cartesian coordinates of spacetime with new quantum coordinates, say u and v, that are *anticommuting*: $u \times v = -v \times u$. Supersymmetry can then be thought of as translations in this quantum-mechanically augmented form of spacetime.

5. For the reader interested in more details of this technical issue we note the following. In note 6 of Chapter 6 we mentioned that the standard model invokes a "mass-giving particle"—the Higgs boson—to endow the particles of Tables 1.1 and 1.2 with their observed masses. For this procedure to work, the Higgs particle itself cannot be too heavy; studies show that its mass should certainly be no greater than about 1,000 times the mass of a proton. But it turns out that quantum fluctuations tend to contribute substantially to the mass of the Higgs particle, potentially driving its mass all the way to the Planck scale. Theorists have found, however, that this outcome, which would uncover a major defect in the standard model, can be avoided if certain parameters in the standard model (most notably, the so-called bare mass of the Higgs particle) are finely tuned to better than 1 part in 10^{15} to cancel the effects of these

quantum fluctuations on the Higgs particle's mass.

6. One subtle point to note about Figure 7.1 is that the strength of the weak force is shown to be between that of the strong and electromagnetic forces, whereas we have previously said that it is weaker than both. The reason for this lies in Table 1.2, in which we see that the messenger particles of the weak force are quite massive, whereas those of the strong and electromagnetic forces are massless. Intrinsically, the strength of the weak force (as measured by its coupling constant—an idea we will come upon in Chapter 12) is as shown in Figure 7.1, but its massive messenger particles are sluggish conveyers of its influence and diminish its effects. In Chapter 14 we will see how the gravitational force fits into Figure 7.1.

7. Edward Witten, lecture at the Heinz Pagels Memorial Lecture Series, Aspen, Colorado, 1997.

8. For an in-depth discussion of these and related ideas, see Steven Weinberg, *Dreams of a Final Theory*.

Chapter 8

1. This is a simple idea, but since the imprecision of common language can sometimes lead to confusion, two clarifying remarks are in order. First, we are assuming that the ant is constrained to live on the *surface* of the garden hose. If, on the contrary, the ant could burrow into the *interior* of the hose—if it could penetrate into the rubber material of the hose—we would need three numbers to specify its position, since we would need to also specify how deeply it had burrowed. But if the ant lives only on the hose's surface, its location can be specified with just two numbers. This leads to our second point. Even with the ant living on the hose's surface, we could, if we so chose, specify its location with three numbers: the ordinary left-right, back-forth, and up-down positions in our familiar three-dimensional space. But once we know that the ant lives on the surface of the hose, the two numbers referred to in the text give the *minimal* data that uniquely specify the ant's position. This is what we mean by saying that the surface of the hose is two-dimensional.

2. Surprisingly, the physicists Savas Dimopoulos, Nima Arkani-Hamed, and Gia Dvali, building on earlier insights of Ignatios Antoniadis and Joseph Lykken, have pointed out that even if an extra curled-up dimension were as large as a millimeter in size, it is possible that it would not yet have been detected experimentally. The reason is that particle accelerators probe the microworld by utilizing the strong, weak, and electromagnetic forces. The gravitational force, being incredibly feeble at technologically accessible energies, is generally ignored. But Dimopoulos and his collaborators note that if the extra curled-up dimension has an impact predominantly on the gravitational force (something, it turns out, that is quite plausible in string theory), all extant experiments could well have overlooked it. New, highly sensitive gravitational

experiments will look for such "large" curled-up dimensions in the near future. A positive result would be one of the greatest discoveries of all time.

3. Edwin Abbott, *Flatland* (Princeton: Princeton University Press, 1991).

4. A. Einstein in letter to T. Kaluza as quoted in Abraham Pais, *"Subtle is the Lord": The Science and the Life of Albert Einstein* (Oxford: Oxford University Press, 1982), p. 330.

5. A. Einstein in letter to T. Kaluza as quoted in D. Freedman and P. van Nieuwenhuizen, "The Hidden Dimensions of Spacetime," *Scientific American* 252 (1985), 62.

6. Ibid.

7. Physicists found that the most difficult feature of the standard model to incorporate through a higher-dimensional formulation is something known as *chirality*. So as not to overburden the discussion we have not covered this concept in the main text, but for readers who are interested we do so briefly here. Imagine that someone shows you a film of some particular scientific experiment and confronts you with the unusual challenge of determining whether the film shot the experiment directly or whether it shot the experiment by looking at its reflection in a mirror. As the cinematographer was quite expert, there are no telltale signs of a mirror being involved. Is this a challenge you can meet? In the mid-1950s, the theoretical insights of T. D. Lee and C. N. Yang, and the experimental results of C. S. Wu and collaborators, showed that you *can* meet the challenge, so long as an appropriate experiment had been filmed. Namely, their work established that the laws of the universe are not perfectly mirror symmetric in the sense that the mirror-reflected version of certain processes—those directly dependent on the weak force—*cannot happen in our world,* even though the original process can. And so, as you watch the film if you see one of these forbidden processes occur, you will know that you are watching a mirror-reflected image of the experiment, as opposed to the experiment itself. Since mirrors interchange left and right, the work of Lee, Yang, and Wu established that the universe is not perfectly left-right symmetric—in the language of the field, the universe is *chiral.* It is this feature of the standard model (the weak force, in particular) that physicists found nearly impossible to incorporate into a higher-dimensional supergravity framework. To avoid confusion, we note that in Chapter 10 we will discuss a concept in string theory known as "mirror symmetry," but the use of the word "mirror" in that context is completely different from its use here.

8. For the mathematically inclined reader, we note that a Calabi-Yau manifold is a complex Kähler manifold with vanishing first Chern class. In 1957 Calabi conjectured that every such manifold admits a Ricci-flat metric, and in 1977 Yau proved this to be true.

9. This illustration is courtesy of Andrew Hanson of Indiana University, and was made using the *Mathematica* 3-D graphing package.

10. For the mathematically inclined reader we note that this particular Calabi-Yau space is a real three-dimensional slice through the quintic hypersurface in complex projective four-space.

Chapter 9

1. Edward Witten, "Reflections on the Fate of Spacetime" *Physics Today,* April 1996, p. 24.

2. Interview with Edward Witten, May 11, 1998.

3. Sheldon Glashow and Paul Ginsparg, "Desperately Seeking Superstrings?" *Physics Today,* May 1986, p. 7.

4. Sheldon Glashow, in *The Superworld I,* ed. A. Zichichi (New York: Plenum, 1990), p. 250.

5. Sheldon Glashow, *Interactions* (New York: Warner Books, 1988), p. 335.

6. Richard Feynman, in *Superstrings: A Theory of Everything?* ed. Paul Davies and Julian Brown (Cambridge, Eng: Cambridge University Press, 1988).

7. Howard Georgi, in *The New Physics,* ed. Paul Davies (Cambridge: Cambridge University Press 1989), p. 446.

8. Interview with Edward Witten, March 4, 1998.

9. Interview with Cumrun Vafa, January 12, 1998.

10. Murray Gell-Mann, as quoted in Robert P. Crease and Charles C. Mann, *The Second Creation* (New Brunswick, N.J.: Rutgers University Press), 1996, p. 414.

11. Interview with Sheldon Glashow, December 28, 1997.

12. Interview with Sheldon Glashow, December 28, 1997.

13. Interview with Howard Georgi, December 28, 1997. During the interview, Georgi also noted that the experimental refutation of the prediction of proton decay that emerged from his and Glashow's first proposed grand unified theory (see Chapter 7) played a significant part in his reluctance to embrace superstring theory. He noted poignantly that his grand unified theory invoked a vastly higher energy realm than any theory previously considered, and when its prediction was proved wrong— when it resulted in his "being slapped down by nature"—his attitude toward studying extremely high energy physics abruptly changed. When I asked him whether experimental confirmation of his grand unified theory might have inspired him to lead the charge to the Planck scale, he responded, "Yes, it likely would have."

14. David Gross, "Superstrings and Unification," in *Proceedings of the XXIV International Conference on High Energy Physics,* ed. R. Kotthaus and J. Kühn (Berlin: Springer-Verlag, 1988), p. 329.

15. Having said this, it's worth bearing in mind the long-shot possibility, pointed out in endnote 8 of Chapter 6, that strings just *might* be significantly longer than originally thought and therefore might be subject to direct experimental observation by accelerators within a few decades.

16. For the mathematically inclined reader we note that the more precise mathematical statement is that the number of families is half the absolute value of the Euler number of the Calabi-Yau space. The Euler number itself is the alternating sum of the dimensions of the manifold's homology groups—the latter being what we loosely refer to as multidimensional holes. So, three families emerge from Calabi-Yau spaces whose Euler number is ± 6.

17. Interview with John Schwarz, December 23, 1997.

18. For the mathematically inclined reader we note that we are referring to Calabi-Yau manifolds with a finite, nontrivial fundamental group, the order of which, in certain cases, determines the fractional charge denominators.

19. Interview with Edward Witten, March 4, 1998.

20. For the expert we note that some of these processes violate lepton number conservation as well as charge-parity-time (CPT) reversal symmetry.

Chapter 10

1. For completeness, we note that although much of what we have covered to this point in the book applies equally well to open strings (a string with loose ends) or closed-string loops (the strings on which we have focused), the topic discussed here is one in which the two kinds of strings would appear to have different properties. After all, an open string will not get entangled by looping around a circular dimension. Nevertheless, through work that ultimately has played a pivotal part in the second superstring revolution, in 1989 Joe Polchinski from the University of California at Santa Barbara and two of his students, Jian-Hui Dai and Robert Leigh, showed how open strings fit perfectly into the conclusions we find in this chapter.

2. In case you are wondering why the possible uniform vibrational energies are *whole number* multiples of $1/R$, you need only think back to the discussion of quantum mechanics—the warehouse in particular—from Chapter 4. There we learned that quantum mechanics implies that energy, like money, comes in discrete lumps: whole number multiples of various energy denominations. In the case of uniform vibrational string motion in the Garden-hose universe, this energy denomination is precisely $1/R$, as we demonstrated in the text using the uncertainty principle. Thus the uniform vibrational energies are whole number multiples of $1/R$.

3. Mathematically, the identity between the string energies in a universe with a circular dimension whose radius is either R or $1/R$ arises from the fact that the energies are of the form $v/R + wR$, where v is the vibration number and w is the winding number. This equation is invariant under the simultaneous interchange of v and w as well as R and $1/R$—i.e., under the interchange of vibration and winding numbers and inversion of the radius. In our discussion we are working in Planck units, but we can work in more conventional units by rewriting the energy formula in terms of $\sqrt{\alpha'}$—

the so-called string scale—whose value is about the Planck length, 10^{-33} centimeter. We can then express string energies as $v/R + wR/\alpha'$, which is invariant under interchange of v and w as well as R and α'/R, where the latter two are now expressed in terms of conventional units of distance.

4. You may be wondering how it's possible for a string that stretches all the way around a circular dimension of radius R to nevertheless measure the radius to be $1/R$. Although a thoroughly justifiable concern, its resolution actually lies in the imprecise phrasing of the question itself. You see, when we say that the string is wrapped around a circle of radius R, we are by necessity invoking a definition of distance (so that the phrase "radius R" has meaning). But *this* definition of distance is the one relevant for the unwound string modes—that is, the vibration modes. From the point of view of this definition of distance—and only this definition—the winding string configurations appear to stretch around the circular part of space. However, from the second definition of distance, the one that caters to the wound-string configurations, they are every bit as localized in space as are the vibration modes from the viewpoint of the first definition of distance, and the radius they "see" is $1/R$, as discussed in the text.

This description gives some sense of why wound and unwound strings measure distances that are inversely related. But as the point is quite subtle, it is perhaps worth noting the underlying technical analysis for the mathematically inclined reader. In ordinary point-particle quantum mechanics, distance and momentum (essentially energy) are related by Fourier transform. That is, a position eigenstate $|x>$ on a circle of radius R can be defined by $|x>=\Sigma_v e^{ixp}|p>$ where $p = v/R$ and $|p>$ is a momentum eigenstate (the direct analog of what we have called a uniform-vibration mode of a string—overall motion without change in shape). In string theory, though, there is a second notion of position eigenstate $|\tilde{x}>$ defined by making use of the winding string states: $|\tilde{x}>= \Sigma_w e^{i\tilde{x}\tilde{p}}|\tilde{p}>$ where $|\tilde{p}>$ is a winding eigenstate with $\tilde{p} = wR$. From these definitions we immediately see that x is periodic with period $2\pi R$ while $\tilde{x}$ is periodic with period $2\pi/R$, showing that x is a position coordinate on a circle of radius R while $\tilde{x}$ is the position coordinate on a circle of radius $1/R$. Even more explicitly, we can now imagine taking the two wavepackets $|x>$ and $|\tilde{x}>$, both starting say, at the origin, and allowing them to evolve in time to carry out our operational approach for defining distance. The radius of the circle, as measured by either probe, is then proportional to the required time lapse for the packet to return to its initial configuration. Since a state with energy E evolves with a phase factor involving Et, we see that the time lapse, and hence the radius, is $t \sim 1/E \sim R$ for the vibration modes and $t \sim 1/E \sim 1/R$ for the winding modes.

5. For the mathematically inclined reader, we note that, more precisely, the number of families of string vibrations is one-half the absolute value of the Euler characteristic of the Calabi-Yau space, as mentioned in note 16 of Chapter 9. This is given

by the absolute value of *difference* between $h^{2,1}$ and $h^{1,1}$, where $h^{p,q}$ denotes the (p,q) Hodge number. Up to a numerical shift, these count the number of nontrivial homology three-cycles ("three-dimensional holes") and the number of homology two-cycles ("two-dimensional holes"). And so, whereas we speak of the total number of holes in the main text, the more precise analysis shows that the number of families depends on the absolute value of difference between the odd- and even-dimensional holes. The conclusion, however, is the same. For instance, if two Calabi-Yau spaces differ by the interchange of their respective $h^{2,1}$ and $h^{1,1}$ Hodge numbers, the number of particle families—and the total number of "holes"—will not change.

6. The name comes from the fact that the "Hodge diamonds"—a mathematical summary of the holes of various dimensions in a Calabi-Yau space—for each Calabi-Yau space of a mirror pair are mirror reflections of one another.

7. The term *mirror symmetry* is also used in other, completely different contexts in physics, such as in the question of chirality—that is, whether the universe is left-right symmetric—as discussed in note 7 of Chapter 8.

Chapter 11

1. The mathematically inclined reader will recognize that we are asking whether the topology of space is dynamical—that is, whether it can change. We note that although we will often use the language of dynamical topology change, in practice we are usually considering a one-parameter family of *spacetimes* whose topology changes as a function of the parameter. Technically speaking, this parameter is not time, but in certain limits can essentially be identified with time.

2. For the mathematically inclined reader, the procedure involves blowing down rational curves on a Calabi-Yau manifold and then making use of the fact that, under certain circumstances, the resulting singularity can be repaired by distinct small resolutions.

3. K. C. Cole, *New York Times Magazine,* October 18, 1987, p. 20.

Chapter 12

1. Albert Einstein, as quoted in John D. Barrow, *Theories of Everything* (New York: Fawcett-Columbine, 1992), p. 13.

2. Let's briefly summarize the differences between the five string theories. To do so, we note that vibrational disturbances along a loop of string can travel clockwise or counterclockwise. The Type IIA and Type IIB strings differ in that in the latter theory, these clockwise/counterclockwise vibrations are identical, while in the former, they are exactly opposite in form. *Opposite* has a precise mathematical meaning in this context, but it's easiest to think about in terms of the spins of the resulting vibrational patterns in each theory. In the Type IIB theory, it turns out that all particles spin in the

same direction (they have the same chirality), whereas in the Type IIA theory, they spin in both directions (they have both chiralities). Nevertheless, each theory incorporates supersymmetry. The two heterotic theories differ in a similar but more dramatic way. Each of their clockwise string vibrations looks like those of the Type II string (when focusing on just the clockwise vibrations, the Type IIA and Type IIB theories are the same), but their counterclockwise vibrations are those of the original bosonic string theory. Although the bosonic string has insurmountable problems when chosen for both clockwise and counterclockwise string vibrations, in 1985 David Gross, Jeffrey Harvey, Emil Martinec, and Ryan Rhom (all then at Princeton University and dubbed the "Princeton String Quartet") showed that a perfectly sensible theory emerges if it is used in combination with the Type II string. The really odd feature of this union is that it has been known since the work of Claude Lovelace of Rutgers University in 1971 and the work of Richard Brower of Boston University, Peter Goddard of Cambridge University, and Charles Thorn of the University of Florida at Gainesville in 1972 that the bosonic string requires a 26-dimensional spacetime, whereas the superstring, as we have discussed, requires a 10-dimensional one. So the heterotic string constructions are a strange hybrid—a *heterosis*—in which counterclockwise vibrational patterns live in 26 dimensions and clockwise patterns live in 10 dimensions! Before you get caught up in trying to make sense of this perplexing union, Gross and his collaborators showed that the extra 16 dimensions on the bosonic side must be curled up into one of two very special higher-dimensional doughnutlike shapes, giving rise to the Heterotic-O and Heterotic-E theories. Since the extra 16 dimensions on the bosonic side are rigidly curled up, each of these theories behaves as though it really has 10 dimensions, just as in the Type II case. Again, both heterotic theories incorporate a version of supersymmetry. Finally, the Type I theory is a close cousin of the Type IIB string except that, in addition to the closed loops of string we have discussed in previous chapters, it also has strings with unconnected ends—so-called *open strings*.

3. When we speak of "exact" answers in this chapter, such as the "exact" motion of the earth, what we really mean is the exact prediction for some physical quantity *within some chosen theoretical framework*. Until we truly have the final theory—perhaps we now do, perhaps we never will—all of our theories will themselves be approximations to reality. But this notion of approximate has nothing to do with our discussion in this chapter. Here we are concerned with the fact that within a chosen theory, it is often difficult, if not impossible, to extract the exact predictions that the theory makes. Instead, we have to extract such predictions using approximation methods based on a perturbative approach.

4. These diagrams are string theory versions of the so-called Feynman diagrams, invented by Richard Feynman for performing perturbative calculations in point-particle quantum field theory.

5. More precisely, every virtual string pair, that is, every loop in a given diagram, contributes—among other more complicated terms—a multiplicative factor of the string coupling constant. More loops translate into more factors of the string coupling constant. If the string coupling constant is less than 1, repeated multiplications make the overall contribution ever smaller; if it is 1 or larger, repeated multiplications yield a contribution with the same or larger magnitude.

6. For the mathematically inclined reader, we note that the equation states that spacetime must admit a Ricci-flat metric. If we split spacetime into a Cartesian product of four-dimensional Minkowski spacetime and a six-dimensional compact Kähler space, Ricci-flatness is equivalent to the latter being a Calabi-Yau manifold. This is why Calabi-Yau spaces play such a prominent role in string theory.

7. Of course, nothing absolutely ensures that these indirect approaches are justified. For example, just as some faces are not left-right symmetric, it *might* be that the laws of physics are different in other far-flung regions of the universe, as we will discuss briefly in Chapter 14.

8. The expert reader will recognize that these statements require so-called N=2 supersymmetry.

9. To be a little more precise, if we call the Heterotic-O coupling constant g_{HO} and the Type I coupling constant g_I, then the relation between the two theories states that they are physically identical so long as $g_{HO} = 1/g_I$, which is equivalent to $g_I = 1/g_{HO}$. When one coupling constant is big the other is small.

10. This is a close analog of the R, $1/R$ duality discussed previously. If we call the Type IIB string coupling constant g_{IIB} then the statement that appears to be true is that the values g_{IIB} and $1/g_{IIB}$ describe the same physics. If g_{IIB} is big, $1/g_{IIB}$ is small, and vice versa.

11. If all but four dimensions are curled up, a theory with more than eleven total dimensions necessarily gives rise to massless particles with spin greater than 2, something that both theoretical and experimental considerations rule out.

12. A notable exception is the important 1987 work of Duff, Paul Howe, Takeo Inami, and Kelley Stelle in which they drew on earlier insights of Eric Bergshoeff, Ergin Sezgin, and Townsend to argue that ten-dimensional string theory should have a deep eleven-dimensional connection.

13. More precisely, this diagram should be interpreted as saying that we have a single theory that depends on a number of parameters. The parameters include coupling constants as well as geometrical size and shape parameters. In principle, we should be able to use the theory to calculate particular values for all of these parameters—a particular value for its coupling constant and a particular form for the spacetime geometry—but within our current theoretical understanding, we do not know how to accomplish this. And so, to understand the theory better string theo-

rists study its properties as the values of these parameters are varied over all possibilities. If the parameter values are chosen to lie in any of the six peninsular regions of Figure 12.11, the theory has the properties inherent to one of the five string theories, or to eleven-dimensional supergravity, as marked. If the parameter values are chosen to lie in the central region, the physics is governed by the still mysterious M-theory.

14. We should note, though, that even in the peninsular regions there are some exotic ways in which branes can have an effect on familiar physics. For example, it has been suggested that our three extended spatial dimensions might *themselves* be a three-brane that is large and unfurled. If so, as we go about our daily business we would be gliding through the interior of a three-dimensional membrane. Investigations of such possibilities are now being undertaken.

15. Interview with Edward Witten, May 11, 1998.

Chapter 13

1. The expert reader will recognize that under mirror symmetry, a collapsing three-dimensional sphere on one Calabi-Yau space gets mapped to a collapsing two-dimensional sphere on the mirror Calabi-Yau space—apparently putting us back in the situation of flops discussed in Chapter 11. The difference, however, is that a mirror rephrasing of this sort results in the antisymmetric tensor field $B_{\mu\nu}$—the real part of the complexified Kähler form on the mirror Calabi-Yau space—vanishing, and this is a far more drastic sort of singularity than that discussed in Chapter 11.

2. More precisely, these are examples of *extremal* black holes: black holes that have the minimum mass consistent with the force charges they carry, just like the BPS states in Chapter 12. Similar black holes will also play a pivotal role in the following discussion on black hole entropy.

3. The radiation emitted from a black hole should be just like that emitted from a hot oven—the very problem, discussed at the outset of Chapter 4, that played such a pivotal role in the development of quantum mechanics.

4. It turns out that because the black holes involved in space-tearing conifold transitions are extremal, they do not Hawking radiate, regardless of how light they become.

5. Stephen Hawking, lecture at Amsterdam Symposium on Gravity, Black Holes, and Strings, June 21, 1996.

6. In their initial calculation, Strominger and Vafa found that the mathematics was made easier by working with five—not four—extended spacetime dimensions. Surprisingly, after completing their calculation of the entropy of such a five-dimensional black hole they realized that no theoretician had as yet constructed such hypothetical extremal black holes in the setting of five-dimensional general relativity. Since only

by comparing their answer to the area of the event horizon of such a hypothetical black hole could they confirm their results, Strominger and Vafa then set out to mathematically construct such a five-dimensional black hole. They succeeded. It was then a simple matter to show that the microscopic string theory calculation of the entropy was in agreement with what Hawking would have predicted based on the area of the black hole's event horizon. But it is interesting to realize that because the black hole solution was found later, Strominger and Vafa did not know the answer they were shooting for while undertaking their entropy calculation. Since their work, numerous researchers, led most notably by Princeton physicist Curtis Callan, have succeeded in extending the entropy calculations to the more familiar setting of four extended spacetime dimensions, and all are in agreement with Hawking's predictions.

7. Interview with Sheldon Glashow, December 29, 1997.

8. Laplace, *Philosophical Essay on Probabilities,* trans. Andrew I. Dale (New York: Springer-Verlag, 1995).

9. Stephen Hawking, in Hawking and Roger Penrose, *The Nature of Space and Time* (Princeton: Princeton University Press, 1995), p. 41.

10. Stephen Hawking, lecture at the Amsterdam Symposium on Gravity, Black Holes, and Strings, June 21, 1997.

11. Interview with Andrew Strominger, December 29, 1997.

12. Interview with Cumrun Vafa, January 12, 1998.

13. Stephen Hawking, lecture at the Amsterdam Symposium on Gravity, Black Holes, and Strings, June 21, 1997.

14. This issue also has some bearing on the information-loss question, as some physicists have speculated over the years that there might be a central "nugget" embedded in the depths of a black hole that stores all of the information carried by matter that gets trapped within the hole's horizon.

15. In fact, the space-tearing conifold transitions discussed in this chapter involve black holes and hence might seem to be tied up with the question of their singularities. But recall that the conifold tear occurs just as the black hole has shed all its mass, and is therefore not directly related to questions concerning black hole singularities.

Chapter 14

1. More precisely, the universe should be filled with photons conforming to the radiation thermally emitted by a perfectly absorbent body—a "black-body" in the language of thermodynamics—with the stated temperature range. This is the same radiation spectrum emitted quantum mechanically by black holes, as explained by Hawking, and by a hot oven, as explained by Planck.

2. The discussion conveys the spirit of the issues involved although we are glossing over some subtle features having to do with the motion of light in an expanding

universe that affect the detailed numerics. In particular, although special relativity declares that nothing can travel faster than the speed of light, this does *not* preclude two photons carried along on the expanding spatial fabric from receeding from one another at a speed exceeding that of light. For example, at the time the universe first became transparent, about 300,000 years ATB, locations in the heavens that were about 900,000 light-years apart would have been able to have influenced each other, even though the distance between them exceeds 300,000 light-years. The extra factor of three comes from the expansion of the spatial fabric. This means that as we run the cosmic film backward in time, by the time we get to 300,000 years ATB, two points in the heavens need only be less than 900,000 light-years apart to have had a chance to influence each other's temperature. These detailed numerics do not change the qualitative features of the issues discussed.

3. For a detailed and lively discussion of the discovery of the inflationary cosmological model and the problems it resolves, see Alan Guth, *The Inflationary Universe* (Reading, Mass: Addison-Wesley, 1997).

4. For the mathematically inclined reader, we note that the idea underlying this conclusion is the following: If the sum of the spacetime dimensions of the paths swept out by each of two objects is greater than or equal to the spacetime dimension of the arena through which they are moving then they will generically intersect. For instance, point particles sweep out one-dimensional spacetime paths—the sum of the spacetime dimensions for two such particle paths is therefore two. The spacetime dimension of Lineland is also two, and hence their paths will generally intersect (assuming their velocities have not been finely tuned to be exactly equal). Similarly, strings sweep out two-dimensional spacetime paths (their world-sheets); for two strings the sum in question is therefore four. This means that strings moving in four spacetime dimensions (three space and one time) will generally intersect.

5. With the discovery of M-theory and the recognition of an eleventh dimension, string theorists have begun studying ways of curling up all *seven* extra dimensions in a manner that puts them all on more or less equal footing. The possible choices for such seven-dimensional manifolds are known as *Joyce* manifolds, after Domenic Joyce of Oxford University, who is credited with finding the first techniques for their mathematical construction.

6. Interview with Cumrun Vafa, January 12, 1998.

7. The expert reader will note that our description is taking place in the so-called string frame of reference, in which increasing curvature during the pre–big bang arises from (a dilaton-driven) increase in the strength of the gravitational force. In the so-called Einstein frame, the evolution would be described as an accelerating contraction phase.

8. Interview with Gabriele Veneziano, May 19, 1998.

9. Smolin's ideas are discussed in his book *The Life of the Cosmos* (New York: Oxford University Press, 1997).

10. Within string theory, for example, this evolution could be driven by small changes to the shape of the curled-up dimensions from one universe to its offspring. From our results on space-tearing conifold transitions, we know that a sufficiently long sequence of such small changes can take us from one Calabi-Yau to any other, allowing the multiverse to sample the reproductive efficiency of all universes based on strings. After the multiverse has passed through sufficiently many stages of reproduction, Smolin's hypothesis would lead us to expect that the typical universe will have a Calabi-Yau component that is optimized for fertility.

Chapter 15

1. Interview with Edward Witten, March 4, 1998.

2. Some theorists see a hint of this idea in the *holographic principle,* a concept originated by Susskind and the renowned Dutch physicist Gerard 't Hooft. Just as a hologram can reproduce a *three*-dimensional visual image from a specially designed *two*-dimensional film, Susskind and 't Hooft have suggested that all of the physical happenings we encounter may actually be encoded fully through equations defined in a *lower*-dimensional world. Although this may sound as strange as trying to draw someone's portrait by viewing only their shadow, we can get a sense of what it means, and understand part of Susskind's and 't Hooft's motivation, by thinking about black hole entropy as discussed in Chapter 13. Recall that the entropy of a black hole is determined by the *surface area* of its event horizon—and *not* by the total volume of space that the event horizon bounds. Therefore, the disorder of a black hole, and correspondingly the information it can embody, is encoded in the *two*-dimensional data of surface area. It is almost as if the event horizon of the black hole acts like a hologram by capturing all the information content of the black hole's three-dimensional interior. Susskind and 't Hooft have generalized this idea to the whole universe by suggesting that everything that occurs in the "interior" of the universe is merely a reflection of data and equations defined on a distant, bounding surface. Recently, work by the Harvard physicist Juan Maldacena, together with important subsequent work by Witten and of Princeton physicists Steven Gubser, Igor Klebanov, and Alexander Polyakov, has shown that, at least in certain cases, *string theory embodies the holographic principle.* In a manner that is currently being investigated vigorously, it appears that the physics of a universe governed by string theory has an equivalent description that involves only physics that takes place on such a bounding surface—a surface necessarily of lower dimensionality than the interior. Some string theorists have suggested that fully understanding the holographic principle and its role in string theory may well lead to the third superstring revolution.

3. *Sir Isaac Newton's Mathematical Principles of Natural Philosophy and His System of the World,* trans. Motte and Cajori (Berkeley: University of California Press, 1962), Vol. I, p. 6.

4. If you are familiar with linear algebra, one simple and relevant way of thinking about noncommutative geometry is to replace conventional Cartesian coordinates, which commute under multiplication, with matrices, which do not.

5. Interview with Cumrun Vafa, January 12, 1998.

6. Interview with Edward Witten, May 11, 1998.

7. Quoted in Banesh Hoffman with Helen Dukas, *Albert Einstein, Creator and Rebel* (New York: Viking, 1972), p. 18.

8. Martin J. Klein, "Einstein: The Life and Times, by R. W. Clark," (book review) *Science* 174, pp. 1315–16.

9. Jacob Bronkowski, *The Ascent of Man* (Boston: Little, Brown, 1973), p. 20.

Glossary of Scientific Terms

Absolute zero. The lowest possible temperature, about −273 degrees Celsius, or 0 on the Kelvin scale.

Acceleration. A change in an object's speed or direction. See also *velocity.*

Accelerator. See *particle accelerator.*

Amplitude. The maximum height of a wave peak or the maximum depth of a wave trough.

Anthropic principle. Doctrine that one explanation for why the universe has the properties we observe is that, were the properties different, it is likely that life would not form and therefore we would not be here to observe the changes.

Antimatter. Matter that has the same gravitational properties as ordinary matter, but that has an opposite electric charge as well as opposite nuclear *force charges.*

Antiparticle. A particle of *antimatter.*

ATB. Acronym for "after the bang"; usually used in reference to time elapsed since the *big bang.*

Atom. Fundamental building block of matter, consisting of a *nucleus* (comprising *protons* and *neutrons*) and an orbiting swarm of *electrons.*

Big bang. Currently accepted theory that the expanding universe began some 15 billion years ago from a state of enormous energy, density, and compression.

Big crunch. One hypothesized future for the universe in which the current expansion stops, reverses, and results in all space and all matter collapsing together; a reversal of the *big bang.*

Black hole. An object whose immense gravitational *field* entraps anything, even light, that gets too close (closer than the black hole's *event horizon*).

413

Black-hole entropy The *entropy* embodied within a *black hole*.

Boson. A particle, or pattern of *string* vibration, with a whole number amount of *spin;* typically a *messenger particle*.

Bosonic string theory. First known string theory; contains *vibrational patterns* that are all *bosons*.

BPS states. Configurations in a *supersymmetric* theory whose properties can be determined exactly by arguments rooted in *symmetry*.

Brane. Any of the extended objects that arise in *string theory*. A one-brane is a *string*, a two-brane is a membrane, a three-brane has three extended dimensions, etc. More generally, a *p*-brane has *p* spatial dimensions.

Calabi-Yau space, Calabi-Yau shape. A space (shape) into which the extra spatial dimensions required by *string theory* can be *curled up,* consistent with the equations of the theory.

Charge. See *force charge*.

Chiral, Chirality. Feature of fundamental particle physics that distinguishes left-from right-handed, showing that the universe is not fully left-right symmetric.

Closed string. A type of *string* that is in the shape of a loop.

Conifold transition. Evolution of the *Calabi-Yau* portion of space in which its fabric rips and repairs itself, yet with mild and acceptable physical consequences in the context of *string theory*. The tears involved are more severe than those in a *flop transition*.

Cosmic microwave background radiation. Microwave radiation suffusing the universe, produced during the *big bang* and subsequently thinned and cooled as the universe expanded.

Cosmological constant. A modification of *general relativity*'s original equations, allowing for a static universe; interpretable as a constant energy density of the vacuum.

Coupling constant. See *string coupling constant*.

Curled-up dimension. A spatial *dimension* that does not have an observably large spatial extent; a spatial dimension that is crumpled, wrapped, or curled up into a tiny size, thereby evading direct detection.

Curvature. The deviation of an object or of space or of *spacetime* from a *flat* form and therefore from the rules of geometry codified by Euclid.

Dimension. An independent axis or direction in space or *spacetime*. The familiar space around us has three dimensions (left-right, back-forth, up-down) and the familiar *spacetime* has four (the previous three axes plus the past-future axis). *Superstring theory* requires the universe to have additional spatial dimensions.

Dual, Duality, Duality symmetries. Situation in which two or more theories appear to be completely different, yet actually give rise to identical physical consequences.

Electromagnetic field. Force field of the *electromagnetic force,* consisting of electric and magnetic lines of force at each point in space.

Electromagnetic force. One of the four fundamental forces, a union of the electric and magnetic forces.

Electromagnetic gauge symmetry. *Gauge symmetry* underlying *quantum electrodynamics.*

Electromagnetic radiation. The energy carried by an *electromagnetic wave.*

Electromagnetic wave. A wavelike disturbance in an *electromagnetic field;* all such waves travel at the speed of light. Visible light, X rays, microwaves, and infrared radiation are examples.

Electron. Negatively charged particle, typically found orbiting the nucleus of an *atom.*

Electroweak theory. *Relativistic quantum field theory* describing the *weak force* and the *electromagnetic force* in one unified framework.

Eleven-dimensional supergravity. Promising higher-dimensional *supergravity* theory developed in the 1970s, subsequently ignored, and more recently shown to be an important part of *string theory.*

Entropy. A measure of the disorder of a physical system; the number of rearrangements of the ingredients of a system that leave its overall appearance intact.

Equivalence principle. See *principle of equivalence.*

Event horizon. The one-way surface of a *black hole;* once penetrated, the laws of gravity ensure that there is no turning back, no escaping the powerful gravitational grip of the black hole.

Extended dimension. A space (and *spacetime) dimension* that is large and directly apparent; a dimension with which we are ordinarily familiar, as opposed to a *curled-up dimension.*

Extremal black holes. *Black holes* endowed with the maximal amount of *force charge* possible for a given total mass.

Families. Organization of matter particles into three groups, with each group being known as a family. The particles in each successive family differ from those in the previous by being heavier, but carry the same electric and nuclear *force charges.*

Fermion. A particle, or pattern of *string* vibration, with half a whole odd number amount of *spin;* typically a matter particle.

Feynman sum-over-paths. See *sum-over-paths.*

Field, Force field. From a *macroscopic* perspective, the means by which a force communicates its influence; described by a collection of numbers at each point in space that reflect the strength and direction of the force at that point.

Flat. Subject to the rules of geometry codified by Euclid; a shape, like the surface of a perfectly smooth tabletop, and its higher-dimensional generalizations.

Flop transition. Evolution of the *Calabi-Yau* portion of space in which its fabric rips and repairs itself, yet with mild and acceptable physical consequences in the context of string theory.

Foam. See *spacetime foam*.

Force charge. A property of a particle that determines how it responds to a particular force. For instance, the electric charge of a particle determines how it responds to the *electromagnetic force*.

Frequency. The number of complete wave cycles a wave completes each second.

Gauge symmetry. *Symmetry* principle underlying the quantum-mechanical description of the three nongravitational forces; the symmetry involves the invariance of a physical system under various shifts in the values of *force charges*, shifts that can change from place to place and from moment to moment.

General relativity. Einstein's formulation of gravity, which shows that space and time communicate the gravitational force through their *curvature*.

Gluon. Smallest bundle of the *strong force field*; *messenger particle* of the strong force.

Grand unification. Class of theories that merge all three nongravitational forces into a single theoretical framework.

Gravitational force. The weakest of the four fundamental forces of nature. Described by Newton's universal theory of gravity, and subsequently by Einstein's *general relativity*.

Graviton. Smallest bundle of the *gravitational force field*; *messenger particle* for the gravitational force.

Heterotic-E string theory (Heterotic $E_8 \times E_8$ string theory). One of the five *superstring theories*; involves closed strings whose right-moving vibrations resemble those of the *Type II string* and whose left-moving vibrations involve those of the *bosonic string*. Differs in important but subtle ways from the *Heterotic-O string theory*.

Heterotic-O string theory (Heterotic O(32) string theory). One of the five *superstring theories*; involves closed strings whose right-moving vibrations resemble those of the *Type II string* and whose left-moving vibrations involve those of the *bosonic string*. Differs in important but subtle ways from the *Heterotic-E string theory*.

Higher-dimensional supergravity. Class of *supergravity* theories in more than four *spacetime* dimensions.

Horizon problem. Cosmological puzzle associated with the fact that regions of the universe that are separated by vast distances nevertheless have nearly identical properties such as temperature. *Inflationary cosmology* offers a solution.

Infinities. Typical nonsensical answer emerging from calculations that involve *general relativity* and *quantum mechanics* in a point-particle framework.

Inflation, Inflationary cosmology. Modification to the earliest moments of the standard *big bang* cosmology in which universe undergoes a brief burst of enormous expansion.

Initial conditions. Data describing the beginning state of a physical system.

Interference pattern. Wave pattern that emerges from the overlap and the intermingling of waves emitted from different locations.

Kaluza-Klein theory. Class of theories incorporating extra *curled-up dimensions,* together with *quantum mechanics.*

Kelvin. A temperature scale in which temperatures are quoted relative to *absolute zero.*

Klein-Gordon equation. A fundamental equation of *relativistic quantum field theory.*

Laplacian determinism. Clockwork conception of the universe in which complete knowledge of the state of the universe at one moment completely determines its state at all future and past moments.

Light clock. A hypothetical clock that measures elapsed time by counting the number of round-trip journeys completed by a single *photon* between two mirrors.

Lorentz contraction. Feature emerging from *special relativity,* in which a moving object appears shortened along its direction of motion.

Macroscopic. Refers to scales typically encountered in the everyday world and larger; roughly the opposite of microscopic.

Massless black hole. In string theory, a particular kind of *black hole* that may have large mass initially, but that becomes ever lighter as a piece of the *Calabi-Yau* portion of space shrinks. When the portion of space has shrunk down to a point, the initially massive black hole has no remaining mass—it is massless. In this state, it no longer manifests such usual black hole properties as an *event horizon.*

Maxwell's theory, Maxwell's electromagnetic theory. Theory uniting electricity and magnetism, based on the concept of the *electromagnetic field,* devised by Maxwell in the 1880s; shows that visible light is an example of an *electromagnetic wave.*

Messenger particle. Smallest bundle of a *force field;* microscopic conveyer of a force.

Mirror symmetry. In the context of *string theory,* a *symmetry* showing that two different *Calabi-Yau shapes,* known as a mirror pair, give rise to identical physics when chosen for the *curled-up dimensions* of *string theory.*

M-theory. Theory emerging from the *second superstring revolution* that unites the previous five *superstring theories* within a single overarching framework. M-theory appears to be a theory involving eleven *spacetime dimensions,* although many of its detailed properties have yet to be understood.

Multidimensional hole. A generalization of the hole found in a doughnut to higher-dimensional versions.

Multi-doughnut, Multi-handled doughnut. A generalization of a doughnut shape (a torus) that has more than one hole.

Multiverse. Hypothetical enlargement of the cosmos in which our universe is but one of an enormous number of separate and distinct universes.

Neutrino. Electrically neutral particle, subject only to the *weak force.*

Neutron. Electrically neutral particle, typically found in the nucleus of an *atom,* consisting of three *quarks* (two down-quarks, one up-quark).

Newton's laws of motion. Laws describing the motion of bodies based on the conception of an absolute and immutable space and time; these laws held sway until Einstein's discovery of *special relativity.*

Newton's universal theory of gravity. Theory of gravity declaring that the force of attraction between two bodies is proportional to the product of their masses and inversely proportional to the square of the distance between them. Subsequently supplanted by Einstein's *general relativity.*

Nonperturbative. Feature of a theory whose validity is not dependent on approximate, *perturbative* calculations; an exact feature of a theory.

Nucleus. The core of an *atom,* consisting of *protons* and *neutrons.*

Observer. Idealized person or piece of equipment, often hypothetical, that measures relevant properties of a physical system.

One-loop process. Contribution to a calculation in *perturbation theory* in which one virtual pair of *strings* (or particles in a point-particle theory) is involved.

Open string. A type of *string* with two free ends.

Oscillatory pattern. See *vibrational pattern.*

Particle accelerator. Machine for boosting particles to nearly light speed and slamming them together in order to probe the structure of matter.

Perturbation theory. Framework for simplifying a difficult problem by finding an approximate solution that is subsequently refined as more details, initially ignored, are systematically included.

Perturbative approach, Perturbative method. See *perturbation theory.*

Phase. When used in reference to matter, describes its possible states: solid phase, liquid phase, gas phase. More generally, refers to the possible descriptions of a physical system as features on which it depends (temperature, *string coupling constant* values, form of *spacetime,* etc.) are varied.

Phase transition. Evolution of a physical system from one *phase* to another.

Photoelectric effect. Phenomenon in which *electrons* are ejected from a metallic surface when light is shone upon it.

Photon. Smallest packet of the *electromagnetic force field; messenger particle* of the *electromagnetic force;* smallest bundle of light.

Planck energy. About 1,000 kilowatt hours. The energy necessary to probe to distances as small as the *Planck length.* The typical energy of a vibrating *string* in *string theory.*

Planck length. About 10^{-33} centimeters. The scale below which *quantum fluctuations* in the fabric of *spacetime* would become enormous. The size of a typical *string* in *string theory.*

Planck mass. About ten billion billion times the mass of a *proton;* about one-hundredth of a thousandth of a gram; about the mass of a small grain of dust. The typical mass equivalent of a vibrating *string* in *string theory.*

Planck's constant. Denoted by the symbol $\hbar$, Planck's constant is a fundamental parameter in *quantum mechanics.* It determines the size of the discrete units of energy, mass, *spin,* etc. into which the microscopic world is partitioned. Its value is 1.05×10^{-27} grams-cm/sec.

Planck tension. About 10^{39} tons. The tension on a typical *string* in *string theory.*

Planck time. About 10^{-43} seconds. Time at which the size of the universe was roughly the *Planck length;* more precisely, time it takes light to travel the *Planck length.*

Primordial nucleosynthesis. Production of atomic nuclei occurring during the first three minutes after the *big bang.*

Principle of equivalence. Core principle of *general relativity* declaring the indistinguishability of accelerated motion and immersion in a gravitational field (over small enough regions of observation). Generalizes the *principle of relativity* by showing that all observers, regardless of their state of motion, can claim to be at rest, so long as they acknowledge the presence of a suitable gravitational field.

Principle of relativity. Core principle of *special relativity* declaring that all constant-*velocity observers* are subject to an identical set of physical laws and that, therefore, every constant-velocity observer is justified in claiming that he or she is at rest. This principle is generalized by the *principle of equivalence.*

Product. The result of multiplying two numbers.

Proton. Positively charged particle, typically found in the nucleus of an *atom,* consisting of three *quarks* (two up-quarks and one down-quark).

Quanta. The smallest physical units into which something can be partitioned, according to the laws of quantum mechanics. For instance, *photons* are the quanta of the electromagnetic field.

Quantum chromodynamics (QCD). *Relativistic quantum field theory* of the *strong force* and *quarks,* incorporating *special relativity.*

Quantum claustrophobia. See *quantum fluctuations.*

Quantum determinism. Property of *quantum mechanics* that knowledge of the quantum state of a system at one moment completely determines its quantum state at future and past moments. Knowledge of the quantum state, however, determines only the probability that one or another future will actually ensue.

Quantum electrodynamics (QED). *Relativistic quantum field theory* of the *electromagnetic force* and *electrons,* incorporating *special relativity.*

Quantum electroweak theory. See *electroweak theory.*

Quantum field theory. See *relativistic quantum field theory.*

Quantum fluctuation. Turbulent behavior of a system on microscopic scales due to the *uncertainty principle.*

Quantum foam. See *spacetime foam.*

Quantum geometry. Modification of *Riemannian geometry* required to describe accurately the physics of space on *ultramicroscopic* scales, where quantum effects become important.

Quantum gravity. A theory that successfully mergers *quantum mechanics* and *general relativity,* possibly involving modifications of one or both. *String theory* is an example of a theory of quantum gravity.

Quantum mechanics. Framework of laws governing the universe whose unfamiliar features such as *uncertainty, quantum fluctuations,* and *wave-particle duality* become most apparent on the microscopic scales of *atoms* and subnuclear particles.

Quantum tunneling. Feature of *quantum mechanics* showing that objects can pass through barriers that should be impenetrable according to Newton's classical laws of physics.

Quark. A particle that is acted upon by the *strong force.* Quarks exist in six varieties (up, down, charm, strange, top, bottom) and three "colors" (red, green, blue).

Radiation. The energy carried by waves or particles.

Reciprocal. The inverse of a number; for example, the reciprocal of 3 is ⅓, the reciprocal of ½ is 2.

Relativistic quantum field theory. Quantum-mechanical theory of fields, such as the *electromagnetic field,* that incorporates *special relativity.*

Resonance. One of the natural states of oscillation of a physical system.

Riemannian geometry. Mathematical framework for describing curved shapes of any dimension. Plays a central role in Einstein's description of *spacetime* in *general relativity.*

Schrödinger equation. Equation governing the evolution of probability waves in *quantum mechanics.*

Schwarzschild solution. Solution to the equations of *general relativity* for a spherical distribution of matter; one implication of this solution is the possible existence of *black holes.*

Second law of thermodynamics. Law stating that total *entropy* always increases.

Second superstring revolution. Period in the development of *string theory* beginning around 1995 in which some *nonperturbative* aspects of the theory began to be understood.

Singularity. Location where the fabric of space or *spacetime* suffers a devastating rupture.

Smooth, Smooth space. A spatial region in which the fabric of space is flat or gently curved, with no pinches, ruptures, or creases of any kind.

Space-tearing flop transition. See *flop transition.*

Spacetime. A union of space and time originally emerging from *special relativity.* Can be viewed as the "fabric" out of which the universe is fashioned; it constitutes the dynamical arena within which the events of the universe take place.

Spacetime foam. Frothy, writhing, tumultuous character of the *spacetime* fabric on *ultramicroscopic* scales, according to a conventional point-particle perspective. An essential reason for the incompatibility of *quantum mechanics* and *general relativity* prior to *string theory.*

Special relativity. Einstein's laws of space and time in the absence of gravity (see also *general relativity*).

Sphere. The outer surface of a ball. The surface of a familiar three-dimensional ball has two dimensions (which can be labeled by two numbers such as "latitude" and "longitude," as on the surface of the earth). The concept of a sphere, though, applies more generally to balls and hence their surfaces, in any number of dimensions. A one-dimensional sphere is a fancy name for a circle; a zero-dimensional sphere is two points (as explained in the text). A three-dimensional sphere is harder to picture; it is the surface of a four-dimensional ball.

Spin. A quantum-mechanical version of the familiar notion of the same name; particles have an intrinsic amount of spin that is either a whole number or half a whole number (in multiples of *Planck's constant*), and which never changes.

Standard model of cosmology. *Big bang* theory together with an understanding of the three nongravitational forces as summarized by the *standard model of particle physics*.

Standard model of particle physics, Standard model, Standard theory. An enormously successful theory of the three nongravitational forces and their action on matter. Effectively the union of *quantum chromodynamics* and the *electroweak theory*.

String. Fundamental one-dimensional object that is the essential ingredient in *string theory*.

String coupling constant. A (positive) number that governs how likely it is for a given *string* to split apart into two strings or for two strings to join together into one—the basic processes in *string theory*. Each *string theory* has its own string coupling constant, the value of which should be determined by an equation; currently such equations are not understood well enough to yield any useful information. Coupling constants less than 1 imply that *perturbative methods* are valid.

String mode. A possible configuration (*vibrational pattern, winding configuration*) that a *string* can assume.

String theory. *Unified theory* of the universe postulating that fundamental ingredients of nature are not zero-dimensional point particles but tiny one-dimensional filaments called *strings*. String theory harmoniously unites *quantum mechanics* and *general relativity*, the previously known laws of the small and the large, that are otherwise incompatible. Often short for *superstring theory*.

Strong force, Strong nuclear force. Strongest of the four fundamental forces, responsible for keeping *quarks* locked inside *protons* and *neutrons* and for keeping protons and neutrons crammed inside of atomic nuclei.

Strong force symmetry. *Gauge symmetry* underlying the *strong force*, associated with invariance of a physical system under shifts in the color charges of *quarks*.

Strongly coupled. Theory whose *string coupling constant* is larger than 1.

Strong-weak duality. Situation in which a *strongly coupled* theory is *dual*—physically identical—to a different, *weakly coupled* theory.

Sum-over-paths. Formulation of *quantum mechanics* in which particles are envisioned to travel from one point to another along all possible paths between them.

Supergravity. Class of point-particle theories combining *general relativity* and *supersymmetry*.

Superpartners. Particles whose *spins* differ by 1/2 unit and that are paired by *supersymmetry*.

Superstring theory. *String theory* that incorporates *supersymmetry*.

Supersymmetric quantum field theory. *Quantum field theory* incorporating *supersymmetry*.

Supersymmetric standard model. Generalization of the *standard model of particle physics* to incorporate *supersymmetry*. Entails a doubling of the known elementary particle species.

Supersymmetry. A *symmetry* principle that relates the properties of particles with a whole number amount of *spin (bosons)* to those with half a whole (odd) number amount of *spin (fermions)*.

Symmetry. A property of a physical system that does not change when the system is transformed in some manner. For instance, a *sphere* is rotationally symmetrical since its appearance does not change if it is rotated.

Symmetry breaking. A reduction in the amount of *symmetry* a system appears to have, usually associated with a *phase transition*.

Tachyon. Particle whose mass (squared) is negative; its presence in a theory generally yields inconsistencies.

Thermodynamics. Laws developed in the nineteenth century to describe aspects of heat, work, energy, *entropy*, and their mutual evolution in a physical system.

Three-brane. See *brane*.

Three-dimensional sphere. See *sphere*.

Time dilation. Feature emerging from *special relativity*, in which the flow of time slows down for an *observer* in motion.

T.O.E. (Theory of Everything). A quantum-mechanical theory that encompasses all forces and all matter.

Topologically distinct. Two shapes that cannot be deformed into one another without tearing their structure in some manner.

Topology. Classification of shapes into groups that can be deformed into one another without ripping or tearing their structure in any way.

Topology-changing transition. Evolution of spatial fabric that involves rips or tears, thereby changing the *topology* of space.

Torus. The two-dimensional surface of a doughnut.

Two-brane. See *brane*.

Two-dimensional sphere. See *sphere*.

Type I string theory. One of the five *superstring theories*; involves both *open* and *closed strings*.

Type IIA string theory. One of the five *superstring theories*; involves *closed strings* with left-right symmetric *vibrational patterns*.

Type IIB string theory. One of the five *superstring theories*; involves *closed strings* with left-right asymmetric *vibrational patterns*.

Ultramicroscopic. Length scales shorter than the *Planck length* (and also time scales shorter than the *Planck time*).

Uncertainty principle. Principle of *quantum mechanics,* discovered by Heisenberg, that there are features of the universe, like the position and *velocity* of a particle, that cannot be known with complete precision. Such uncertain aspects of the microscopic world become ever more severe as the distance and time scales on which they are considered become ever smaller. Particles and fields undulate and jump between all possible values consistent with the quantum uncertainty. This implies that the microscopic realm is a roiling frenzy, awash in a violent sea of *quantum fluctuations.*

Unified theory, Unified field theory. Any theory that describes all four forces and all of matter within a single, all-encompassing framework.

Uniform vibration. The overall motion of a *string* in which it moves without changes in shape.

Velocity. The speed and the direction of an object's motion.

Vibrational mode. See *vibrational pattern.*

Vibrational pattern. The precise number of peaks and troughs as well as their amplitude as a *string* oscillates.

Vibration number. Whole number describing the energy in the *uniform vibrational* motion of a *string;* the energy in its overall motion as opposed to that associated with changes in its shape.

Virtual particles. Particles that erupt from the vacuum momentarily; they exist on borrowed energy, consistent with the *uncertainty principle,* and rapidly annihilate, thereby repaying the energy loan.

Wave function. Probability waves upon which *quantum mechanics* is founded.

Wavelength. The distance between successive peaks or troughs of a wave.

Wave-particle duality. Basic feature of *quantum mechanics* that objects manifest both wavelike and particle-like properties.

W bosons. See *weak gauge boson.*

Weak force, Weak nuclear force. One of the four fundamental forces, best known for mediating radioactive decay.

Weak gauge boson. Smallest bundle of the *weak force field; messenger particle* of the *weak force;* called W or Z boson.

Weak gauge symmetry. *Gauge symmetry* underlying the *weak force.*

Weakly coupled. Theory whose *string coupling constant* is less than 1.

Winding energy. The energy embodied by a *string* wound around a circular *dimension* of space.

Winding mode. A *string* configuration that wraps around a circular spatial *dimension.*

Winding number. The number of times a *string* is wound around a circular spatial dimension.

World-sheet. Two-dimensional surface swept out by a *string* as it moves.

Wormhole. A tube-like region of space connecting one region of the universe to another.

Z boson. See *weak gauge boson*.

Zero-dimensional sphere. See *sphere*.

References and Suggestions for
Further Reading

Abbot, Edwin A. *Flatland: A Romance of Many Dimensions*. Princeton: Princeton University Press, 1991.

Barrow, John D. *Theories of Everything*. New York: Fawcett-Columbine, 1992.

Bronowski, Jacob. *The Ascent of Man*. Boston: Little, Brown, 1973.

Clark, Ronald W. *Einstein, The Life and Times*. New York: Avon, 1984.

Crease, Robert P., and Charles C. Mann. *The Second Creation*. New Brunswick, N.J.: Rutgers University Press, 1996.

Davies, P. C. W. *Superforce*. New York: Simon & Schuster, 1984.

Davies, P. C. W., and J. Brown, eds. *Superstrings: A Theory of Everything?* Cambridge, Eng.: Cambridge University Press, 1988.

Deutsch, David. *The Fabric of Reality*. New York: Allen Lane, 1997.

Einstein, Albert. *The Meaning of Relativity*. Princeton: Princeton University Press, 1988.

———. *Relativity*. New York: Crown, 1961.

Ferris, Timothy. *Coming of Age in the Milky Way*. New York: Anchor, 1989.

———. *The Whole Shebang*. New York: Simon & Schuster, 1997.

Fölsing, Albrecht. *Albert Einstein*. New York: Viking, 1997.

Feynman, Richard. *The Character of Physical Law*. Cambridge, Mass.: MIT Press, 1995.

Gamow, George. *Mr. Tompkins in Paperback*. Cambridge, Eng.: Cambridge University Press, 1993.

Gell-Mann, Murray. *The Quark and the Jaguar*. New York: Freeman, 1994.

Glashow, Sheldon. *Interactions*. New York: Time-Warner Books, 1988.

Guth, Alan H. *The Inflationary Universe*. Reading, Mass.: Addison-Wesley, 1997.

Hawking, Stephen. *A Brief History of Time*. New York: Bantam Books, 1988.

Hawking, Stephen, and Roger Penrose. *The Nature of Space and Time*. Princeton: Princeton University Press, 1996.

Hey, Tony, and Patrick Walters. *Einstein's Mirror*. Cambridge, Eng.: Cambridge University Press, 1997.

Kaku, Michio. *Beyond Einstein*. New York: Anchor, 1987.

———. *Hyperspace*. New York: Oxford University Press, 1994.

Lederman, Leon, with Dick Teresi. *The God Particle*. Boston: Houghton Mifflin, 1993.

Lindley, David. *The End of Physics*. New York: Basic Books, 1993.

———. *Where Does the Weirdness Go?* New York: Basic Books, 1996.

Overbye, Dennis, *Lonely Hearts of the Cosmos*. New York: HarperCollins, 1991.

Pais, Abraham. *Subtle Is the Lord: The Science and the Life of Albert Einstein*. New York: Oxford University Press, 1982.

Penrose, Roger. *The Emperor's New Mind*. Oxford, Eng.: Oxford University Press, 1989.

Rees, Martin J. *Before the Beginning*. Reading, Mass.: Addison-Wesley, 1997.

Smolin, Lee. *The Life of the Cosmos*. New York: Oxford University Press, 1997.

Thorne, Kip. *Black Holes and Time Warps*. New York: Norton, 1994.

Weinberg, Steven. *The First Three Minutes*. New York: Basic Books, 1993.

———. *Dreams of a Final Theory*. New York: Pantheon, 1992.

Wheeler, John A. *A Journey into Gravity and Spacetime*. New York: Scientific American Library, 1990.

Index

Page numbers in *italics* refer to illustrations and tables.